Danil Prokhorov (Ed.)

Computational Intelligence in Automotive Applications

Studies in Computational Intelligence, Volume 132

Editor-in-Chief
Prof. Janusz Kacprzyk
Systems Research Institute
Polish Academy of Sciences
ul. Newelska 6
01-447 Warsaw
Poland
E-mail: kacprzyk@ibspan.waw.pl

Further volumes of this series can be found on our homepage:
springer.com

Vol. 111. David Elmakias (Ed.)
New Computational Methods in Power System Reliability, 2008
ISBN 978-3-540-77810-3

Vol. 112. Edgar N. Sanchez, Alma Y. Alanís and Alexander G. Loukianov
Discrete-Time High Order Neural Control: Trained with Kalman Filtering, 2008
ISBN 978-3-540-78288-9

Vol. 113. Gemma Bel-Enguix, M. Dolores Jiménez-López and Carlos Martín-Vide (Eds.)
New Developments in Formal Languages and Applications, 2008
ISBN 978-3-540-78290-2

Vol. 114. Christian Blum, Maria José Blesa Aguilera, Andrea Roli and Michael Sampels (Eds.)
Hybrid Metaheuristics, 2008
ISBN 978-3-540-78294-0

Vol. 115. John Fulcher and Lakhmi C. Jain (Eds.)
Computational Intelligence: A Compendium, 2008
ISBN 978-3-540-78292-6

Vol. 116. Ying Liu, Aixin Sun, Han Tong Loh, Wen Feng Lu and Ee-Peng Lim (Eds.)
Advances of Computational Intelligence in Industrial Systems, 2008
ISBN 978-3-540-78296-4

Vol. 117. Da Ruan, Frank Hardeman and Klaas van der Meer (Eds.)
Intelligent Decision and Policy Making Support Systems, 2008
ISBN 978-3-540-78306-0

Vol. 118. Tsau Young Lin, Ying Xie, Anita Wasilewska and Churn-Jung Liau (Eds.)
Data Mining: Foundations and Practice, 2008
ISBN 978-3-540-78487-2

Vol. 119. Slawomir Wiak, Andrzej Krawczyk and Ivo Dolezel (Eds.)
Intelligent Computer Techniques in Applied Electromagnetics, 2008
ISBN 978-3-540-78489-0

Vol. 120. George A. Tsihrintzis and Lakhmi C. Jain (Eds.)
Multimedia Interactive Services in Intelligent Environments, 2008
ISBN 978-3-540-78491-3

Vol. 121. Nadia Nedjah, Leandro dos Santos Coelho and Luiza de Macedo Mourelle (Eds.)
Quantum Inspired Intelligent Systems, 2008
ISBN 978-3-540-78531-6

Vol. 122. Tomasz G. Smolinski, Mariofanna G. Milanova and Aboul-Ella Hassanien (Eds.)
Applications of Computational Intelligence in Biology, 2008
ISBN 978-3-540-78533-0

Vol. 123. Shuichi Iwata, Yukio Ohsawa, Shusaku Tsumoto, Ning Zhong, Yong Shi and Lorenzo Magnani (Eds.)
Communications and Discoveries from Multidisciplinary Data, 2008
ISBN 978-3-540-78732-7

Vol. 124. Ricardo Zavala Yoe
Modelling and Control of Dynamical Systems: Numerical Implementation in a Behavioral Framework, 2008
ISBN 978-3-540-78734-1

Vol. 125. Larry Bull, Bernadó-Mansilla Ester and John Holmes (Eds.)
Learning Classifier Systems in Data Mining, 2008
ISBN 978-3-540-78978-9

Vol. 126. Oleg Okun and Giorgio Valentini (Eds.)
Supervised and Unsupervised Ensemble Methods and their Applications, 2008
ISBN 978-3-540-78980-2

Vol. 127. Régie Gras, Einoshin Suzuki, Fabrice Guillet and Filippo Spagnolo (Eds.)
Statistical Implicative Analysis, 2008
ISBN 978-3-540-78982-6

Vol. 128. Fatos Xhafa and Ajith Abraham (Eds.)
Metaheuristics for Scheduling in Industrial and Manufacturing Applications, 2008
ISBN 978-3-540-78984-0

Vol. 129. Natalio Krasnogor, Giuseppe Nicosia, Mario Pavone and David Pelta (Eds.)
Nature Inspired Cooperative Strategies for Optimization (NICSO 2007), 2008
ISBN 978-3-540-78986-4

Vol. 130. Richi Nayak, Nikhil Ichalkaranje and Lakhmi C. Jain (Eds.)
Evolution of the Web in Artificial Intelligence Environments, 2008
ISBN 978-3-540-79139-3

Vol. 131. Roger Lee and Haeng-Kon Kim (Eds.)
Computer and Information Science, 2008
ISBN 978-3-540-79186-7

Vol. 132. Danil Prokhorov (Ed.)
Computational Intelligence in Automotive Applications, 2008
ISBN 978-3-540-79256-7

Danil Prokhorov
(Ed.)

Computational Intelligence in Automotive Applications

Corrected Second Printing

Springer

Danil Prokhorov
Toyota Technical Center – A Division
of Toyota Motor Engineering
and Manufacturing (TEMA)
Ann Arbor, MI 48105
USA
Email: dvprokhorov@gmail.com

ISBN 978-3-642-42425-0   ISBN 978-3-540-79257-4 (eBook)

Studies in Computational Intelligence   ISSN 1860-949X

© 2008 Springer-Verlag Berlin Heidelberg, Corrected Second Printing 2008

Softcover re-print of the Hardcover 1st edition 2008
This work is subject to copyright. All rights are reserved, whether the whole or part of the material is concerned, specifically the rights of translation, reprinting, reuse of illustrations, recitation, broadcasting, reproduction on microfilm or in any other way, and storage in data banks. Duplication of this publication or parts thereof is permitted only under the provisions of the German Copyright Law of September 9, 1965, in its current version, and permission for use must always be obtained from Springer. Violations are liable to prosecution under the German Copyright Law.

The use of general descriptive names, registered names, trademarks, etc. in this publication does not imply, even in the absence of a specific statement, that such names are exempt from the relevant protective laws and regulations and therefore free for general use.

Typesetting: Scientific Publishing Services Pvt. Ltd., Chennai, India.
Cover Design: Deblik, Berlin, Germany.

Printed in acid-free paper

9 8 7 6 5 4 3 2 1

springer.com

Editor

Danil V. Prokhorov

About the Editor

Danil V. Prokhorov began his technical career in St. Petersburg, Russia, after graduating with Honors from Saint Petersburg State University of Aerospace Instrumentation in 1992 (MS in Robotics). He worked as a research engineer in St. Petersburg Institute for Informatics and Automation, one of the instututes of the Russian Academy of Sciences. He came to US in late 1993 to study for Ph.D. in neurocomputing. He got involved with automotive research in 1995 when he was a Summer intern at Ford Scientific Research Lab in Dearborn, MI. Upon his graduation from the EE Department of Texas Tech University, Lubbock, in 1997, he joined Ford to pursue application-driven research on neural networks and other machine learning algorithms. While at Ford, he took part in several production-bound projects including neural network based engine misfire detection. Since 2005 he is with Toyota Technical Center, Ann Arbor, MI, overseeing important mid- and long-term research projects in computational intelligence. In addition to contributing with his numerous technical papers and patents, he has been helping research community with reviewing for many conferences, journals, and for the US funding agencies.

Preface

What is computational intelligence (CI)? Traditionally, CI is understood as a collection of methods from the fields of neural networks (NN), fuzzy logic and evolutionary computation. Various definitions and opinions exist, but what belongs to CI is still being debated; see, e.g., [1], [2], [3]. More recently there has been a proposal to define the CI not in terms of the tools but in terms of challenging problems to be solved [4].

With this edited volume I made an attempt to give a representative sample of contemporary CI activities in automotive applications to illustrate the state of the art. While CI research and achievements have been illustrated in the book form for other specialized fields (see, e.g., [5] and [6]), this is the first volume of its kind dedicated to automotive technology. As if reflecting the general lack of consensus on what constitutes the field of CI, this volume illustrates automotive applications of not only neural and fuzzy computations[1] which are considered to be the "standard" CI topics but also others, such as decision trees, graphical models, Support Vector Machines (SVM), multi-agent systems, etc.

This book is neither an introductory text, nor a comprehensive overview of all CI research in this area. Hopefully, as a broad and representative sample of CI activities in automotive applications it will be worthy of reading for both professionals and students. When the amount of details appears insufficient, the reader is encouraged to consult other relevant sources provided by the chapter authors.

Chapter "Learning-based Driver Workload Estimation" discusses research on estimation of driver cognitive workload and proposes a new methodology to design driver workload estimation systems. The methodology is based on decision-tree learning. It derives optimized models to assess the time-varying workload level from data which includes not only measurements from various sensors but also subjective workload level ratings.

---

[1] Another "standard" CI topic called evolutionary computation (EC) is not represented in this volume in the form of a separate chapter, although some EC elements are mentioned or referenced throughout the book. Relevant publications on EC for automotive applications are available (e.g., [7]), but unfortunately were not available as contributors of this volume.

Chapter "Visual Monitoring of Driver Inattention" introduces a prototype computer vision system for real-time detection of driver fatigue. The system includes an image acquisition module with an infrared illuminator, pupil detection and tracking module, and algorithms for detecting appropriate visual behaviors and monitoring six parameters which may characterize the fatigue level of a driver. To increase effectiveness of monitoring, a fuzzy classifier is implemented to fuse all these parameters into a single gauge of driver inattentiveness. The system tested on real data from different drivers operates with high accuracy and robustly at night.

Chapter "Understanding Driving Activity Using Ensemble Methods" complements Chapter "Visual Monitoring of Driver Inattention" by discussing whether driver inattention can be detected without eye and head tracking systems. Instead of limiting themselves to working with just a few signals from preselected sensors, the authors chose to operate on hundreds of signals reflecting real-time environment both outside and inside the vehicle. The discovery of relationships in the data useful for driver activity classification, as well as ranking signals in terms of their importance for classification, is entrusted to an approach called random forest, which turned out to be more effective than either hidden Markov models or SVM.

Chapter "Computer Vision and Machine Learning for Enhancing Pedestrian Safety" overviews methods for pedestrian detection which use information from on-board and infrastructure based sensors. Many of the discussed methods are sufficiently generic to be useful for object detection, classification and motion prediction in general.

Chapter "Application of Graphical Models in the Automotive Industry" describes briefly how graphical models, such as Bayesian and Markov networks, are used at Volkswagen and Daimler. Production planning at Volkswagen and demand prediction benefit significantly from the graphical model based system developed. Another data mining system is developed for Daimler to help assessing the quality of vehicles and identifying causes of troubles when the vehicles have already spent some time in service. It should be noted that other automotive companies are also pursuing data mining research (see, e.g., [8]).

Chapter "Extraction of Maximum Support Rules for the Root Cause Analysis" discusses extraction of rules from manufacturing data for root cause analysis and process optimization. An alternative approach to traditional methods of root cause analysis is proposed. This new approach employs branch-and-bound principles, and it associates process parameters with results of measurements which is helpful in identification of the main drivers for quality variations of an automotive manufacturing process.

Chapter "Neural networks in automotive applications" provides an overview of neural network technology, concentrating on three main roles of neural networks: models, virtual or soft sensors and controllers. Training of NN is also discussed, followed by a simple example illustrating importance of recurrent NN.

Chapter "On learning machines for engine control" deals with modeling for control of turbocharged spark ignition engines with variable camshaft timing.

Two examples are considered: 1) estimation of the in-cylinder air mass in which open loop neural estimators are combined with a dynamic polytopic observer, and 2) modeling an in-cylinder residual gas fraction by a linear programming support vector regression method. The authors argue that models based on first principles ("white boxes") and neural or other "black box" models must be combined and utilized in the "grey box" approach to obtain results which are not just superior than any alternatives but also more acceptable to automotive engineers.

Chapter "Recurrent neural networks for AFR estimation and control in spark ignition automotive engines" complements Chapter "On learning machines for engine control" by discussing specifics of the air-fuel ratio (AFR) control. Recurrent NN are trained off-line and employed as both the AFR virtual sensor and the inverse model controller. The authors also provide a comparison with a conventional control strategy on a real engine.

Chapter "Intelligent Vehicle Power Management - an overview" presents four case studies: a conventional vehicle power controller and three different approaches for a parallel HEV power controller. They include controllers based on dynamic programming and neural networks, and fuzzy logic controllers one of which incorporates predictions of driving environment and driving patterns.

Chapter "Integrated Diagnostic Process for Automotive Systems" provides an overview of model-based and data-driven diagnostic methods applicable to complex systems. Selected methods are applied to three automotive examples, one of them being a hardware-in-the-loop system, in which the methods are put to work together to solve diagnostic and prognostic problems. It should be noted that integration of different approaches is an important theme for automotive research spanning the entire product life cycle (see, e.g., [9]).

Chapter "Automotive manufacturing: intelligent resistance welding" introduces a real-time control system for resistance spot welding. The control system is built on the basis of neural networks and fuzzy logic. It includes a learning vector quantization NN for assessing the quality of weld nuggets and a fuzzy logic process controller. Experimental results indicate substantial quality improvement over a conventional controller.

Chapter "Intelligent control of mobility systems" (ICMS) overviews projects of the ICMS Program at the National Institute of Standards and Technology (NIST). The program provides architecture, interface and data standards, performance test methods and infrastructure technology available to manufacturing industry and government agencies in developing and applying intelligent control technology to mobility systems. A common theme among these projects is autonomy and the four dimensional/real-time control systems (4D/RCS) control architecture for intelligent systems proposed and developed in the NIST Intelligent Systems Division.

Unlike the book's index, each chapter has its own bibliography for convenience of the reader, with little overlap among references of different chapters.

This volume highlights important challenges facing CI in the automotive domain. Better vehicle diagnostics/vehicle system safety, improved control of vehicular systems and manufacturing processes to save resources and minimize impact on the environment, better driver state monitoring, improved safety of pedestrians, making vehicles more intelligent on the road - these are important directions where the CI technology can and should make the impact. All of these are consistent with Toyota vision [10]:

Toyota's vision is to balance "Zeronize" and "Maximize". "Zeronize" symbolizes the vision and philosophy of our persistent efforts in minimizing negative aspects vehicles have such as environmental impact, traffic congestion and traffic accidents, while "Maximize" symbolizes the vision and philosophy of our persistent efforts in maximizing the positive aspects vehicles have such as fun, delight, excitement and comfort, that people seek in automobiles.

I am very thankful to all the contributors of this edited volume for their willingness to participate in this project, their patience and valuable time. I am also grateful to Prof. Janusz Kacprzyk, the Springer Series Editor, for his encouragement to organize and edit this volume, as well as Thomas Ditzinger, the Springer production editor for his support of this project.

January 2008
Danil V. Prokhorov

References

[1] http://en.wikipedia.org/wiki/Computational_intelligence
[2] Bezdek, J.C.: What is computational intelligence? In: Zurada, M., Robinson (eds.) Computational Intelligence: Imitating Life, pp. 1–12. IEEE Press, New York (1994)
[3] Marks II, R.J.: Intelligence: Computational Versus Artificial. IEEE Transactions on Neural Networks 4(5), 737–739 (1993)
[4] Duch, W.: What is computational intelligence and what could it become? In: Duch, W., Mandziuk, J. (eds.) Challenges for Computational Intelligence. Studies in Computational Intelligence (J. Kacprzyk Series Editor), vol. 63. Springer, Heidelberg (2007), http://cogprints.org/5358/
[5] Ruano, A.: Intelligent Control Systems Using Computational Intelligence Techniques (IEE Control Series). IEE (2005)
[6] Begg, R., Lai, D.T.H., Palaniswami, M.: Computational Intelligence in Biomedical Engineering. CRC Press, Taylor & Francis Books Inc. (2007)
[7] Laumanns, M., Laumanns, N.: Evolutionary Multiobjective Design in Automotive Development. Applied Intelligence 23, 55–70 (2005)
[8] Montgomery, T.A.: Text Mining on a Budget: Reduce, Reuse, Recycle. In: Michigan Leadership Summit on Business Intelligence and Advanced Analytics, Troy, MI (March 8, 2007), http://www.cmurc.com/bi-PreviousEvents.htm
[9] Struss, P., Price, C.: Model-Based Systems in the Automotive Industry. AI Magazine 24(4), 17–34 (Winter 2003)
[10] Toyota ITS vision, http://www.toyota.co.jp/en/tech/its/vision/

Contents

Learning-Based Driver Workload Estimation
Yilu Zhang, Yuri Owechko, Jing Zhang 1

Visual Monitoring of Driver Inattention
*Luis M. Bergasa, Jesús Nuevo, Miguel A. Sotelo, Rafael Barea,
Elena Lopez* 25

Understanding Driving Activity Using Ensemble Methods
*Kari Torkkola, Mike Gardner, Chris Schreiner, Keshu Zhang,
Bob Leivian, Harry Zhang, John Summers* 53

Computer Vision and Machine Learning for Enhancing Pedestrian Safety
Tarak Gandhi, Mohan Manubhai Trivedi 79

Application of Graphical Models in the Automotive Industry
Matthias Steinbrecher, Frank Rügheimer, Rudolf Kruse 103

Extraction of Maximum Support Rules for the Root Cause Analysis
Tomas Hrycej, Christian Manuel Strobel 117

Neural Networks in Automotive Applications
Danil Prokhorov 133

On Learning Machines for Engine Control
Gérard Bloch, Fabien Lauer, Guillaume Colin 165

Recurrent Neural Networks for AFR Estimation and Control in Spark Ignition Automotive Engines
Ivan Arsie, Cesare Pianese, Marco Sorrentino 191

Intelligent Vehicle Power Management – An Overview
Yi L. Murphey 223

An Integrated Diagnostic Process for Automotive Systems
*Pattipati Krishna, Kodali Anuradha, Luo Jianhui, Choi Kihoon,
Singh Satnam, Sankavaram Chaitanya, Mandal Suvasri, Donat William,
Namburu Setu Madhavi, Chigusa Shunsuke, Qiao Liu* 253

Automotive Manufacturing: Intelligent Resistance Welding
Mahmoud El-Banna, Dimitar Filev, Ratna Babu Chinnam 291

Intelligent Control of Mobility Systems
*James Albus, Roger Bostelman, Raj Madhavan, Harry Scott,
Tony Barbera, Sandor Szabo, Tsai Hong, Tommy Chang,
Will Shackleford, Michael Shneier, Stephen Balakirsky, Craig Schlenoff,
Hui-Min Huang, Fred Proctor* 315

Index 363

Author Index 367

List of Contributors

James Albus, Roger Bostelman, Raj Madhavan, Harry Scott, Tony Barbera, Sandor Szabo, Tsai Hong, Tommy Chang, Will Shackleford, Michael Shneier, Stephen Balakirsky, Craig Schlenoff, Hui-Min Huang, Fred Proctor
Intelligent Systems Division, National Institute of Standards and Technology (NIST), 100 Bureau Drive, Mail Stop 8230, Gaithersburg, MD 20899-8230, USA
raj.madhavan@nist.gov,
roger.bostelman@nist.gov

Ivan Arsie
Department of Mechanical Engineering, University of Salerno, 84084 Fisciano (SA), Italy
iarsie@unisa.it

Rafael Barea
Department of Electronics, University of Alcala, CAMPUS. 28805 Alcal de Henares (Madrid) Spain
barea@depeca.uah.es

Luis M. Bergasa
Department of Electronics, University of Alcala, CAMPUS. 28805 Alcal de Henares (Madrid) Spain
bergasa@depeca.uah.es

Gérard Bloch
Centre de Recherche en Automatique de Nancy (CRAN), Nancy-University, CNRS, CRAN-ESSTIN, 2 rue Jean Lamour, 54519 Vandoeuvre lès Nancy, France
gerard.bloch@esstin.uhp-nancy.fr

Ratna Babu Chinnam
Wayne State University, Detroit, MI 48202, USA
r_chinnam@wayne.edu

Guillaume Colin
Laboratoire de Mécanique et d'Energétique (LME), University of Orléans, 8 rue Léonard de Vinci, 45072 Orléans Cedex 2, France
guillaume.colin@univ-orleans.fr

Mahmoud El-Banna
University of Jordan, Amman 11942, Jordan
m.albanna@ju.edu.jo

Dimitar Filev
Ford Motor Company, Dearborn, MI
48121, USA
dfilev@ford.com

Tarak Gandhi
Laboratory for Safe and Intelligent
Vehicles
(LISA), University of California San
Diego, La Jolla, CA 92093
tgandhi@ucsd.edu

Tomas Hrycej
formerly with DaimlerChrysler
Research, Ulm,
Germany
tomas_hrycej@yahoo.de

Rudolf Kruse
Department of Knowledge Processing
and
Language Engineering, Otto-von-
Guericke University of Magdeburg,
Universittsplatz 2, 39106 Magdeburg,
Germany
kruse@iws.cs.uni-magdeburg.de

Fabien Lauer
Centre de Recherche en Automatique
de Nancy
(CRAN), Nancy-University, CNRS,
CRAN-ESSTIN, 2 rue Jean Lamour,
54519 Vandoeuvre lès Nancy, France
fabien.lauer@esstin.uhp-nancy.fr

Elena Lopez
Department of Electronics,
University of Alcala, CAMPUS. 28805
Alcal de Henares (Madrid)
Spain
elena@depeca.uah.es

Jianhui Luo
Qualtech Systems, Inc., Putnam Park,
Suite 603,
100 Great Meadow Road,
Wethersfield, CT 06109, USA

Yi L. Murphey
Department of Electrical and
Computer
Engineering, University of
Michigan-Dearborn, Dearborn, MI
48128

Setu Madhavi Namburu,
Shunsuke Chigusa, Liu Qiao
Toyota
Technical Center - a division of Toyota
Motor Engineering and
Manufacturing (TEMA), 1555
Woodridge Rd., Ann Arbor, MI 48105
liu.qiao@tema.toyota.com

Jess Nuevo
Department of Electronics,
University of Alcala, CAMPUS. 28805
Alcal de Henares (Madrid)
Spain
jnuevo@depeca.uah.es

Yuri Owechko
HRL Laboratories, LLC., 3011 Malibu
Canyon
Road, Malibu, CA, USA
yowechko@hrl.com

Krishna Pattipati,
Anuradha Kodali,
Kihoon Choi, Satnam Singh,
Chaitanya Sankavaram,
Suvasri Mandal, William Donat
University of Connecticut, Storrs, CT,
06268, USA
krishna@engr.uconn.edu

Cesare Pianese
Department of Mechanical
Engineering,
University of Salerno, 84084 Fisciano
(SA), Italy
pianese@unisa.it

Danil V. Prokhorov
Toyota Technical Center - a division of Toyota Motor Engineering and Manufacturing (TEMA), Ann Arbor, MI
48105, USA
dvprokhorov@gmail.com

Frank Rgheimer
Department of Knowledge Processing and Language Engineering,
Otto-von-Guericke University of Magdeburg,
Universittsplatz 2, 39106 Magdeburg, Germany
ruegheim@iws.cs.uni-magdeburg.de

Marco Sorrentino
Department of Mechanical Engineering,
University of Salerno, 84084 Fisciano (SA), Italy
msorrentino@unisa.it

Miguel A. Sotelo
Department of Electronics,
University of Alcala, CAMPUS. 28805 Alcal de Henares (Madrid)
Spain
sotelo@depeca.uah.es

Matthias Steinbrecher
Department of Knowledge Processing and Language Engineering,
Otto-von-Guericke University of Magdeburg,
Universittsplatz 2, 39106 Magdeburg, Germany
msteinbr@iws.cs.uni-magdeburg.de

Christian Manuel Strobel
University of Karlsruhe (TH),
Germany
mstrobel@statistik.uni-karlsruhe.de

Kari Torkkola, Mike Gardner, Chris Schreiner, Keshu Zhang, Bob Leivian, Harry Zhang, John Summers
Motorola Labs, USA
Kari.Torkkola@motorola.com

Mohan Manubhai Trivedi
Laboratory for Safe and Intelligent Vehicles (LISA), University of California San Diego, La Jolla, CA 92093
mtrivedi@ucsd.edu

Jing Zhang
R&D Center, General Motors Cooperation, 30500
Mound Road, Warren, MI, USA
jing.zhang@gm.com

Yilu Zhang
R&D Center, General Motors Cooperation, 30500
Mound Road, Warren, MI, USA
yilu.zhang@gm.com

Learning-Based Driver Workload Estimation

Yilu Zhang[1], Yuri Owechko[2], and Jing Zhang[3]

[1] R&D Center, General Motors Cooperation, 30500 Mound Road, Warren, MI
  yilu.zhang@gm.com
[2] HRL Laboratories, LLC., 3011 Malibu Canyon Road, Malibu, CA
  yowechko@hrl.com
[3] R&D Center, General Motors Cooperation, 30500 Mound Road, Warren, MI
  jing.zhang@gm.com

A popular definition of workload is given by O'Donnell and Eggmeir, which states that "The term workload refers to that portion of the operator's limited capacity actually required to perform a particular task" [1]. In the vehicle environment, the "particular task" refers to both the vehicle control, which is the primary task, and other secondary activities such as listening to the radio. Three major types of driver workload are usually studied, namely, visual, manual, and cognitive. Auditory workload is not treated as a major type of workload in the driving context because the auditory perception is not considered as a major requirement to perform a driving task. Even when there is an activity that involves audition, the driver is mostly affected cognitively.

Lately, the advanced computer and telecommunication technology is introducing many new in-vehicle information systems (IVISs), which give drivers more convenient and pleasant driving experiences. Active research is being conducted to provide IVISs with both high functionality and high usability. On the usability side, driver's workload is a heated topic advancing in at least two major directions. One is the offline assessment of the workload imposed by IVISs, which can be used to improve the design of IVISs. The other effort is the online workload estimation, based on which IVISs can provide appropriate service at appropriate time, which is usually termed as *Workload Management* . For example, the incoming phone call may be delayed if the driver is engaged in a demanding maneuver.

Among the three major types of driver workload, cognitive workload is the most difficult to measure. For example, withdrawing hands from the steering wheel to reach for a coffee cup requires extra manual workload. It also may require extra visual workload in that the position of the cup may need to be located. Both types of workload are directly measurable through such observations as hands-off-wheel and eyes-off-road time. On the other hand, engaging in thinking (the so-called minds-off-road phenomenon) is difficult to detect. Since the cognitive workload level is internal to the driver, it can only be inferred based on the information that is observable.

In this chapter, we report some of our research results on driver's cognitive workload estimation[1]. After the discussion of the existing practices, we propose a new methodology to design driver workload estimation systems, that is, using machine-learning techniques to derive optimized models to index workload. The advantage of this methodology will be discussed, followed by the presentation of some experimental results. This chapter concludes with discussion of future work.

1 Background

Driver Workload Estimation (DWE) refers to the activities of monitoring the driver, the vehicle, and the driving environment in real-time, and acquiring the knowledge of driver's workload level continuously. A typical DWE system takes sensory information of the driver, the vehicle and the driving environment as inputs, and generates an index to the driver's workload level as shown in Fig. 1. The central issue of DWE is to design the driver workload estimation algorithm that generates the workload index with high accuracy.

A practical DWE system fulfills the following three requirements in order to identify driver's cognitive status while the driver is engaged in naturalistic driving practice.

- Continuously measurable: A DWE system has to be able to continuously measure workload while the driver is driving the vehicle so that workload management can be conducted appropriately to avoid overloading the driver.
- Residual capacity sensitive: The residual capacity of the driver refers to the amount of spare capacity of the driver while he/she is performing the primary task of maneuvering the vehicle, and, if applicable, engaging in secondary tasks such as drinking a cup of coffee or operating an IVIS. If the residual capacity is high, the driver is typically doing well in vehicle control and may be able to be engaging in even more secondary activities. Residual capacity is the primary interest for DWE.
- Highly non-intrusive: A DWE system should not interfere with the driver by any means.

Before the discussion of existing DWE methodologies in the next section, it is helpful to give a brief introduction of cognitive workload assessment methods. There exist four major categories of cognitive workload assessment methods in present-day practice [2], namely primary-task performance measures, secondary-task performance measures, subjective measures, and physiological measures.

The primary-task performance measures evaluate cognitive workload based on driving performance, such as lane-position deviation, lane exceedences, brake pressure, and vehicle headway. These measures are usually direct and continuous.

In the secondary-task approach, the driver is required to perform one or multiple secondary tasks, e.g., pushing a button when flashing LED light is detected

---

[1] To simplify the terminology, we use "workload" interchangeably with "cognitive workload" in this chapter.

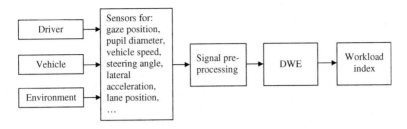

Fig. 1. The working process of a driver workload estimation system

in the peripheral vision. The secondary-task performance such as reaction time is measured as the index of driver's cognitive workload level. Secondary-task performance may introduce tasks unnecessary to regular driving and is intrusive to the driver.

With the subjective measure approach, driver's personal opinion on his/her operative experience is elicited. After a trip or an experiment, the subject is asked to describe or rate several dimensions of effort required to perform the driving task. If the driver and the experimenter establish clear mutual understanding of the rating scale, the subjective measure can be very reliable. Examples of popular subjective workload rating index are NASA Task Load Index (TLX) [3] and Subjective Workload Assessment Technique (SWAT) [4].

Physiological measures include brain activities such as event-related potential (ERP) and Electroencephalogram (EEG), cardiac activities such as heart rate variance, as well as ocular activities such as eye closure duration and pupil diameter changes. Physiological measures are continuous and residual-capacity sensitive. The challenge, however, lies in reliably acquiring and interpreting the physiological data sets, in addition to user acceptance issues.

Among the four workload assessment methods, primary-task performance measures and physiological measures (obtained by non-intrusive sensors) fulfill the above-discussed DWE requirements and are generally appropriate for DWE applications. Although not directly suitable for real-time DWE, the secondary-task performance measures and the subjective measures are still valuable in developing DWE since they provide ways to calibrate the index generated by a DWE system.

2 Existing Practice and Its Challenges

Most existing researches on DWE follow this pattern. First, analyze the correlation between various features, such as lane position deviation, and driver's workload. The ground truth of driver's workload is usually assessed by subjective measures, secondary-task performance, or the analysis of the task. The features are usually selected according to the prior understanding of human behaviors and then tested using well-designed experiments. While there are attempts reported to analyze the features simultaneously [5], usually the analysis is done

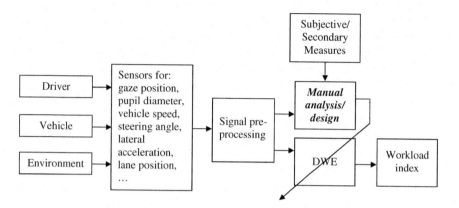

Fig. 2. The existing DWE system design process

on individual features [6] [7]. Second, models are designed to generate workload index by combining features that have high correlation with driver workload. We refer to the above methodology as *manual analysis and modeling*. The manual DWE design process is illustrated in Fig. 2. Research along this line has achieved encouraging success. The well-known existing models include the steering entropy [8] and the SEEV model [9]. However, there are yet difficulties in developing a robust cognitive workload estimator for practical applications, the reasons of which are discussed below.

First, the existing data analysis methods very much rely on the domain knowledge in the field of human behavior. Although many studies have been conducted and many theories have been proposed to explain the way that human beings manage resources and workload [10] [11] [12] [2], the relationship between overt human behavior and cognitive activities is by and large unclear to the scientific community. It is extremely difficult to design the workload estimation models based on this incomplete domain knowledge.

Second, manual data analysis and modeling are not efficient. Until now, a large number of features related to driver's cognitive workload have been studied. A short list of them includes: lane position deviation, the number of lane departure, lane departure duration, speed deviation, lateral deviation, steering hold, zero-crossing and steering reversal rate, brake pressure, the number of brake presses, and vehicle headway. With the fast advancing sensing technology, the list is quickly expanding. It has been realized that while each individual feature may not index workload well under various scenarios, the fusion of multiple features tends to provide better overall performance. However, in the course of modeling the data to estimate workload, the models tend to be either relatively simple, such as the linear regression models, or narrowly scoped by covering a small number of features. It is usually expensive and time-consuming to iteratively design models over a large number of features and validate models on a huge data set.

Third, most researchers choose workload inference features by analyzing the *correlation* between the observations of driver's behavior and driver's workload

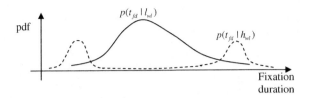

Fig. 3. The probability distribution function of fixation duration under high and low workload

level. This analysis requires the assumption of unimode Gaussian distribution, which is very likely to be violated in reality. In addition, a feature showing low correlation with the workload levels is not necessarily a bad workload indicator.

For example, driver's eye fixation duration is one of the extensively studied features for workload estimation. However, studies show contradictory findings in the relation between workload level and fixation duration. Some of them show positive correlation [13] [14] while others show negative correlation [15] [16]. Does this mean fixation duration is not a good workload indicator? Not necessarily. The fact, that the average fixation duration may become either longer or shorter when driver's workload is high, implies that the probability distribution function (pdf) of fixation duration under high workload ($p(t_{fd}|h_{wl})$) is multi-modal, as shown in Fig.3. With collected ocular data, one may estimate the conditional pdfs ($p(t_{fd}|h_{wl})$ and $p(t_{fd}|l_{wl})$) and the prior probabilities for high and low workload ($P(h_{wl})$ and $P(l_{wl})$). With this knowledge, standard Bayesian analysis will tell the probability of high workload given the fixation duration,

$$p(h_{wl}|t_{fd}) = \frac{p(t_{fd}|h_{wl})P(h_{wl})}{p(t_{fd}|h_{wl})P(h_{wl}) + p(t_{fd}|l_{wl})P(l_{wl})}.$$

3 The Proposed Approach: Learning-Based DWE

We proposed a learning-based DWE design process a few years ago [17] [18]. Under this framework, instead of manually analyzing the significance of individual features or a small set of features, the whole set of features are considered simultaneously. Machine-learning techniques are used to tune the DWE system, and derive an optimized model to index workload.

Machine learning is concerned with the design of algorithms that encode inductive mechanisms so that solutions to broad classes of problems may be derived from examples. It is essentially data-driven and is fundamentally different from traditional AI such as expert systems where rules are extracted mainly by human experts. Machine learning technology has been proved to be very effective in discovering the underlying structure of data and, subsequently, generate models that are not discovered from domain knowledge. For example, in the automatic speech recognition (ASR) domain, models and algorithms based on machine learning outperform all other approaches that have been attempted to date [19]. Machine learning has found increasing applicability in fields as varied as banking, medicine, marketing, condition monitoring, computer vision, and robotics [20].

Machine learning technology has been implemented in the context of driver behavior modeling. Kraiss [21] showed that a neural network could be trained to emulate an algorithmic vehicle controller and that individual human driving characteristics were identifiable from the input/output relations of a trained network. Forbes et.al. [22] used dynamic probabilistic networks to learn the behavior of vehicle controllers that simulate good drivers. Pentland and Liu [23] demonstrated that human driving actions, such as turning, stopping, and changing lane, could be accurately recognized very soon after the beginning of the action using Markov dynamic model (MDM). Oliver and Pentland [24] reported a hidden Markov model-based framework to predict the most likely maneuvers of human drivers in realistic driving scenarios. Mitrović [25] developed a method to recognize driving events, such as driving on left/right curves and making left/right turns, using hidden Markov models. Simmons et.al. [26] presented a hidden Markov model approach to predict a driver's intended route and destination based on observation of his/her driving habits.

Thanks to the obvious relation between driver behavior and driver workload, our proposal of learning-based DWE is a result of the above progress. Similar ideas were proposed by other researchers [27] [28] around the time frame of our work and many followup works have been reported ever since [29] [30] [31].

3.1 Learning-Based DWE Design Process

The learning-based DWE design process is shown in Fig. 4. Compared to the one shown in Fig. 2, the new process replaces the module of manual analysis/design with a module of a machine learning algorithm, which is the key to learning-based DWE.

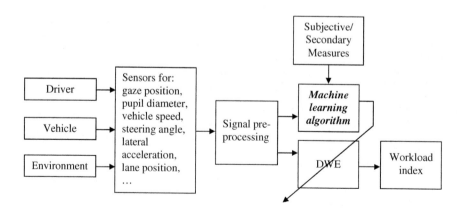

Fig. 4. The learning-based DWE design process

A well-posed machine learning problem requires a task definition, a performance measure, and a set of training data, which are defined as follows for the learning-based DWE:

Task: identify driver's cognitive workload level in a time interval of reasonable length, e.g., every few seconds.
Performance measure: the rate of correctly estimating driver's cognitive workload level.
Training data: recorded driver's behavior including both driving performance and physiological measures together with the corresponding workload levels assessed by subjective measures, secondary task performance, or task analysis.

In order to design the learning-based DWE algorithm, training data need to be collected while subjects drive a vehicle in pre-designed experiments. The data includes the sensory information of the maneuvering of the vehicle (e.g. lane position, which reflects driver's driving performance) and the driver's overt behavior (e.g., eye movement and heart beat), depending on the availability of the sensor on the designated vehicle. The data also includes the subjective workload ratings and/or the secondary-task performance ratings of the subjects. These ratings serve as the training labels.

After some preprocessing on the sensory inputs, such as the computation of mean and standard deviation, the data is fed to a machine-learning algorithm to extract the relationship between the noisy sensory information and the driver's workload level. The computational intelligence algorithm can be decision tree, artificial neural network, support vector machine, or methods based on discriminant analysis. The learned estimator, a mapping from the sensory inputs to the driver's cognitive workload level, can be a set of rules, a look-up table, or a numerical function, depending on the algorithm used.

3.2 Benefits of Learning-Based DWE

Changing from a manual analysis and modeling perspective to a learning-based modeling perspective will gain us much in terms of augmenting domain knowledge, and efficiently and effectively using data.

A learning process is an automatic knowledge extraction process under certain learning criteria. It is very suitable for a problem as complicated as workload estimation. Machine learning techniques are meant for analyzing huge amounts of data, discovering patterns, and extracting relationships. The use of machine-learning techniques can save labor-intensive manual process to derive combined workload index and, therefore, can take full advantage of the availability of various sensors. Finally, most machine learning techniques do not require the assumption of the unimode Gaussian distribution. In addition to the advantages discussed above, this change makes it possible for a DWE system to be adaptive to individual drivers. We will come back to this issue in Section 7.

Having stated the projected advantages, we want to emphasize that the learning-based approach benefits from the prior studies on workload estimation, which have identified a set of salient features, such as fixation duration, pupil diameter, and lane position deviation. We utilize the known salient features as candidate inputs.

4 Experimental Data

Funded by GM R&D under a contract, researchers from the University of Illinois at Urbana-Champaign conducted a driving simulator study to understand driver's workload. The data collected in the simulator study was used to conduct some preliminary studies on learning-based DWE as presented in this chapter.

The simulator system has two Ethernet-connected PCs running GlobalSim's Vection Simulation Software version 1.4.1, a product currently offered by DriveSafety Inc. (www.drivesafety.com). One of the two computers (the subject computer) generates the graphical dynamic driving scenes on a standard 21-in monitor with the resolution of 1024×768 (Fig. 5). Subjects use a non-force feedback Microsoft Sidewinder USB steering wheel together with the accelerator and brake pedals to drive the simulator. The second computer (the experimental computer) is used by an experimenter to create driving scenarios, and collect and store the simulated vehicle data, such as vehicle speed, acceleration, steering angle, lateral acceleration, lane position, etc. To monitor the driver's behavior closely, a gaze tracking system is installed on the subject computer and running at the same time as the driving simulation software. The gaze tracking system is an Applied Science Lab remote monocular eye tracker, Model 504 with pan/tilt optics and a head tracker. It measures the pupil diameter and the point of gaze at 60 Hz with an advertised tracking accuracy of about ±0.5 degree [32]. The gaze data is also streamed to and logged by the experimenter computer. A complete data table is shown in Table 1.

Twelve students participated in the experiment. Each participant drove the simulator in three different driving scenarios, namely, highway, urban, and rural (Fig. 5). There were two sessions of driving for each scenario, each lasting about 8-10 minutes. In each session, the participants were asked to perform secondary

Table 1. The vehicle and gaze data collected by the simulator system

Vehicle	velocity, lane position, speed limit, steer angle, acceleration, brake, gear, horn, vehicle heading, vehicle pitch, vehicle roll, vehicle X, vehicle Y, vehicle Z, turn signal status, latitudinal acceleration, longitudinal acceleration, collision, vehicle ahead or not, headway time, headway distance, time to collide, terrain type, slip
Gaze	gaze vertical position, gaze horizontal position, pupil diameter, eye to scene point of gaze distance, head x position, head y position, head z position, head azimuth, head elevation, head roll

Fig. 5. The screen shots of the driving scene created by the GlobalSim Vection Simulation Software in three different driving scenarios: (a) urban, (b) highway, and (c) rural

tasks (two verbal tasks and two spatial-imagery tasks) during four different 30-second periods called *critical periods*. In the verbal task, the subjects were asked to name words starting with a designated letter. In the spatial-imagery task, the subjects were asked to imagine the letters from A to Z with one of the following characteristics: a) remaining unchanged when flipped sideways, b) remaining unchanged when flipped upside down, c) containing a close part such as "A", d) having no enclosed part, e) containing a horizontal line, f) containing a vertical line. Another four 30-second critical periods were identified as control sessions in each session, during which no secondary tasks were introduced. In the following analysis, we concentrate on the data during the critical periods. In total, there were 12 *subjects* × 3 *scenarios* × 2 *sessions* × 8 *critical periods/session* = 576 critical periods. Because of some technical difficulties during the experiment, the data from some of the critical periods were missing, which ended up with a total of 535 critical periods for use. The total number of data entries is 535 *critical periods* × 30 *seconds* × 60 Hz = 1,036,800.

In the simulator study, there was no direct evidence of driver's workload level, such as the subjective workload assessment. However, workload level for each data entry was needed in order to evaluate the idea of learning-based DWE. We made an assumption that drivers bear more workload when engaging in the secondary tasks. As we know, the primary driving task includes vehicle control (maintaining the vehicle in a safe location with an appropriate speed), hazard awareness (detecting hazards and handling the elicited problems), and navigation (recognizing landmarks and taking actions to reach destination) [33]. The visual perception, spatial cognitive processing, and manual responses involved in these subtasks all require brain resources. In various previous studies, many secondary mental tasks, such as verbal and spatial-imagery tasks, have been shown to compete for the limited brain resources with the primary driving task. The secondary mental tasks affect the drivers by reducing their hazard detection capability and delaying the decision-making time [13] [14] [34]. Although a driver may respond to the multi-tasks by changing resource allocation strategy to make the cooperation more efficient, in general, the more tasks a driver is conducting at a time, the more resources he/she is consuming and, therefore, the higher workload he/she is bearing. Based on this assumption, we labeled all the sensor inputs falling into the dual-task critical periods with high workload. The sensor inputs falling into the control critical periods were labeled with low workload. We understand that driver's workload may fluctuate during a critical period depending on the driving condition and her actual involvement in the secondary task.

5 Experimental Process

We preprocessed the raw measurements from the sensors and generated vectors of features over the fixed-size rolling time windows as shown in Fig. 6. Table 2 lists all the features we used. The "regions" in Table 2 refer to the eight regions of driver's front view as shown in Fig. 7. It is desirable to estimate the workload

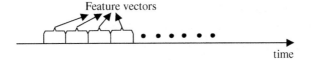

Fig. 6. The rolling time windows for computing the feature vectors

Table 2. The features used to estimate driver's workload

Feature number	Features
1	m_{spd}: mean vehicle velocity
2	V_{spd}: standard deviation of vehicle velocity
3	m_{lp}: mean lane position
4	V_{lp}: standard deviation of vehicle lane position
5	m_{str}: mean steering angle
6	V_{str}: standard deviation of steering angle
7	m_{acc}: mean vehicle acceleration
8	V_{acc}: standard deviation of vehicle acceleration
9	m_{pd}: mean pupil diameter
10	V_{pd}: standard deviation of pupil diameter
11-18	n_i: number of entries for the gaze moving into region i, $i = 1, 2, ..., 8$
19-26	ts_i: portion of time the gaze stayed in region i, $i = 1, 2, ..., 8$
27-34	tv_i: mean visit time for region i, $i = 1, 2, ..., 8$

at a frequency as high as possible. However, it is not necessary to assess it at a frequency of 60 Hz because the driver's cognitive status does not change at that high rate. In practice, we tried different time window sizes, of which the largest was 30 seconds, which equals the duration of a critical period.

While many learning methods can be implemented, such as Bayesian learning, artificial neural networks, hidden Markov models, case based reasoning, and genetic algorithms, we used decision tree learning , one of the most widely used methods for inductive inference, to show the concept.

A decision tree is a hierarchical structure, in which each node corresponds to one attribute of the input attribute vector. If the attribute is categorical, each arc branching from the node represents a possible value of that attribute. If the attribute is numerical, each arc represents an interval of that attribute. The leaves of the tree specify the expected output values corresponding to the attribute vectors. The path from the root to a leaf describes a sequence of decisions made to generate the output value corresponding to an attribute vector. The goal of decision-tree learning is to find out the attribute and the splitting value for each node of the decision tree. The learning criterion can be to reduce entropy [35] or to maximize t-statistics [36], among many others.

For the proof-of-concept purpose, we used the decision-tree learning software, See5, developed by Quinlan [35]. In a See5 tree, the attribute associated with each

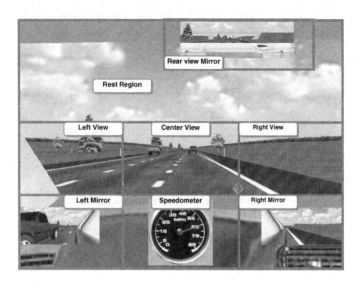

Fig. 7. The screen of the driving scene is divided into eight regions in order to count the gaze entries in each region. The region boundaries were not shown on the screen during the experiment.

node is the most informative one among the attributes not yet considered in the path from the root. The significance of finding the most informative attribute is that making the decision on the most informative attribute can reduce the uncertainty about the ultimate output value to the highest extent.

In information theory, the uncertainty metric of a data set S is defined by entropy $H(S)$,

$$H(S) = -\Sigma_{i=1}^{c} P_i log_2(P_i),$$

where, S is a set of data, c is the number of categories in S, and P_i is the proportion of category i in S. The uncertainty about a data set S when the value of a particular attribute A is known is given by the conditional entropy $H(S|A)$,

$$H(S|A) = -\Sigma_{v \in Value(A)} P(A=v) H(S|A=v),$$

where $Value(A)$ is the set of all possible values for attribute A, and $P(A=v)$ is the proportion of data in S, whose attribute A has the value v. If we use S_v to represent the subset of S for which attribute A has the value v, the conditional entropy $H(S|A)$ can be rewritten as,

$$H(S|A) = -\Sigma_{v \in Value(A)} \frac{|S_v|}{|S|} H(S_v),$$

where $|\bullet|$ is the number of data points in the respective data set. As a result, the information gain of knowing the value of attribute A is defined as,

$$Gain(S,A) = H(S) - \Sigma_{v \in Value(A)} \frac{|S_v|}{|S|} H(S_v),$$

So, the most informative attribute A_{mi} is determined by,

$$A_{mi} = \arg\max_{for\ all\ As} Gain(S, A).$$

To improve the performance, a popular training algorithm called adaptive boosting or AdaBoost was used. The AdaBoost algorithm [37] [38] is an interactive learning process which combines the outputs of a set of N "weak" classifiers trained with different weightings of the data in order to create a "strong" composite classifier. A "weak" learning algorithm can generate a hypothesis that is slightly better than random for any data distribution. A "strong" learning algorithm can generate a hypothesis with an arbitrarily low error rate, given sufficient training data. In each successive round of weak classifier training, greater weight is placed on those data points that were mis-classified in the previous round. After completion of N rounds, the N weak classifiers are combined in a weighted sum to form the final strong classifier.

Freund and Shapire have proven that if each weak classifier performs slightly better than a random classifier, then the training error will decrease exponentially with N. In addition, they showed that the test set or generalization error is bounded with high probability by the training set error plus a term that is proportional to the square root of N/M, where M is the number of training data. These results show that, initially, AdaBoost will improve generalization performance as N is increased, due to the fact that the training set error decreases more than the increase of the N/M term. However, if N is increased too much, the N/M term increases faster than the decrease of the training set error. That is when overfitting occurs and the reduction in generalization performance will follow. The optimum value of N and the maximum performance can be increased by using more training data. AdaBoost has been validated in a large number of classification applications. See5 incorporates AdaBoost as a training option. We utilized boosting with $N = 10$ to obtain our DWE prediction results. Larger values of N did not improve performance significantly.

6 Experimental Results

The researchers from the University of Illinois at Urbana-Champaign reported the effect of secondary tasks on driver's behavior in terms of the following features,

- portion of gaze time in different regions of driver's front view;
- mean pupil size;
- mean and standard deviation of lane position;
- mean and standard deviation of the vehicle speed.

The significance of the effect was based on the analysis of variance (ANOVA) [39] with respect to each of these features individually. The general conclusion was that the effect of secondary tasks on some features was significant, such as speed deviation and lane position. However, there was an interaction effect of driving environments and tasks on these features, which means the significance of the task effect was not consistent over different driving environments [40].

It should be noted that ANOVA assumes Gaussian statistics and does not take into account the possible multi-modal nature of the feature probability distributions. Even for those features showing significant difference with respect to secondary tasks, a serious drawback of ANOVA is that this analysis only tells the significance on average. We can not tell from this analysis how robust an estimator can be if we use the features on a moment-by-moment basis.

In the study presented in this chapter, we followed two strategies when conducting the learning process, namely driver-independent and driver-dependent. In the first strategy, we built models over all of the available data. Depending on how the data were allocated to the training and testing sets, we performed training experiments for two different objectives: subject-level and segment-level training, the details of which are presented in the following subsection. In the second training strategy, we treated individual subjects' data separately. That is, we used part of one subject's data for training and tested the learned estimator on the rest data of the same subject. This is a driver-dependent case. The consideration here is that since the workload level is driver-sensitive, individual difference makes a difference in the estimation performance.

The performance of the estimator was assessed with the cross-validation scheme, which is widely adopted by the machine learning community. Specifically, we divided all the data into subsets, called *folds*, of equal sizes. All the folds except one were used to train the estimator while the left-out fold was used for performance evaluation. Given a data from the left-out fold, if the estimation of the learned decision tree was the same as the label of the data, we counted it as a success. Otherwise, it was an error. The correct estimation rate, r_c, was given by,

$$r_c = \frac{n_c}{n_{tot}},$$

where n_c was the total number of successes and n_{tot} was the total number of data entries. This process rotated through each fold and the average performance on the left-out folds was recorded. A cross validation process involves ten folds (ten subsets) is called a *ten-fold cross validation*. Since the estimator is always evaluated on the data disjoint from the training data, the performance evaluated through the cross validation scheme correctly reflects the actual generalization capability of the derived estimator.

6.1 Driver-Independent Training

The structure of the dataset is illustrated schematically in the upper part of Fig. 8. The data can be organized into a hierarchy where individual subjects are at the top. Each subject experienced eight critical periods with single or dual tasks under urban, highway, and rural driving scenarios. Each critical period can be divided into short time windows or segments to compute the vectors of features. For clarity we do not show all subjects, scenarios, and critical periods in Fig. 8.

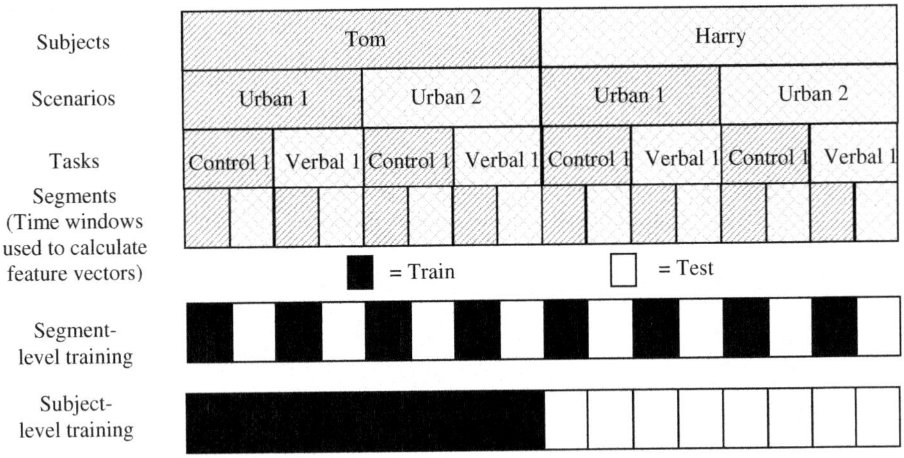

Fig. 8. Allocation of time windows or segments to training and testing sets for two training objectives

Table 3. The test set confusion table for the *subject-level* driver-independent training with time window size of 30s. In the table, dual-task refers to the cases when the subjects were engaged in both primary and secondary tasks and, therefore, bore high workload. Similarly, single-task refers to the cases when the subjects were engaged only in primary driving task and bore low workload.

Workload estimation	Dual-task	Single-task	Total
High workload	186	93	-
Low workload	84	172	-
Correct estimation rate (%)	69	65	67

In the subject-level training, all of the segments for one subject were allocated to either the training or testing set. The data for any individual subject did not appear in both the training and testing sets. The subject-level training was used for estimating the workload of a subject never seen before by the system. It is the most challenging workload estimation problem. In the segment-level training, segments from each critical period were allocated disjointly to both the training and testing sets. This learning problem corresponds to estimating workload for individuals who are available to train the system beforehand. It is an easier problem than the subject-level training.

The test set confusion table for the subject-level training with 30-second time window is shown in Table 3. We were able to achieve an overall correct estimation rate of 67% for new subjects that were not in the training set using segments equal in length to the critical periods (30s). The rules converted from the learned decision tree are shown in Fig. 9. Reducing the segment length increased the

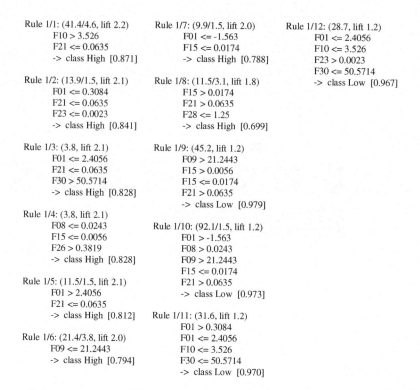

Fig. 9. The rules converted from the driver-independent decision tree. Each rule is characterized by the statistics (N/E, lift L), where N is the number of training cases covered by the rule, E (if shown) is the number of them that do not belong to the rule's class, and L is the estimated accuracy of the rule (the number in square brackets, e.g. [0.871]) divided by the prior probability of the rule's class. The reason that N and E may be non-integral numbers is that when the value of an attribute in the tree is not known, See5 splits the case and sends a fraction down each branch. Please refer to Table 2 for the attribute indices. For example, F01 refers to feature 1 in Table 2.

Table 4. The correct estimation rates in the *subject-level* driver-independent training with different time window sizes

Time window size (s)	1.9	7.5	30
Correct estimation rate (%)	61	63	67

number of feature vectors in the training set but the variance was also increased due to the shorter averaging period. The overall effect of reducing the segment length on subject-level training was a degradation in performance (See Table 4).

The test set confusion table for segment-level training is shown in Table 5 for segments that were equal to the critical period (30s). In this case, the data

Table 5. The test set confusion table for the *segment-level* driver-independent training with time window size of 30s. In the table, dual-task refers to the cases when the subjects were engaged in both primary and secondary tasks and, therefore, bore high workload. Similarly, single-task refers to the cases when the subjects were engaged only in primary driving task and bore low workload.

Workload estimation	Dual-task	Single-task	Total
High workload	219	50	-
Low workload	51	215	-
Correct estimation rate (%)	81	81	81

Table 6. The correct estimation rates in the *segment-level* driver-independent training with different time window sizes

Time window size (s)	1.9	7.5	30
Correct estimation rate (%)	71	78	81

from all subjects was disjointly distributed in either the training or testing sets. Compared to the results for subject-level training, the correct estimation rate was improved from 67% to 81%. Similar to the case of subject-level training, the overall effect of reducing the segment length on segment-level training was a degradation in performance (See Table 6).

Note that, considering the significant reduction of segment length, the degradation of estimation performance shown in both Table 4 and Table 6 seems to be quite acceptable.

6.2 Driver-Dependent Training

The cognitive capacity required to perform the same tasks varies from person to person [2]. So does the workload level for different drivers. This may violate one of the assumptions for any learning-based estimators, i.e., the distribution of the training data is the same as the distribution of the testing data. As a result, the workload estimator obtained from the data of some drivers may not yield good estimation performance when applied to other drivers' data. The driver-dependent training strategy is adopted to evaluate and address this issue.

In driver-dependent training strategy, the data from a single subject was divided into disjoint training and test sets for ten-fold cross validation. The ten correct estimation rates for each subject was averaged to represent the estimation performance over that subject's data. The standard deviation of the ten correct estimation rates was also calculated to evaluate the robustness of the estimation performance. Since there was only limited data from each subject, we used relatively short time-window size in order to make sure there was enough data to train the decision tree. Table 7 shows the performance of the learned

Table 7. The correct estimation rates of driver-dependent training with 0.5 second time window under ten-fold cross validation

Subjects	1	2	3	4	5	6	7	8	9	10	11	12	Average
Correct estimation rate (%)	85.7	80.6	86.4	81.8	77.8	82.4	88.0	87.9	81.3	93.4	90.8	88.4	85.4
Standard deviation (%)	0.7	0.8	0.6	0.8	0.9	0.9	0.7	0.7	0.8	0.4	0.5	0.5	0.69

Table 8. The average correct estimation rates of driver-dependent training with different time window sizes under ten-fold cross-validation

Time window size (s)	0.5	2	5
Correct estimation rate (%)	85.4	80.1	78.2

estimator for each of the twelve subjects when the size of the time window was 0.5 second. The highest correct estimation rate reached 93.4% and the average performance was 85.4%. Because of the limited amount of data, we were only able to label the data with a temporal resolution of 30 seconds (the duration of a critical period). This labeling strategy failed to reflect the possible workload fluctuation within each critical period. As a result, it introduced errors to both training and evaluation. In addition, the level of fluctuation differed for different subjects, which is probably the major reason for the performance variance among subjects. However, given that the standard deviation of the performance for each subject is under one percentage point, the estimator design is highly robust.

We tried different window sizes for the feature vector computation and the average correct estimation rates are listed in Table 8. As the windows size got larger, the amount of training data was reduced, which contributed to the degraded performance of the estimator.

6.3 Feature Combination

To understand the contribution of different subsets of the features to the estimation performance, we trained See 5 decision trees using various combinations of the features with the segment-level training. The time window size was set to be 30 seconds. The performance for various groupings of features is shown in Table 9. The best performance we were able to obtain was 81% using all of features. The eyegaze-related features were more predictive than the driving-performance features since removing all other features (the driving-performance features) from the training set reduced performance by only 1 percentage point from 81% to 80%. Removing the eyegaze-related features (feature 9-34), on the other hand, reduced performance from 81% to 60%. All of the driving-performance and eyegaze-related features, however, contributed to workload detection accuracy, albeit in varying degrees.

Table 9. The correct estimation rates in the 30-second segment-level driver independent training with various feature combinations. Please refer to Table 2 for the feature indices.

Feature Combination	Number of features	Correct estimation Rate (%)
All features (F01-F34)	34	81
Eyegaze-related features (F09-F34)	26	80
All but pupil-diameter features (All but F09-F10)	32	70
Pupil-diameter features only (F09-F10)	2	61
Driving-performance features (F01-F08)	8	60

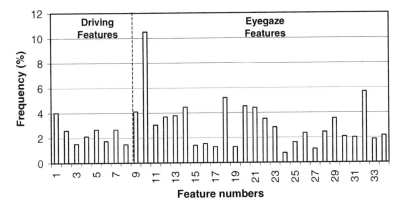

Fig. 10. Histogram of feature frequency in the workload estimator trained by See5. Please refer to Table 2 for the features.

From past experience we have found that there is a high correlation between feature frequency in a decision tree and the predictive power of the feature. We did an analysis on the learned See5 trees by counting how many times each of the feature appeared. A histogram of feature usage in the learned rule sets is shown in Fig. 10. The feature with the highest frequency is the standard deviation of pupil diameter, which is consistent with the known correlation between pupil changes and workload [41], especially in the controlled illumination conditions of the driving simulator.

Among the driving-performance features, vehicle speed (F1) and speed deviation (F2) have the highest frequency of occurrence. The selection of F2 by the learning algorithm is consistent with the ANOVA results done by the University of Illinois at Urbana-Champaign. The ANOVA analysis did not include vehicle speed since the vehicle speed was largely determined by the driving scenarios such as the speed limit of the road and was not considered a good predictor of workload. However, it is understandable that a driver under high workload would tend to have higher speed deviation when the vehicle speed is high, compared to the case when the vehicle speed is low. In other words, under high workload, speed deviation correlates with vehicle speed. This correlation together with the correlation between speed deviation and workload may have made the decision-tree algorithm to pick vehicle speed relatively frequently in order to determine the workload.

It is noteworthy, however, that all of the features are used at least to some extent by the workload estimator, which indicates that all of the feature have some predictive power. The best estimation performance should be obtained when all of the features are used together.

7 Conclusions and Future Work

Compared to the existing practice, the proposed learning-based DWE does not require sophisticated domain knowledge or labor-intensive manual analysis and modeling efforts. With the help of mature machine learning techniques, the proposed method provides an efficient way to derive models over a large number of sensory inputs, which is critical for a problem as complicated as cognitive workload estimation.

The preliminary experimental results of learning-based DWE show that a driver's cognitive status could be estimated with an average correct rate of more than 85% for driver-dependent training. Although this performance is still a long way from a practical DWE system, it is very encouraging especially considering that the estimation was conducted at a rate of twice a second. As we anticipated, the estimation performance was not as good in the driver-independent case as in the driver-dependent case, which restates the importance that a DWE system should capture individual difference.

Recall that our labeling strategy was that all the sensor inputs in the dual-task critical periods had high workload label and those in the control critical periods were labeled with low workload. Within each critical period, the driver might switch their attention between the driving and secondary tasks and, thus, change their actual workload level, which may have contributed to the error. In addition, in the conducted experiments, we used a general machine learning package. Usually a customized algorithm has a better performance than a general-purpose one. It will be interesting to see how much improvement we can achieve with a fine tuned learning algorithm dedicated to our specific problem.

The significance of learning-based DWE is not limited to quickly getting good estimation results. With the manual analysis and modeling methodology, a DWE system works in a static mode. That is, the rules for workload estimation and the thresholds specified in the rules are usually determined through certain human factors studies before the vehicles are sold to the customers. Whether the induced rules represent the customer population very much depends on how the sample subjects are selected and the sample size of the studies. It is also questionable whether one set of rules fit customers with different genders, different ages, different driving experiences, different education background, and etc. Ideally, the rules for DWE should be tailored to each individual driver in order to capture individual difference.

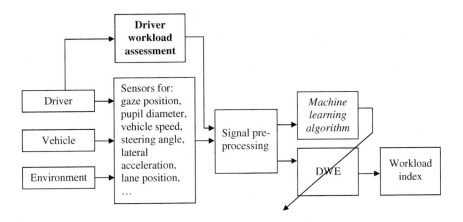

Fig. 11. A possible implementation of adaptive DWE

On the other hand, the learning-based method provides a possibility to automate the adaptation process. Fig. 11 illustrates one possible implementation of adaptive DWE. Instead of collecting the training data in a pre-designed experiments, adaptive DWE collects data when the end customer is driving the vehicle. Compared to Fig. 4, Fig. 11 replaces the Subjective/Secondary Measure module with a Driver Workload Assessment module. The Driver Workload Assessment module allows the driver, while driving, to occasionally submit the subjective assessment of workload level (e.g. 1, 2, ...10, or high, medium, low) through an driver-vehicle interaction mechanism, such as a push button or voice user interface. Such action triggers the collection of a running cache of streaming sensor data. With both the sensory readings and the subjective assessment (the label of workload level) in place, the machine-learning algorithm module can update the DWE module without manual interference, using pre-specified learning mechanisms, e.g. decision tree learning with reducing entropy as the learning criterion.

Having stated the advantages that the learning-based method may deliver, it is important to understand that not all problems are solved. For example, there is still a bottleneck in the learning-based DWE design process, i.e., the subject/secondary task measures module that provides the training data with the labels. We need to address the automatic label generation issue in order to reach a full automatic DWE design process.

Acknowledgments

The authors would like to thank X.S. Zheng, G. McConkie, and Y.-C Tai for collecting the data under a contract with GM, which made this research possible.

References

1. O'Donnel, R.D., Eggemeier, F.T.: Workload assessment methodology. In: Boff, K.R., Kaufman, L., Thomas, J.P. (eds.) Handbook of Perception and Human Performance, Cognitive Processes and Performance, vol. II. Wiley, New York (1986)
2. Wickens, C.D., Hollands, J.G.: Engineering Psychology and Human Performance, 3rd edn. Prentice-Hall Inc., Upper Saddle River (2000)
3. Hart, S.G., Seaveland, L.E.: Development of nasa-tls (task load index): results of emperical and theoretical research. In: Hancock, P.A., Meshkati, N. (eds.) Human Mental Workload. North Holland, Amsterdam (1988)
4. Reid, G.B., Nygren, T.E.: The subjective workload assessment technique: a scaling procedure for measuring mental workload. In: Hancock, P.A., Meshkati, N. (eds.) Human Mental Workload. North Holland, Amsterdam (1988)
5. Angell, L.S., Young, R.A., Hankey, J.M., Dingus, T.A.: An evaluation of alternative methods for assessing driver workload in the early development of in-vehicle information systems. In: Proceedings of SAE Government/Industry Meeting, Washington, DC (May 2002)
6. Hurwitz, J., Wheatley, D.J.: Using driver performance measures to estimate workload. In: Proceedings of the Human Factors and Ergonomics Society 46th Annual Meeting, Santa Monica, CA, pp. 1804–1808 (2002)
7. Mayser, C., Piechulla, W., Weiss, K.-E., König, W.: Driver workload monitoring. In: Proceedings of the Internationale Ergonomie-Konferenz der GfA, ISOES und FEES, München, Germany, May 7-9 (2003)
8. Nakayama, O., Futami, T., Nakamura, T., Boer, E.R.: Development of a steering entropy method for evaluating driver workload. In: Proceedings of SAE International Congress and Exposition Detroit, Detroit, MI (March 1999)
9. Wickens, C.D., Helleberg, J., Goh, J., Xu, X., Horrey, B.: Pilot task management: testing an attentional expected value model of visual scanning. Tech. Rep. ARL-01-14/NASA-01-7, University of Illinois, Aviation Research Lab, Savoy, IL (2001)
10. Card, S.K., Moran, T.P., Newell, A.: The model human processor: an engineering model of human performance. In: Handbook of Perception and Human Performance, pp. 1–35. John Wiley and Sons, New York (1986)
11. Kieras, D.E., Meyer, D.E.: An overview of the epic architecture for cognition and performance with application to human-computer interaction. Human-Computer Interaction 12, 391–438 (1997)
12. Anderson, J.R., Lebiere, C.: The Atomic Components of Thought. Erlbaum, Mahwah (1998)
13. Recarte, M.A., Nunes, L.M.: Effects of verbal and spatial-imagery task on eye fixations while driving. Journal of Experimental Psychology: Applied 6(1), 31–43 (2000)
14. Harbluk, J.L., Noy, Y.I.: The impact of cognitive distraction on driver visual behaviour and vehicle control. Tech. Rep., Transport Canada, (February 2002), http://www.tc.gc.ca/roadsafety/tp/tp13889/en/menu.htm
15. Rahimi, M., Briggs, R.P., Thom, D.R.: A field evaluation of driver eye and head movement strategies toward environmental targets and distracters. Applied Ergonomiecs 21(4), 267–274 (1990)
16. Chapman, P.R., Underwood, G.: Visual search of dynamic scenes: event types and the role of experience in viewing driving situations. In: Underwood, G. (ed.) Eye Guidance in Reading and Scence Perception, pp. 369–393. Elsevier, Amsterdam (1998)

[17] Zhang, Y., Owechko, Y., Zhang, J.: Machine learning-based driver workload estimation. In: Proc. of the 7th International Symposium on Advanced Vehicle Control, Arnhem, Netherlands, August 23-27 (2004)
[18] Zhang, Y., Owechko, Y., Zhang, J.: Driver cognitive workload estimation: a data-driven perspective. In: Proc. of 7th International IEEE Conference on Intelligent Transportation Systems, Washington D.C., October 3-6 (2004)
[19] Mitchell, T.M.: Machine Learning. WCB/McGraw-Hill, Boston (1997)
[20] Bhagat, P.: Pattern Recognition in Industry. Elsevier Science, Oxford (2005)
[21] Kraiss, K.-F.: Implementation of user-adaptive assistants with neural operator models. Control Engineering Practice 3(2), 249–256 (1995)
[22] Forbes, J., Huang, T., Kanazawa, K., Russell, S.: Batmobile: Towards a bayesian automated taxi. In: Proc. Int'l Joint Conf. on Artificial Intelligence, Montreal, Canada, pp. 1878–1885 (1995)
[23] Pentland, A.P., Liu, A.: Modeling and prediction of human. Behavior Neural Computation 11(2) (1999)
[24] Oliver, N., Pentland, A.P.: A graphical models for driver behavior recognition in a smartcar. In: Proc. of Intelligent Vehicles, Detroit, Michigan (October 2000)
[25] Mitrovic, D.: Reliable method for driving events recognition. IEEE Transactions on Intelligent Transportation Systems 6(2), 198–205 (2005)
[26] Simmons, R., Browning, B., Zhang, Y., Sadekar, V.: Learning to predict driver route and destination intent. In: Proc. of the 9th International IEEE Conference on Intelligent Transportation Systems, Toronto, Canada, September 17-20 (2006)
[27] Rakotonirainy, A., Tay, R.: Driver inattention detection through intelligent analysis of readily available sensors. In: Proceedings of The 7th International IEEE Conference on Intelligent Transportation Systems, Washingtong D.C., October 3-6 (2004)
[28] Rakotonirainy, A., Tay, R.: In-vehicle ambient intelligent transport systems (i-vaits): towards an integrated research. In: Proceedings of The 7th International IEEE Conference on Intelligent Transportation Systems, Washingtong D.C., October 3-6 (2004)
[29] Brooks, C., Rakotonirainy1, A., Maire, F.: Reducing driver distraction through software. In: Proceedings Australasian Road Safety Research Policing Education Conference, Wellington, New Zealand (November 2005)
[30] Bergasa, L.M., Nuevo, J., Sotelo, M.A., Barea, R., Lopez, E.: Visual Monitoring of Driver Inattention. In: Prokhorov, D. (ed.) Comput. Intel. in Automotive Applications, ch.2, vol. 132. Springer, Heidelberg (2009)
[31] Liang, Y., Reyes, M.L., Lee, J.D.: Real-time detection of driver cognitive distraction using support vector machines. IEEE Transaction On Intelligent Transaction Systems 8(2), 340–350 (2007)
[32] Applied Science Lab Inc., Eye Tracking System Instruction Manual - Model 504, Applied Science Group Inc., Bedford, MA (2001)
[33] Wickens, C.D., Gordon, S.E., Liu, Y.: An Introduction to Human Factors Engineering. Addison-Wesley, New York (1998)
[34] Recarte, M.A., Nunes, L.M.: Mental workload while driving: effects on visual search, discrimination, and decision making. Journal of Experimental Psychology: Applied 9(2), 119–137 (2003)
[35] Quinlan, J.: C4.5: Programs for Machine Learning. Morgan Kaufmann, San Mateo (1993)

[36] Chapman, D., Kaelbling, L.P.: Input generalization in delayed reinforcement learning: An algorithm and performance comparisons. In: Proceedings of the Twelfth International Joint Conference on Artificial Intelligence (IJCAI 1991), Sydney, Australia, pp. 726–731 (August 1991)
[37] Freund, Y., Shapire, R.E.: A decision-theoretic generalization of on-line learning and an application to boosting. Journal of Computer and System Sciences 55(1), 119–139 (1997)
[38] Freund, Y., Shapire, R.E.: A short introduction to boosting. Journal of Japanese Society for Artificial Intelligence 14(5), 771–780 (1999)
[39] Montgomery, D.C., Runger, G.C.: Applied statistics and probability for engineers, 4th edn. Wiley, Chichester (2006)
[40] Zheng, X.S., McConkie, G.W., Tai, Y.-C.: The effect of secondary tasks on drivers' scanning behavior. In: Proceedings of the 47nd Annual Meeting of the Human Factors and Ergonomics Society, pp. 1900–1903 (October 2003)
[41] Kahneman, D.: Attention and Effort. Prentice-Hall, Englewood Cliffs (1973)

Visual Monitoring of Driver Inattention

Luis M. Bergasa, Jesús Nuevo, Miguel A. Sotelo,
Rafael Barea, and Elena Lopez

Department of Electronics, University of Alcala, CAMPUS,
28805 Alcalá de Henares (Madrid) Spain
Tel.: +34-918856569; Fax: +34-918856591
{bergasa,jnuevo,sotelo,barea,elena}@depeca.uah.es

1 Introduction

The increasing number of traffic accidents due to driver inattention has become a serious problem for society. Every year, about 45,000 people die and 1.5 million people are injured in traffic accidents in Europe. These figures imply that one person out of every 200 European citizens is injured in a traffic accident every year and that around one out 80 European citizens dies 40 years short of the life expectancy. It is known that the great majority of road accidents (about 90-95%) are caused by human error. More recent data has identified inattention (including distraction and falling asleep at the wheel) as the primary cause of accidents, accounting for at least 25% of the crashes [15]. Road safety is thus a major European health problem. In the "White Paper on European Transport Policy for 2010", the European Commission declares the ambitious objective of reducing by 50% the number of fatal accidents on European roads by 2010 (European Commission, 2001).

According to the U.S. National Highway Traffic Safety Administration (NHTSA), falling asleep while driving is responsible for at least 100,000 automobile crashes annually. An annual average of roughly 70,000 nonfatal injuries and 1,550 fatalities results from these crashes [32; 33]. These figures only cover crashes happening between midnight and 6 am, involving a single vehicle and a sober driver traveling alone, including the car departing from the roadway without any attempt to avoid the crash. These figures underestimate the true level of the involvement of drowsiness because they do not include crashes at daytime hours involving multiple vehicles, alcohol, passengers or evasive maneuvers. These statistics do not deal with crashes caused by driver distraction either, which is believed to be a larger problem. Between 13% and 50% of crashes are attributed to distraction, resulting in as many as 5000 fatalities per year. Increasing use of in-vehicle information systems (IVISs) such as cell phones, GPS navigation systems, satellite radios and DVDs has exacerbated the problem by introducing additional sources of distraction. That is, the more IVISs the more sources of distraction from the most basic task at hand, i.e. driving the vehicle. Enabling drivers to benefit from IVISs without diminishing safety is an important challenge.

This chapter presents an original system for monitoring driver inattention and alerting the driver when he is not paying adequate attention to the road in order to prevent accidents. According to [40] the driver inattention status can be divided into two main categories: distraction detection and identifying sleepiness. Likewise, distraction can be divided in two main types: visual and cognitive. Visual distraction is straightforward, occurring when drivers look away from the roadway (e.g., to adjust a radio). Cognitive distraction occurs when drivers think about something not directly related to the current vehicle control task (e.g., conversing on a hands-free cell phone or route planning). Cognitive distraction impairs the ability of drivers to detect targets across the entire visual scene and causes gaze to be concentrated in the center of the driving scene. This work is focused in the sleepiness category. However, sleepiness and cognitive distraction partially overlap since the context awareness of the driver is related to both, which represent mental occurrences in humans [26].

The rest of the chapter is structured as follows. In section 2 we present a review of the main previous work in this direction. Section 3 describes the general system architecture, explaining its main parts. Experimental results are shown in section 4. Section 5 discusses weaknesses and improvements of the system. Finally, in section 6 we present the conclusions and future work.

2 Previous Work

In the last few years many researchers have been working on systems for driver inattention detection using different techniques. The most accurate techniques are based on physiological measures like brain waves, heart rate, pulse rate, respiration, etc. These techniques are intrusive, since they need to attach some electrodes on the drivers, causing annoyance to them. Daimler-Chrysler has developed a driver alertness system, Distronic [12], which evaluates the EEG (electroencephalographic) patterns of the driver under stress. In ASV (Advanced Safety Vehicle) project by Toyota (see in [24]), the driver must wear a wristband in order to measure his heart rate. Others techniques monitor eyes and gaze movements using a helmet or special contact lens [3]. These techniques, though less intrusive, are still not acceptable in practice.

A driver's state of vigilance can also be characterized by indirect vehicle behaviors like lateral position, steering wheel movements, and time to line crossing. Although these techniques are not intrusive they are subject to several limitations such as vehicle type, driver experience, geometric characteristics, state of the road, etc. On the other hand, these procedures require a considerable amount of time to analyze user behaviors and thereby they do not work with the so-called micro-sleeps: when a drowsy driver falls asleep for some seconds on a very straight road section without changing the lateral position of the vehicle [21]. To this end we can find different experimental prototypes, but at this moment none of them has been commercialized. Toyota uses steering wheel sensors (steering wheel variability) and pulse sensor to record the heart rate, as mentioned above [24]. Mitsubishi has reported the use of steering wheel sensors and measures of vehicle behavior (such as lateral position of the car) to detect driver drowsiness in their advanced safety vehicle system [24]. Daimler Chrysler has developed a system based on vehicle speed, steering angle and vehicle position relative to road delimitation (recorded by a camera) to detect if the vehicle is about to leave the road [11]. Volvo Cars recently announced its Driver Alert Control system [39], that will be available on its high-end models from 2008. This system uses a camera, a number of sensors and a central unit to monitor the movements of the car within the road lane and to assess whether the driver is drowsy.

People in fatigue show some visual behaviors easily observable from changes in their facial features like eyes, head, and face. Typical visual characteristics observable from the images of a person with reduced alertness level include longer blink duration, slow eyelid movement, smaller degree of eye opening (or even closed), frequent nodding, yawning, gaze (narrowness in the line of sight), sluggish facial expression, and drooping posture. Computer vision can be a natural and non-intrusive technique for detecting visual characteristics that typically characterize a driver's vigilance from the images taken by a camera placed in front of the user. Many researches have been reported in the literature on developing image-based driver alertness using computer vision techniques. Some of them are primarily focused on head and eye tracking techniques using two cameras [27], [38]. In [34] a system called FaceLAB developed by the company Seeing Machines is presented. The 3D pose of the head and the eye-gaze direction are calculated accurately. FaceLAB also monitors the eyelids, to determine eye opening and blink rates. With this information the system estimates the driver's fatigue level. According to FaceLab information, the system operates day and night but at night the performance of the system decreases. All systems explained above rely on manual initialization of feature points. The systems appear to be robust but the manual initialization is a limitation, although it simplifies the whole problem of tracking and pose estimation.

There are other proposals that use only a camera. In [6] we can find a 2D pupil monocular tracking system based on the differences in color and reflectivity between the pupil and iris. The system monitors driving vigilance by studying the eyelid movement. Another successful system of head/eye monitoring and tracking for drowsiness detection using one camera, which is based on color predicates, is presented in [37]. This system is based on passive vision techniques and its functioning can be problematical in poor or very bright lighting conditions. Moreover, it does not work at night, when the monitoring is more important.

In order to work at nights some researches use active illumination based on infrared LED. In [36] a system using 3D vision techniques to estimate and track the 3D line of sight of a person using multiple cameras is proposed. The method relies on a simplified eye model, and it uses the Purkinje images of an infrared light source to determine eye location. With this information, the gaze direction is estimated. Nothing about monitoring driver vigilance is presented. In [23] a system with active infrared LED illumination and a camera is implemented. Because of the LED illumination, the method can easily find the eyes and based on them, the system locates the rest of the facial features. They propose to analytically estimate the local gaze direction based on pupil location. They calculate eyelid movement and face orientation to estimate driver fatigue. Almost all the active systems reported in the literature have been tested in simulated environments but not in real moving vehicles. A moving vehicle presents new challenges like variable lighting, changing background and vibrations that must be taken into account in real systems. In [19] an industrial prototype called *Copilot* is presented. This system uses infrared LED illumination to find the eyes and it has been tested with truck's drivers in real environments. It uses a simple subtraction process to find the eyes and it only calculates a validated parameter called PERCLOS (percent eye closure), in order to measure driver's drowsiness. This system currently works under low light conditions. Recently, Seeing Machines has presented a commercial system called DSS (Driver State Sensor) [35] for driver fatigue detection in transportation operation. The system utilizes a camera and LEDs for night working. It calculates PERCLOS and obtains a 93% correlation with drowsiness.

Systems relying on a single visual cue may encounter difficulties when the required visual features cannot be acquired accurately or reliably, as happens in real conditions. Furthermore, a single visual cue may not always be indicative of the overall mental condition [23]. The use of multiple visual cues reduces the uncertainty and the ambiguity present in the information from a single source. The most recent research in this direction use this hypothesis. The European project AWAKE (2001-2004) [22] proposes a multi-sensor system adapted to the driver, the vehicle, and the environment in an integrated way. This system merges, via an artificial intelligent algorithm, data from on-board driver monitoring sensors (such as an eyelid camera and a steering grip sensor) as well as driver behavior data (i.e. from lane tracking sensor, gas/brake and steering wheel positioning). The system must be personalized for each driver during a learning phase. Another European project, SENSATION(2004-2007) [16] is been currently founded to continue research of the AWAKE project in order to obtain a commercial system. The European project AIDE -Adaptive Integrated Driver-vehicle InterfacE (2004-2008) [15] works in this direction as well.

This chapter describes a real-time prototype system based on computer vision for monitoring driver vigilance using active infrared illumination and a single camera placed on the car dashboard. We have employed this technique because our goal is to monitor a driver in real conditions (vehicle moving) and in a very robust and accurate way mainly at nights (when the probability to crash due to drowsiness is the highest). The proposed system does not need manual initialization and monitors several visual behaviors that typically characterize a person's level of alertness while driving. In a different fashion than other previous works, we have fused different visual cues from one camera using a fuzzy classifier instead of different cues from different sensors. We have analyzed different visual behaviors that characterize a drowsy driver and we have studied the best fusion for optimal detection. Moreover, we have tested our system during several hours in a car moving on a motorway and with different users. The basics of this work were presented by the authors in [5].

3 System Architecture

The general architecture of our system is shown in figure 1. It consists of four major modules: 1) Image acquisition, 2) Pupil detection and tracking, 3) Visual behaviors and 4) Driver monitoring. Image acquisition is based on a low-cost CCD micro-camera sensitive to near-IR. The pupil detection and tracking stage is responsible for segmentation and image processing. Pupil detection is simplified by the "bright pupil" effect, similar to the red eye effect in photography. Then, we use two Kalman filters in order to track the pupils robustly in real-time. In the visual behaviors stage we calculate some parameters from the images in order to detect some visual behaviors easily observable in people in fatigue: slow eyelid movement, smaller degree of eye opening, frequent nodding, blink frequency, and face pose. Finally, in the driver monitoring stage we fuse all the individual parameters obtained in the previous stage using a fuzzy system, yielding the driver inattentiveness level. An alarm is activated if this level is over a certain threshold.

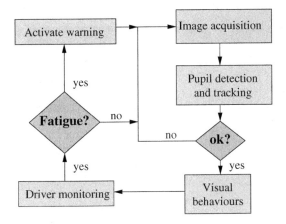

Fig. 1. General architecture

3.1 Image Acquisition System

The purpose of this stage is to acquire the video images of the driver's face. In this application the acquired images should be relatively invariant to light conditions and should facilitate the eye detection and tracking (good performance is necessary). The use of near-IR illuminator to brighten the driver's face serves these goals [25]. First, it minimizes the impact of changes in the ambient light. Second, the near-IR illumination is not detected by the driver, and then, this does not suppose an interference with the user's driving. Third, it produces the bright pupil effect, which constitutes the foundation of our detection and tracking system. A bright pupil is obtained if the eyes are illuminated with an IR illuminator beaming light along the camera optical axis. At the IR wavelength,

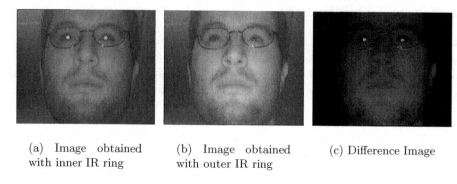

(a) Image obtained with inner IR ring   (b) Image obtained with outer IR ring   (c) Difference Image

Fig. 2. Captured images and their subtraction

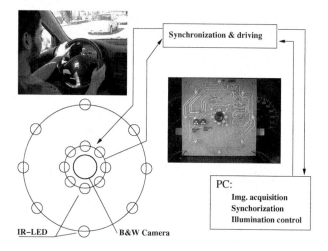

Fig. 3. Block diagram of the prototype

the retina reflects almost all the IR light received along the path back to the camera, and a bright pupil effect will be produced in the image. If illuminated off the camera optical axis, the pupils appear dark since the reflected light of the retina will not enter the camera lens. An example of the bright/dark pupil effect can be seen in figure 2. This pupil effect is clear with and without glasses, with contact lenses and it even works to some extent with sunglasses.

Figure 3 shows the image acquisition system configuration. It is composed by a miniature CCD camera sensitive to near-IR and located on the dashboard of the vehicle. This camera focuses on the driver's head for detecting the multiple visual behaviors. The IR illuminator is composed by two sets of IR LEDs distributed symmetrically along two concentric and circular rings. An embedded PC with a low cost frame-grabber is used for video signal acquisition and signal processing. Image acquiring from the camera and LED excitation is synchronized. The LED

(a) Out-of-the-road lights effect (b) Vehicle lights effect (c) Sunlight effect (d) Sunlight effect with filter

Fig. 4. Effects of external lights in the acquisition system

rings illuminate the driver's face alternatively, one for each image, providing different lighting conditions for almost the same scene.

Ring sizes has been empirically calculated in order to obtain a dark pupil image if the outer ring is turned on and a bright pupil image if the inner ring is turned on. LEDs in the inner ring are as close as possible to the camera, in order to maximize the "bright pupil" effect. The value of the outer ring radius is a compromise between the resulting illumination, that improves as it is increased, and the available space in the car's dashboard. The symmetric position of the LEDs in the rings, around the camera optical axis, cancels shadows generated by LEDs. The inner ring configuration obtains the bright pupil effect because the center of the ring coincides with the camera optical axis, working as if there were an only LED located on the optical axis of the lens. The outer ring provides ambient illumination that is used for contrast enhancing. In spite of those LEDs producing the dark pupil effect, a glint can be observed on each pupil.

The explained acquisition system works very well under controlled light conditions, but real scenarios present new challenges that must be taken into account. Lighting conditions were one of the most important problems to solve in real tests. As our system is based on the reflection of the light emitted by the IR LEDs, external light sources are the main source of noise. Three main sources can be considered, as are depicted in figure 4: artificial light from elements just outside the road (such as light bulbs), vehicle lights, and sun light. The effect of lights from elements outside the road mainly appears in the lower part of the image (figure 4(a)) because they are situated above the height of the car and the beam enters the car with a considerable angle. Then, this noise can be easily filtered. On the other hand, when driving on a double direction road, vehicle lights directly illuminate the driver, increasing the pixels level quickly and causing the pupil effect to disappear (figure 4(b)). Once the car has passed, the light level reduces very fast. Only after a few frames, the AGC (Automatic Gain Controller) integrated in the camera compensates the changes, so very light and dark images are obtained, affecting the performance of the inner illumination system.

Regarding the sun light, it only affects at day time but its effect changes as function of the weather (sunny, cloudy, rainy, etc) and the time of the day. With the exception of the sunset, dawn and cloudy days, sun light hides the inner infrared illumination and then the pupil effect disappears (figure 4(c)). For minimizing interference from light sources beyond the IR light emitted by our LEDs, a narrow band-pass filter, centered at the LED wavelength, has been attached between the CCD and the lens. This filter solved the problem of artificial lights and vehicle light almost completely, but it adds a new drawback for it reduces the intensity of the image, and then the noise is considerably amplified by the AGC. The filter does not eliminate the sun light interference, except for cases when the light intensity is very low. This is caused by the fact that the power emitted by the sun in the band of the filter is able to hide the inner illumination. An image of this case, taken by the sunset, is depicted in figure 4(d). A possible solution for this problem could be the integration of IR filters in the car glasses.

3.2 Pupil Detection and Tracking

This stage starts with pupil detection. As mentioned above, each pair of images contains an image with bright pupil and another one with a dark pupil. The first image is then digitally subtracted from the second to produce the difference image. In this image, pupils appear as the brightest parts in the image as can be seen in figure 2. This method minimizes the ambient light influence by subtracting it in the generation of the difference image. This procedure yields high contrast images where the pupils are easily found. It can be observed that the glint produced by the outer ring of LEDs usually falls close to the pupil, with the same grey level as the bright pupil. The shape of the pupil blob in the difference image is not a perfect ellipse because the glint cuts the blob, affecting the modeling of the pupil blobs and, consequently, the calculation depending on it, as will be explained later. This is the reason why the system only uses subtracted images during initialization, and when light conditions are poor (this initialization time varies depending on the driver and light conditions, but it was below 5 seconds for all test). In other cases, only the image obtained with the inner ring is processed, increasing accuracy and reducing computation time.

Pupils are detected on the resulting image, by searching the entire image to locate two bright blobs that satisfy certain constraints. The image is binarized, using an adaptive threshold, for detecting the brighter blobs in the image.

A standard 8-connected components analysis [18] is then applied to the binarized difference image to identify binary blobs that satisfy certain size and shape constraints. The blobs that are out of some size constraints are removed, and for the others an ellipse model is fit to each one. Depending on their size, intensity, position and distance, best candidates are selected, and all the possible pairs between them are evaluated. The pair with the highest qualification is chosen as the detected pupils, and its centroids are returned as the pupil positions.

One of the main characteristics of this stage is that it is applicable to any user without any supervised initialization. Nevertheless, the reflection of the IR in the pupils under the same conditions varies from one driver to another. Even for the same driver, the intensity depends on the gaze point, head position and the opening of the eye. Apart from those factors, lighting conditions change with

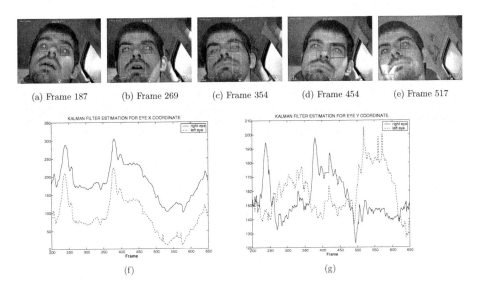

Fig. 5. Tracking results for a sequence

time, which modifies the intensity of the pupils. On the other hand, the size of the pupils also depends on the user, and the distance to the camera. To deal with those differences in order to be generic, our system uses an adaptive threshold in the binarization stage. The parameters of the detected pupils are used to update the statistics that set thresholds and margins in the detection process. Those statistics include size, grey level, position and apparent distance and angle between pupils, calculated over a time window of 2 seconds. The thresholds also get their values modified if the pupils are not found, widening the margins to make more candidates available to the system.

Another question related to illumination that is not usually addressed in the literature is the sensitivity of the eye to IR emission. As the exposure time to the IR source increases, its power has to be reduced in order to avoid damaging the internal tissues of the eye. This imposes a limit on the emission of the near-IR LEDs. To calculate the power of our system, we have followed the recommendations of [1], based on IEC 825-1 and CENELEC 60825-1 infrared norms. With these limitations, no negative effects have been reported in the drivers that collaborated in the tests.

To continuously monitor the driver it is important to track his pupils from frame to frame after locating the eyes in the initial frames. This can be done efficiently by using two Kalman filters, one for each pupil, in order to predict pupil positions in the image. We have used a pupil tracker based on [23] but we have tested it with images obtained from a car moving on a motorway. Kalman filters presented in [23] works reasonably well under frontal face orientation with open eyes. However, it will fail if the pupils are not bright due to oblique face orientations, eye closures, or external illumination interferences. Kalman filter also

Fig. 6. System working with user wearing glasses

fails when a sudden head movement occurs because the assumption of smooth head motion has not been fulfilled. To overcome this limitation we propose a modification consisting on an adaptive search window, which size is determined automatically, based on pupil position, pupil velocity, and location error. This way, if Kalman filtering tracking fails in a frame, the search window progressively increases its size. With this modification, the robustness of the eye tracker is significantly improved, for the eyes can be successfully found under eye closure or oblique face orientation.

The state vector of the filter is represented as $\mathbf{x}_t = (\mathbf{c}_t, \mathbf{r}_t, \mathbf{u}_t, \mathbf{v}_t)$, where $(\mathbf{c}_t, \mathbf{r}_t)$ indicates the pupil pixel position (its centroid) and $(\mathbf{u}_t, \mathbf{v}_t)$ is its velocity at time t in $\mathbf{c}$ and $\mathbf{r}$ directions, respectively. Figure 5 shows an example of the pupil tracker working in a test sequence. Rectangles on the images indicate the search window of the filter, while crosses indicate the locations of the detected pupils. Figures 5(f) and 5(g) draw the estimation of the pupil positions for the sequence under test. The tracker is found to be rather robust for different users without glasses, lighting conditions, face orientations and distances between the camera and the driver. It automatically finds and tracks the pupils even with closed eyes and partially occluded eyes, and can recover from tracking-failures. The system runs at 25 frames per second.

Performance of the tracker gets worse when users wear eyeglasses because different bright blobs appear in the image due to IR reflections in the glasses, as can be seen in figure 6. Although the degree of reflection on the glasses depends on its material and the relative position between the user's head and the illuminator, in the real tests carried out, the reflection of the inner ring of LEDs appears as a filled circle on the glasses, of the same size and intensity as the pupil. The reflection of the outer ring appears as a circumference with bright points around it and with similar intensity to the pupil. Some ideas for improving the tracking with glasses are presented in section 5. The system was also tested with people wearing contact lenses. In this case no differences in the tracking were obtained compared to the drivers not wearing them.

3.3 Visual Behaviors

Eyelid movements and face pose are some of the visual behaviors that reflect a person's level of inattention. There are several ocular measures to

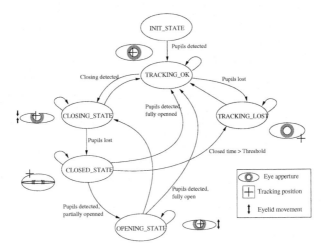

Fig. 7. Finite State Machine for ocular measures

characterize sleepiness such as eye closure duration, blink frequency, fixed gaze, eye closure/opening speed, and the recently developed parameter PERCLOS [14; 41]. This last measure indicates the accumulative eye closure duration over time excluding the time spent on normal eye blinks. It has been found to be the most valid ocular parameter for characterizing driver fatigue [24]. Face pose determination is related to computation of face orientation and position, and detection of head movements. Frequent head tilts indicate the onset of fatigue. Moreover, the nominal face orientation while driving is frontal. If the driver faces in other directions for an extended period of time, it is due to visual distraction. Gaze fixations occur when driver's eyes are nearly stationary. Their fixation position and duration may relate to attention orientation and the amount of information perceived from the fixated location, respectively. This is a characteristic of some fatigue and cognitive distraction behaviors and it can be measured by estimating the fixed gaze. In this work, we have measured all the explained parameters in order to evaluate its performance for the prediction of the driver inattention state, focusing on the fatigue category.

To obtain the ocular measures we continuously track the subject's pupils and fit two ellipses, to each of them, using a modification of the LIN algorithm [17], as implemented in the OpenCV library [7]. The degree of eye opening is characterized by the pupil shape. As eyes close, the pupils start getting occluded by the eyelids and their shapes get more elliptical. So, we can use the ratio of pupil ellipse axes to characterize the degree of eye opening. To obtain a more robust estimation of the ocular measures and, for example, to distinguish between a blink and an error in the tracking of the pupils, we use a Finite State Machine (FSM) as we depict in figure 7. Apart from the *init_state*, five states have been defined: *tracking_ok*, *closing*, *closed*, *opening* and *tracking_lost*. Transitions between states are achieved from frame to frame as a function of the width-height ratio of the pupils.

The system starts at the *init_state*. When the pupils are detected, the FSM passes to the *tracking_ok* state indicating that the pupil's tracking is working correctly. Being in this state, if the pupils are not detected in a frame, a transition to the *tracking_lost* state is produced. The FSM stays in this state until the pupils are correctly detected again. In this moment, the FSM passes to the *tracking_ok* state. If the width-height ratio of the pupil increases above a threshold (20% of the nominal ratio), a closing eye action is detected and the FSM changes to the *closing_state*. Because the width-height ratio may increase due to other reasons, such as segmentation noise, it is possible to return to the *tracking_ok* state if the ratio does not constantly increase.

When the pupil ratio is above the 80% of its nominal size or the pupils are lost, being in *closing_state*, a transition of the FSM to *closed_state* is provoked, which means that the eyes are closed. A new detection of the pupils from the *closed_state* produces a change to *opening_state* or *tracking_ok* state, depending on the degree of opening of the eyelid. If the pupil ratio is between the 20% and the 80% a transition to the *opening_state* is produced, if it is below the 20% the system pass to the *tracking_ok* state. Being in *closed_state*, a transition to the *tracking_lost* state is produced if the closed time goes over a threshold. A transition from opening to closing is possible if the width-height ratio increases again. Being in *opening_state*, if the pupil ratio is below the 20% of the nominal ratio a transition to *tracking_ok* state is produced.

Ocular parameters that characterize eyelid movements have been calculated as a function of the FSM. PERCLOS is calculated from all the states, except from the *tracking_lost* state, analyzing the pupil width-height ratio. We consider that an eye closure occurs when the pupil ratio is above the 80% of its nominal size. Then, the eye closure duration measure is calculated as the time that the system is in the *closed_state*. To obtain a more robust measurement of the PERCLOS, we compute this running average. We compute this parameter by measuring the percentage of eye closure in a 30-second window. Then, PERCLOS measure represents the time percentage that the system is at the *closed_state* evaluated in 30 seconds and excluding the time spent in normal eye blinks. Eye closure/opening speed measures represent the amount of time needed to fully close the eyes or to fully open the eyes. Then, eye closure/opening speed is calculated as the time during which pupil ratio passes from 20% to 80% or from 80% to 20% of the nominal ratio, respectively. In other words, the time that the system is in the *closing_state* or *opening_state*, respectively. Blink frequency measure indicates the number of blinks detected in 30 seconds. A blink action will be detected as a consecutive transition among the following states: closing, closed, and opening, given that this action was carried out in less than a predefined time. Many physiology studies have been carried out on the blinking duration. We have used the recommendation value derived in [31] but this could be easily modified to conform to other recommended value. Respecting the eye nominal size used for the ocular parameters calculation, it varies depending on the driver. To calculate its correct value a histogram of the eyes opening degree for the last 2000 frames not exhibiting drowsiness is obtained. The most frequent value on the histogram is considered to be the nominal size. PERCLOS is computed separately in both eyes and the final value is obtained as the mean of both.

Besides, face pose can be used for detecting fatigue or visual distraction behaviors among the categories defined for inattentive states. The nominal face orientation while driving is frontal. If the driver's face orientation is in other directions for an extended period of time it is due to visual distractions, and if it occurs frequently (in the case of various head tilts), it is a clear symptom of fatigue. In our application, the precise degree of face orientation for detecting this behaviors is not necessary because face poses in both cases are very different from the frontal one. What we are interested in is to detect whether the driver's head deviates too much from its nominal position and orientation for an extended period of time or too frequently (nodding detection).

This work provides a novel solution to the coarse 3D face pose estimation using a single un-calibrated camera, based on the method proposed in [37]. We use a model-based approach for recovering the face pose by establishing the relationship between 3D face model and its two-dimensional (2D) projections. A weak perspective projection is assumed so that face can be approximated as a planar object with facial features, such as eyes, nose and mouth, located symmetrically on the plane. We have performed a robust 2D face tracking based on the pupils and the nostrils detections on the images. Nostrils detection has been carried out in a way similar to that used for the pupils' detection. From these positions the 3D face pose is estimated, and as a function of it, face direction is classified in nine areas, from *upper left* to *lower right*.

This simple technique works fairly well for all the faces we tested, with left and right rotations specifically. A more detailed explanation about our method was presented by the authors in [5]. As the goal is to detect whether the face pose of the driver is not frontal for an extended period of time, this has been computed using only a parameter that gives the percentage of time that the driver has been looking at the front, over a 30-second temporal window.

Nodding is used to quantitatively characterize one's level of fatigue. Several systems have been reported in the literature to calculate this parameter from a precise estimation of the driver's gaze [23][25]. However, these systems have been tested in laboratories but not in real moving vehicles. The noise introduced in real environments makes these systems, based on exhaustive gaze calculation, work improperly. In this work, a new technique based on position and speed data from the Kalman filters used to track the pupils and the FSM is proposed. This parameter measures the number of head tilts detected in the last 2 minutes. We have experimentally observed that when a nodding is taking place, the driver closes his or her eyes and the head goes down to touch the chest or the shoulders. If the driver wakes up in that moment, raising his head, the values of the vertical speed of the Kalman filters will change their sign, as the head rises. If the FSM is in *closed_state* or in *tracking_lost* and the pupils are detected again, the system saves the speeds of the pupils trackers for 10 frames. After that, the data is analyzed to find if it conforms to that of a nodding. If so, the first stored value is saved and used as an indicator of the "magnitude" of the nodding.

Finally, one of the remarkable behaviors that appear in drowsy drivers or cognitively distracted drivers is fixed gaze. A fatigued driver looses the focus of the gaze, not paying attention to any of the elements of the traffic. This loss of concentration is usually correlated with other sleepy behaviors such as a higher blink frequency, a smaller degree of eye opening and nodding. In the case of cognitive distraction, however, fixed gaze is decoupled from other clues. As for the parameters explained above, the existing systems calculate this parameter from a precise estimation of the driver's gaze and, consequently, experience the same problems. In order to develop a method to measure this behavior in a simple and robust way, we present a new technique based on the data from the Kalman filters used to track the pupils.

An attentive driver moves his eyes frequently, focusing to the changing traffic conditions, particularly if the road is busy. This has a clear reflection on the difference between the estimated position from the Kalman filters and the measured ones.

Besides, the movements of the pupils for an inattentive driver present different characteristics. Our system monitors the position on the x coordinate. Coordinate y is not used, as the difference between drowsy and awake driver is not so clear. The fixed gaze parameter is computed locally over a long period of time, allowing for freedom of movement of the pupil over time. We refer here to [5] for further details of the computation of this parameter.

This fixed gaze parameter may suffer from the influence of vehicle vibrations or bumpy roads. Modern cars have reduced vibrations to a point that the effect is legible on the measure. The influence of bumpy roads depends on their particular characteristics. If bumps are occasional, it will only affect few values, making little difference in terms of the overall measure. On the other hand, if bumps are frequent and their magnitude is high enough, the system will probably fail to detect this behavior. Fortunately, the probability for a driver to get distracted or fall asleep is significantly lower in very bumpy roads. The results obtained for all the test sequences with this parameter are encouraging. In spite of using the same a priori threshold for different drivers and situations, the detection was always correct. Even more remarkable was the absence of false positives.

3.4 Driver Monitoring

This section describes the method to determine the driver's visual inattention level from the parameters obtained in the previous section. This process is complicated because several uncertainties may be present. First, fatigue and cognitive distractions are not observable and they can only be inferred from the available information. In fact, this behavior can be regarded as the result of many contextual variables such as environment, health, and sleep history. To effectively monitor it, a system that integrates evidences from multiple sensors is needed. In the present work, several fatigue visual behaviors are subsequently combined to form an inattentiveness parameter that can robustly and accurately characterize one's vigilance level. The fusion of the parameters has been obtained using a fuzzy system. We have chosen this technique for its well known linguistic

Table 1. Fuzzy variables

Variable	Type	Range	Labels	Linguistic terms
PERCLOS	In	[0.0, 1.0]	5	small, medium small, medium, medium large, large
Eye closure duration	In	[1.0 - 30.0]	3	small, medium, large
Blink freq.	In	[1.0 - 30.0]	3	small, medium, large
Nodding freq.	In	[0.0 - 8.0]	3	small, medium, large
Face position	In	[0.0 - 1.0]	5	small, medium small, medium, medium large, large
Fixed gaze	In	[0.0 - 0.5]	5	small, medium small, medium, medium large, large
DIL	Out	[0.0 - 1.0]	5	small, medium small, medium, medium large, large

concept modeling ability. Fuzzy rule expressions are close to expert natural language. Then, a fuzzy system manages uncertain knowledge and infers high level behaviors from the observed data. As an universal approximator, fuzzy inference system can be used for knowledge induction processes. The objective of our fuzzy system is to provide a driver's inattentiveness level (DIL) from the fusion of several ocular and face pose measures, along with the use of expert and induced knowledge. This knowledge has been extracted from the visual observation and the data analysis of the parameters in some simulated fatigue behavior carried out in real conditions (driving a car) with different users. The simulated behaviors have been done according to the physiology study of the US Department of Transportation, presented in [24]. We do not delve into the psychology of driver visual attention, rather we merely demonstrate that with the proposed system, it is possible to collect driver information data and infer whether the driver is attentive or not.

The first step in the expert knowledge extraction process is to define the number and nature of the variables involved in the diagnosis process according to the domain expert experience. The following variables are proposed after appropriate study of our system: PERCLOS, eye closure duration, blink frequency, nodding frequency, fixed gaze and frontal face pose. Eye closing and opening variables are not being used in our input fuzzy set because they mainly depend on factors such as segmentation and correct detection of the eyes, and they take place in the length of time comparable to that of the image acquisition. As a consequence, they are very noisy variables. As our system is adaptive to the user, the ranges of the selected fuzzy inputs are approximately the same for all users. The fuzzy inputs are normalized, and different linguistic terms and its corresponding fuzzy sets are distributed in each of them using induced knowledge based on the hierarchical fuzzy partitioning (HFP) method [20]. Its originality lies in not yielding a single partition, but a hierarchy including partitions with various resolution levels based on automatic clustering data. Analyzing the fuzzy partitions obtained by HFP, we determined that the best suited fuzzy sets and the corresponding linguistic terms for each input variable are those shown in table 2. For the output variable (DIL), the fuzzy set and the linguistic terms were manually chosen. The inattentiveness level range is between 0 and 1, with a normal value up to 0.5. When its value is between 0.5 and 0.75, driver's fatigue is medium, but if the DIL is over 0.75 the driver is considered to be fatigued, and an alarm is activated. Fuzzy sets of triangular shape were chosen, except at the domain edges, where they were semi-trapezoidal.

Based on the above selected variables, experts state different pieces of knowledge (rules) to describe certain situations connecting some symptoms with a certain diagnosis. These rules are of the form *"If condition, Then conclusion"*, where both premise and conclusion use the linguistic terms previously defined, as in the following example:

- **IF** *PERCLOS* is *large* **AND** *Eye Closure Duration* is *large*, **THEN** *DIL* is *large*

In order to improve accuracy and system design, automatic rule generation and its integration in the expert knowledge base were considered. The fuzzy system implementation used the licence-free tool KBCT (Knowledge Base Configuration Tool) [2] developed by the Intelligent Systems Group of the Polytechnics University of Madrid (UPM). A more detailed explanation of this fuzzy system can be found in [5].

4 Experimental Results

The goal of this section is to experimentally demonstrate the validity of our system in order to detect fatigue behaviors in drivers. Firstly, we show some details about the recorded video sequences used for testing, then, we analyze the parameters measured for one of the sequences. Finally, we present the performance of the detection of each one of the parameters, and the overall performance of the system.

4.1 Test Sequences

Ten sequences were recorded in real driving situations over a highway and a two-direction road. Each sequence was obtained for a different user. The images were obtained using the system explained in section 3.1. The drivers simulated some drowsy behaviors according to the physiology study of the US Department of Transportation presented in [24]. Each user drove normally except in one or two intervals where the driver simulated fatigue. Simulating fatigue allows for the system to be tested in a real motorway, with all the sources of noise a deployed system would face. The downside is that there may be differences between an actual drowsy driver and a driver mimicking the standard drowsy behavior, as defined in [24]. We are currently working on testing the system in a truck simulator.

The length of the sequences and the fatigue simulation intervals are shown in table 2. All the sequences were recorded at night except for sequence number 7 that was recorded at day, and sequence number 5 that was recorded at sunset. Sequences were obtained with different drivers not wearing glasses, with the exception of sequence 6, that was recorded for testing the influence of the glasses in real driving conditions.

Table 2. Length of simulated drowsiness sequences

Seq. Num.	Drowsiness Behavior time(sec)	Alertness Behavior time(sec)	Total time(sec)
1	394 (2 intervals:180+214)	516	910
2	90 (1 interval)	210	300
3	0	240	240
4	155 (1 interval)	175	330
5	160 (1 interval)	393	553
6	180 (1 interval)	370	550
7	310 (2 intervals:150+160)	631	941
8	842 (2 intervals:390+452)	765	1607
9	210 (2 intervals:75+135)	255	465
10	673 (2 intervals:310+363)	612	1285

4.2 Parameter Measurement for One of the Test Sequences

The system is currently running on a PC Pentium4 (1,8 Ghz) with Linux kernel 2.6.18 in real time (25 pairs of frames/s) with a resolution of 640x480 pixels. Average processing time per pair of frames is 11.43ms. Figure 8 depicts the parameters measured for sequence number 9. This is a representative test example with a duration of 465 seconds where the user simulates two fatigue behaviors separated by an alertness period. As can be seen, until second 90, and between the seconds 195 and 360, the DIL is below 0.5 indicating an alertness state. In these intervals the PERCLOS is low (below 0.15), eye closure duration is low (below the 200 ms), blink frequency is low (below 2 blinks per 30-second window) and nodding frequency is zero. These ocular parameters indicate a clear alert behavior. The frontal face position parameter is not 1.0, indicating that the

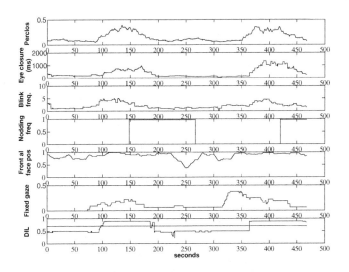

Fig. 8. Parameters measured for the test sequence number 9

predominant position of the head is frontal, but that there are some deviations near the frontal position, typical of a driver with a high vigilance level. The fixed gaze parameter is low because the eyes of the driver are moving caused by a good alert condition. DIL increases over the alert threshold during two intervals (from 90 to 190 and from 360 to 565 seconds) indicating two fatigue behaviors. In both intervals the PERCLOS increases from 0.15 to 0.4, the eye closure duration goes up to 1000 ms, and the blink frequency parameter increases from 2 to 5 blinks. The frontal face position is very close to 1.0 because the head position is fixed and frontal. The fixed gaze parameter increases its value up to 0.4 due to the narrow gaze in the line of sight of the driver. This last variation indicates a typical loss of concentration, and it takes place before other sleepy parameters could indicate increased sleepiness, as can be observed. The nodding is the last fatigue effect to appear. In the two fatigue intervals a nodding occurs after the increase of the other parameters, indicating a low vigilance level. This last parameter is calculated over a temporal window of 2 minutes, so its value remains stable most of the time.

This section described an example of parameter evolution for two simulated fatigue behaviors of one driver. Then, we analyzed the behaviors of other drivers in different circumstances, according to the video tests explained above. The results obtained are similar to those shown for sequence number 9. Overall results of the system are explained in what follows.

4.3 Parameter Performance

The general performance of the measured parameters for a variety of environments with different drivers, according to the test sequences, is presented in table 3. Performance was measured by comparing the algorithm results to results obtained by manually analyzing the recorded sequences on a frame-by-frame basis. Each frame was individually marked with the visual behaviors the driver exhibited, if any. Inaccuracies of this evaluation can be considered negligible for all parameters. Eye closure duration is not easy to evaluate accurately, as the duration of some quick blinks is around 5 to 6 frames at the rate of 25 frames per second (fps), and the starting of the blink can fall between two frames. However, the number of quick blinks is not big enough to make further statistical analysis necessary.

For each parameter the total correct percentage for all sequences excluding sequence number 6 (driver wearing glasses) and sequence number 7 (recorded during the day) is depicted. Then, this column shows the parameter detection performance of the system for optimal situations (driver without glasses driving at night). The performance gets considerably worse by day and it dramatically decreases when drivers wear glasses.

PERCLOS results are quite good, obtaining a total correct percentage of 93.1%. It has been found to be a robust ocular parameter for characterizing driver fatigue. However, it may fail sometimes, for example, when a driver falls asleep without closing her eyes. Eye closure duration performance (84.4%) is a little worse than that of the PERCLOS, because the correct estimation of the

Table 3. Parameter measurement performance

Parameters	Total % correct
PERCLOS	93.1%
Eye closure duration	84.4%
Blink freq.	79.8%
Nodding freq.	72.5%
Face pose	87.5%
Fixed gaze	95.6%

duration is more critical. The variation on the intensity when the eye is partially closed with regard to the intensity when it is open complicates the segmentation and detection. This causes the frame count for this parameter to be usually less than the real one. These frames are considered as closed time. Measured time is slightly over the real time, as a result of delayed detection. Performance of blink frequency parameter is about 80% because some quick blinks are not detected at 25 fps. Then, the three explained parameters are clearly correlated almost linearly, and PERCLOS is the most robust and accurate one.

Nodding frequency results are the worst (72.5%), as the system is not sensible to noddings in which the driver rises her head and then opens her eyes. To reduce false positives, the magnitude of the nodding (i.e., the absolute value of the Kalman filter speed), must be over a threshold. In most of the non-detected noddings, the mentioned situation took place, while the magnitude threshold did not have any influence on any of them. The ground truth for this parameter was obtained manually by localizing the noddings on the recorded video sequences. It is not correlated with the three previous parameters, and it is not robust enough for fatigue detection. Consequently, it can be used as a complementary parameter to confirm the diagnosis established based on other more robust methods.

The evaluation of the face direction provides a measure of alertness related to drowsiness and visual distractions. This parameter is useful for both detecting the pose of the head not facing the front direction and the duration of the displacement. The results can be considered fairly good (87.5%) for a simple model that requires very little computation and no manual initialization. The ground truth in this case was obtained by manually looking for periods in which the driver is not clearly looking in front in the video sequences, and comparing their length to that of the periods detected by the system. There is no a clear correlation between this parameter and the ocular ones for fatigue detection. This would be the most important cue in case of visual distraction detection.

Performance of the fixed gaze monitoring is the best of the measured parameters (95.6%). The maximum values reached by this parameter depend on users' movements and gestures while driving, but a level above 0.05 is always considered to be an indicator of drowsiness. Values greater than 0.15 represent high inattentiveness probability. These values were determined experimentally. This parameter did not have false positives and is largely correlated with the frontal face direction parameter. On the contrary, it is not clearly correlated with the rest of the ocular measurements. For cognitive distraction analysis, this parameter would be the most important cue, as this type of distraction does not normally involve head or eye movements. The ground truth for this parameter was manually obtained by analyzing eye movements frame by frame for the intervals where a fixed gaze behavior was being simulated. We can conclude from these data that fixed gaze and PERCLOS are the most reliable parameters for characterizing driver fatigue, at least for our simulated fatigue study.

All parameters presented in table 3 are fused in the fuzzy system to obtain the DIL for final evaluation of sleepiness. We compared the performance of the system using only the PERCLOS parameter and the DIL(using all of the parameters), in order to test the improvements of our proposal with respect to the most widely used parameter for characterizing driver drowsiness. The system performance was evaluated by comparing the intervals where the PERCLOS/DIL was above a certain threshold to the intervals, manually analyzed over the video sequences, in which the driver simulates fatigue behaviors. This analysis consisted of a subjective estimation of drowsiness by human observers, based on the Wierwille test [41].

As can be seen in table 4, correct detection percentage for DIL is very high (97%). It is higher than the obtained using only PERCLOS, for which the correct detection percentage is about the 90% for our tests. This is due to the fact that fatigue behaviors are not the same for all drivers. Further, parameter evolution and absolute values from the visual cues differ from user to user. Another important fact is the delay between the moment when the driver starts his fatigue behavior simulation and when the fuzzy system detects it. This is a consequence of the window spans used in parameter evaluation. Each parameter responds to a different stage in the fatigue behavior. For example, fixed gaze behavior appears before PERCLOS starts to increase, thus rising the DIL to a value where a noticeable increment of PERCLOS would rise an alarm in few seconds. This is extensible to the other parameters. Using only the PERCLOS would require much more time to activate an alarm (tens of seconds), especially if the PERCLOS increases more slowly for some drivers. Our system provides an accurate characterization of a driver's level of fatigue, using multiple visual parameters to resolve the ambiguity present in the information from a single parameter. Additionally, the system performance is very high in spite of the partial errors associated to each input parameter. This was achieved using redundant information.

Table 4. Sleepiness detection performance

Parameter	Total % correct
PERCLOS	90%
DIL	97%

5 Discussion

It has been shown that the system's weaknesses can be almost completely attributed to the pupil detection strategy, because it is the most sensitive to external interference. As it has been mentioned above, there are a series of situations where the pupils are not detected and tracked robustly enough. Pupil tracking is based on the "bright pupil" effect, and when this effect does not appear clearly enough on the images, the system can not track the eyes. Sunlight intensity occludes the near-IR reflected from the driver's eyes. Fast changes in illumination that the Automatic Gain Control in the camera can not follow produce a similar result. In both cases the "bright pupil" effect is not noticeable in the images, and the eyes can not be located. Pupils are also occluded when the driver's eyes are closed. It is then not possible to track the eyes if the head moves during a blink, and there is an uncertainty of whether the eyes may still be closed or they may have opened and appeared in a position on the image far away from where they were a few frames before. In this situation, the system would progressively extend the search windows and finally locate the pupils, but in this case the measured duration of the blink would not be correct. Drivers wearing glasses pose a different problem. "Bright pupil" effect appears on the images, but so do the reflections of the LEDs from the glasses. These reflections are very similar to the pupil's, making detection of the correct one very difficult.

We are exploring alternative approaches to the problem of pupil detection and tracking, using methods that are able to work 24/7 and in real time, and that yield accurate enough results to be used in other modules of the system. A possible solution is to use an eye or face tracker that does not rely on the "bright pupil" effect. Also, tracking the whole face, or a few parts of it, would make it possible to follow its position when eyes are closed, or occluded.

Face and eye location is an extensive field in computer vision, and multiple techniques have been developed. In recent years, probably the most successful have been texture-based methods and machine learning. A recent survey that compares some of these methods for eye localization can be found in [8]. We have explored the feasibility of using appearance (texture)-based methods, such as Active Appearance Models (AAM)[9]. AAM are generative models, that try to parameterize the contents of an image by generating a synthetic image as close as possible to the given one. The synthetic image is obtained from a model consisting of both appearance and shape. These appearance and shape are learned in a training process, and thus can only represent a constrained range of possible appearances and deformations. They are represented by a series of orthogonal vectors, usually obtained using Principal Component Analysis (PCA), that form a base in the appearance and deformation spaces.

AAMs are linear in both shape and appearance, but are nonlinear in terms of pixel intensities. The shape of the AAM is defined as the coordinates of the v vertices of the shape

$$\mathbf{s} = (x_1, y_1, x_2, y_2, \cdots, x_v, y_v)^t \tag{1}$$

and can be instantiated from the vector base simply as:

$$\mathbf{s} = \mathbf{s_0} + \sum_{i=1}^{n} p_i \cdot \mathbf{s_i} \qquad (2)$$

where $\mathbf{s_0}$ is the *base shape* and $\mathbf{s_i}$ are the *shape vectors*. Appearance is instantiated in the same way

$$A(\mathbf{x}) = A_0(\mathbf{x}) + \sum_{i=1}^{m} \lambda_i \cdot A_i(\mathbf{x}) \qquad (3)$$

where $A_0(\mathbf{x})$ is the *base appearance*, $A_i(\mathbf{x})$ are the *appearance vectors* and λ_i are the weights of these vectors.

The final model instantiation is obtained by warping the appearance $A(\mathbf{x})$, whose shape is $\mathbf{s_0}$, so it conforms to the shape $\mathbf{s}$. This is usually done by triangulating the vertices of the shape, using Delaunay [13] or another triangulation algorithm, as shown in figure 9. The appearance that falls in each triangle is affine warped independently, accordingly to the position of the vertices of the triangle in $\mathbf{s_0}$ and $\mathbf{s}$.

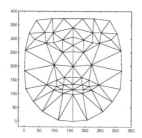

Fig. 9. A triangulated shape

The purpose of fitting the model to a given image is to obtain the parameters that minimize the error between the image I and the model instance:

$$\sum_{\mathbf{x} \in \mathbf{s_0}} \left[A_0(\mathbf{x}) + \sum_{i=1}^{m} \lambda_i A_i(\mathbf{x}) - I(\mathbf{W}(\mathbf{x}; \mathbf{p})) \right]^2 \qquad (4)$$

where $\mathbf{W}(\mathbf{x}; \mathbf{p})$ is a warp defined over the pixel positions $\mathbf{x}$ by the shape parameters $\mathbf{p}$.

These parameters can be then analyzed to gather interesting data, in our case, the position of the eyes and head pose. Minimization is done using the Gauss-Newton method, or some efficient variations, such as the *inverse compositional algorithm* [28; 4].

We tested the performance and robustness of the Active Appearance Models on the same in-car sequences described above. AAMs perform well in sequences

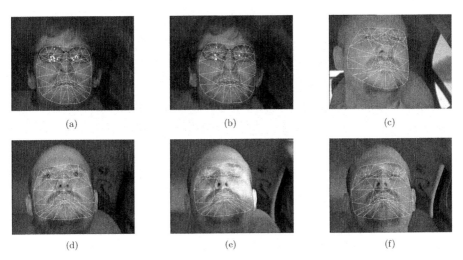

Fig. 10. Fitting results with glasses and sunlight

where the IR-based system did not, such as sequence 6, where the driver is wearing glasses (figures 10(a) and 10(b)), and is able to work with sunlight (10(c)), and track the face under fast illumination changes (10(d) to 10(f)). Also, as the model covers most of the face, the difference between a blink and a tracking loss is clearer, as the model can be fitted when eyes are either open or closed.

On our tests, however, AAM was only fitted correctly when the percentage of occlusion (or self-occlusion, due to head turns) of the face was below 35% of the face. It was also able to fit with low error although the position of the eyes was not determined with the required precision (i.e. the triangles corresponding to the pupil were positioned closer to the corner of the eye than to the pupil). The IR-based system could locate and track an eye when the other eye was occluded, which the AAM-based system is not able to do. More detailed results can be found on [30].

Overall results of face tracking and eye localization with AAM are encouraging, but the mentioned shortcomings indicate that improved robustness is necessary. Constrained Local Models (CLM) are models closely related to AAM, that have shown improved robustness and accuracy [10]. Instead of covering the whole face, CLM only use small rectangular patches placed in specific points that are interesting for its characteristic appearance or high contrast. Constrained Local Models are trained in the same way as AAMs, and both a shape and appearance vector bases are obtained.

Fitting the CLM to an image is done in two steps. First, the same minimization that was used for AAMs is performed, with the difference that now no warping is applied over the rectangles. Those are only displaced over the image. In the second step, the correlation between the patches and the image is maximized, with an iterative algorithm, typically the Nelder-Mead simplex algorithm [29].

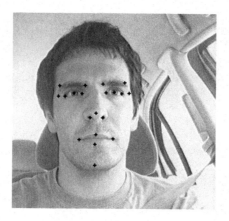

Fig. 11. A Constrained Local Model fitted over a face

The use of small patches and the two-step fitting algorithm make CLM more robust and efficient than AAM. See figure 11 for an example. The CLM is a novel technique that performs well in controlled environments, but that has to be thoroughly tested in challenging operation scenarios.

6 Conclusions and Future Work

We have developed a non-intrusive prototype computer vision system for real-time monitoring of driver's fatigue. It is based on a hardware system for real time acquisition of driver's images using an active IR illuminator and the implementation of software algorithms for real-time monitoring of the six parameters that better characterize the fatigue level of a driver. These visual parameters are PERCLOS, eye closure duration, blink frequency, nodding frequency, face pose and fixed gaze. In an attempt to effectively monitor fatigue, a fuzzy classifier was implemented to merge all these parameters into a single Driver Inattentiveness Level. Monitoring distractions (both visual and cognitive) would be possible using this system. The system development has been discussed. The system is fully autonomous, with automatic (re)initializations if required. It was tested with different sequences recorded in real driving condition with different users during several hours. In each of them, several fatigue behaviors were simulated during the test. The system works robustly at night for users not wearing glasses, yielding accuracy of 97%. Performance of the system decreases during the daytime, especially in bright days, and at the moment it does not work with drivers wearing glasses. A discussion about improvements of the system in order to overcome these weaknesses has been included.

The results and conclusions obtained support our approach to the drowsiness detection problem. In the future the results will be completed with actual drowsiness data. We have the intention of testing the system with more users for long periods of time, to obtain real fatigue behaviors. With this information we will generalize our fuzzy knowledge base. Then, we would like to improve our vision system with some of the techniques mentioned in the previous section, in order to solve the problems of daytime operation and to improve the solution for drivers wearing glasses. We also plan to add two new sensors (a steering wheel sensor and a lane tracking sensor) for fusion with the visual information to achieve correct detection, especially at daytime.

Acknowledgements

This work has been supported by grants TRA2005-08529-C02-01 (MOVICON Project) and PSE-370100-2007-2 (CABINTEC Project) from the Spanish Ministry of Education and Science (MEC). J. Nuevo is also working under a researcher training grant from the Education Department of the Comunidad de Madrid and the European Social Fund.

References

[1] Inc. Agilent Technologies. Application Note 1118: Compliance of Infrared Communication Products to IEC 825-1 and CENELEC EN 60825-1 (1999)
[2] Alonso, J.M., Guillaume, S., Magdalena, L.: KBCT, knowledge base control tool (2003), http://www.mat.upm.es/projects/advocate/en/index.htm
[3] Anon. Perclos and eyetracking: Challenge and opportunity. Technical report, Applied Science Laboratories, Bedford, MA (1999), http://www.a-s-l.com
[4] Baker, S., Matthews, I.: Lucas-Kanade 20 years on: A unifying framework. International Journal of Computer Vision 56(3), 221–255 (2004)
[5] Bergasa, L.M., Nuevo, J., Sotelo, M.A., Barea, R., Lopez, M.E.: Real-time system for monitoring driver vigilance. IEEE Transactions on Intelligent Transportation Systems 7(1), 63–77 (2006)
[6] Boverie, S., Leqellec, J.M., Hirl, A.: Intelligent systems for video monitoring of vehicle cockpit. In: International Congress and Exposition ITS. Advanced Controls and Vehicle Navigation Systems, pp. 1–5 (1998)
[7] Bradski, G., Kaehler, A., Pisarevsky, V.: Learning-based computer vision with intel's open source computer vision library. Intel Technology Journal 09(02) (May 2005)
[8] Campadelli, P., Lanzarotti, R., Lipori, G.: Eye localization: a survey. In: NATO Science Series (2006)
[9] Cootes, T.F., Edwards, G.J., Taylor, C.J.: Active appearance models. IEEE Trans. Pattern Anal. Machine Intell. 23, 681–685 (2001)
[10] Cristinacce, D., Cootes, T.: Feature Detection and Tracking with Constrained Local Models. In: Proc. the British Machine Vision Conf. (2006)
[11] DaimerChrysler, A.G.: The electronic drawbar (June 2001), http://www.daimlerchrysler.com
[12] DaimlerChrysler, A.G.: Driver assistant with an eye for the essentials, http://www.daimlerchrysler.com/dccom
[13] Delaunay, B.: Sur la sphere vide. Izv. Akad. Nauk SSSR, Otdelenie Matematicheskii i Estestvennyka Nauk 7, 793–800 (1934)

[14] Dinges, D., Perclos, F.: A valid psychophysiological measure of alertness as assessed by psychomotor vigilance. Technical Report MCRT-98-006, Federal Highway Administration. Office of motor carriers (1998)
[15] European Project FP6 (IST-1-507674-IP). AIDE - Adaptive Integrated Driver-vehicle Interface (2004–2008), http://www.aide-eu.org/index.html
[16] European Project FP6 (IST-2002-2.3.1.2). Advanced sensor development for attention, stress, vigilance and sleep/wakefulness monitoring (SENSATION) (2004–2007), http://www.sensation-eu.org
[17] Fitzgibbon, A.W., Fisher, R.B.: A buyer's guide to conic fitting. In: Proc. of the 6th British conference on Machine vision, Birmingham, United Kingdom, vol. 2, pp. 513–522 (1995)
[18] Forsyth, D.A., Ponce, J.: Computer Vision: A Modern Approach. Prentice Hall, Englewood Cliffs (2003)
[19] Grace, R.: Drowsy driver monitor and warning system. In: Int. Driving Symposium on Human Factors in Driver Assessment, Training and Vehicle Design (Auguest 2001)
[20] Guillaume, S., Charnomordic, B.: A new method for inducing a set of interpretable fuzzy partitions and fuzzy inference systems from data. Studies in Fuzziness and Soft Computing 128, 148–175 (2003)
[21] Ueno, H., Kaneda, M., Tsukino, M.: Development of drowsiness detection system. In: Proceedings of Vehicle Navigation and Information Systems Conference, pp. 15–20 (1994)
[22] AWAKE Consortium (IST 2000-28062). System for Effective Assessment of Driver Vigilance and Warning According to Traffic Risk Estimation- AWAKE (September 2001-2004), http://www.awake-eu.org
[23] Ji, Q., Yang, X.: Real-time eye, gaze and face pose tracking for monitoring driver vigilance. Real-Time Imaging 8, 357–377 (2002)
[24] Kircher, A., Uddman, M., Sandin, J.: Vehicle control and drowsiness. Technical Report VTI-922A, Swedish National Road and Transport Research Institute (2002)
[25] Koons, D., Flicker, M.: IBM Blue Eyes project (2003), http://almaden.ibm.com/cs/blueeyes
[26] Kutila, M.: Methods for Machine Vision Based Driver Monitoring Applications. PhD thesis, VTT Technical Research Centre of Finland (2006)
[27] Matsumoto, Y., Zelinsky, A.: An algorithm for real-time stereo vision implementation of head pose and gaze direction measurements. In: Procs. IEEE 4th Int. Conf. Face and Gesture Recognition, pp. 499–505 (March 2000)
[28] Matthews, I., Baker, S.: Active appearance models revisited. International Journal of Computer Vision 60(2), 135–164 (2004)
[29] Nelder, J.A., Mead, R.: A simplex method for function minimization. Computer Journal 7(4), 308–313 (1965)
[30] Nuevo, J., Bergasa, L.M., Sotelo, M.A., Ocana, M.: Real-time robust face tracking for driver monitoring. In: Intelligent Transportation Systems Conference, 2006. ITSC 2006, pp. 1346–1351. IEEE, Los Alamitos (2006)
[31] Nunes, L., Recarte, M.A.: Cognitive demands of hands-free phone conversation while driving, ch. F 5, pp. 133–144. Pergamon (2002)
[32] Rau, P.: Drowsy driver detection and warning system for commercial vehicle drivers: Field operational test design, analysis and progress, NHTSA (2005)

[33] Royal, D.: Volume I - Findings; National Survey on Distracted and Driving Attitudes and Behaviours, 2002. Technical Report DOT HS 809 566, The Gallup Organization (March 2003)
[34] Seeing Machines. Facelab transport (August 2006), http://www.seeingmachines.com/transport.htm
[35] Seeing Machines. Driver state sensor (August 2007), http://www.seeingmachines.com/DSS.html
[36] Shih, W., Liu.: A calibration-free gaze tracking technique. In: Proc. 15th Conf. Patterns Recognition, vol. 4, Barcelona, Spain, pp. 201–204 (2000)
[37] Smith, P., Shah, M., Da Vitoria Lobo, N.: Determining driver visual attention with one camera. IEEE Trans. Intell. Transport. Syst. 4(4), 205–218 (2003)
[38] Victor, T., Blomberg, O., Zelinsky, A.: Automating the measurement of driver visual behaviours using passive stereo vision. In: Procs. Int. Conf. Series Vision in Vehicles VIV9, Brisbane, Australia (Auguest 2001)
[39] Volvo Car Corporation. Driver alert control, http://www.volvocars.com
[40] Wierwille, W., Tijerina, L., Kiger, S., Rockwell, T., Lauber, E., Bittne, A.: Final report supplement – task 4: Review of workload and related research. Technical Report DOT HS 808 467(4), USDOT (October 1996)
[41] Wierwille, W., Wreggit, Kirn, Ellsworth, Fairbanks: Research on vehicle-based driver status/performance monitoring; development, validation, and refinement of algorithms for detection of driver drowsiness, final report; technical reports & papers. Technical Report DOT HS 808 247, USDOT (December 1994), www.its.dot.gov

Understanding Driving Activity Using Ensemble Methods

Kari Torkkola, Mike Gardner, Chris Schreiner, Keshu Zhang, Bob Leivian, Harry Zhang, and John Summers

Motorola Labs
Kari.Torkkola@motorola.com

1 Introduction

Motivation for the use of statistical machine learning techniques in the automotive domain arises from our development of context aware intelligent driver assistance systems, specifically, Driver Workload Management systems. Such systems integrate, prioritize, and manage information from the roadway, vehicle, cockpit, driver, infotainment devices, and then deliver it through a multimodal user interface. This could include incoming cell phone calls, email, navigation information, fuel level, and oil pressure to name a very few. In essence, the workload manager attempts to get the right information to the driver at the right time and in the right way in order that driver performance is optimized and distraction is minimized.

In order to do its job, the workload manager system needs to track the wider driving context including the state of the roadway, traffic conditions, and the driver. Current automobiles have a large number of embedded sensors, many of which produce data that are available through the car data bus. The state of many on-board and carried-in devices in the cockpit is also available. New advanced sensors, such as video-based lane departure warning systems and radar-based collision warning systems are currently being deployed in high end car models. All of these could be used to define the driving context [17]. But a number of questions arise:

- What are the range of typical driving maneuvers, as well as near misses and accidents, and how do drivers navigate them?
- Under what conditions does driver performance degrade or driver distraction increase?
- What are the optimal set of sensors and algorithms that can recognize each of these driving conditions near the theoretical limit for accuracy, and what is the sensor set that are accurate yet cost effective?

There are at least two approaches to address these questions. The first is more or less heuristic: experts offer informed opinions on the various matters and sets of sensors are selected accordingly with algorithms coded using rules of thumb.

Individual aspects of the resulting system are tested by using narrow human factors testing controlling all but a few variables. Hundreds of iterations must be completed in order to test the impact on driver performance of the large combinations of driving states, sensor sets, and various algorithms.

The second approach, which we advocate in this chapter and in [28], involves using statistical machine learning techniques that enable the creation of new human factors approaches. Rather than running large numbers of narrowly scoped human factors testing with only a few variables, we chose to invent a "Hypervariate Human Factors Test Methodology" that uses broad naturalistic driving experiences that capture wide conditions resulting in hundreds of variables, but decipherable with machine learning techniques. Rather than pre-selecting our sensor sets we chose to collect data from every sensor we could think of and some not yet invented that might remotely be useful by creating behavioral models overlaid with our vehicle system. Furthermore, all sensor outputs were expanded by standard mathematical transforms that emphasized various aspects of the sensor signals. Again, data relationships are discoverable with machine learning. The final data set consisted of hundreds of driving hours with thousands of variable data outputs which would have been nearly impossible to annotate without machine learning techniques.

In this chapter we describe three major efforts that have employed our machine learning approach. First, we discuss how we have utilized our machine learning approach to detect and classify a wide range of driving maneuvers, and describe a semi-automatic data annotation tool we have created to support our modeling effort. Second, we perform a large scale automotive sensor selection study towards intelligent driver assistance systems. Finally, we turn our attention to creating a system that detects driver inattention by using sensors that are available in the current vehicle fleet (including forwarding looking radar and video-based lane departure system) instead of head and eye tracking systems.

This approach resulted in the creation of two generations of our workload manager system called Driver Advocate that was based on data rather than just expert opinions. The described techniques helped reduce the research cycle times while resulting in broader insight. There was rigorous quantification of theoretical sensor subsystem performance limits and optimal subsystem choices given economic price points. The resulting system performance specs and architecture design created a workload manager that had a positive impact on driver performance [23, 33].

2 Modeling Naturalistic Driving

Having the ability to detect driving maneuvers can be of great benefit in determining a driver's current workload state. For instance, a driving workload manager may decide to delay presenting the driver with non-critical information if the driver was in the middle of a complex driving maneuver. In this section we describe our data-driven approach to classifying driving maneuvers.

Fig. 1. The driving simulator

There are two approaches to collecting large databases of driving sensor data from various driving situations. One can outfit a fleet of cars with sensors and data collection equipment, as has been done in the NHTSA 100-car study [18]. This has the advantage of being as naturalistic as possible. However, the disadvantage is that potentially interesting driving situations will be extremely rare in the collected data. Realistic driving simulators provide much more controlled environments for experimentation and permit the creation of many interesting driving situations within a reasonable time frame. Furthermore, in a driving simulator, it is possible to simulate a large number of potential advanced sensors that would be yet too expensive or impossible to install in a real car. This will also enable us to study what sensors really are necessary for any particular task and what kind of signal processing of those sensors is needed in order to create adequate driving situation models based on those sensors.

We collect data in a driving simulator lab, which is an instrumented car in a surround video virtual world with full visual and audio simulation (although no motion or G-force simulation) of various roads, traffic and pedestrian activity. The driving simulator consists of a fixed based car surrounded by five front and three rear screens (Fig. 1). All driver controls such as the steering wheel, brake, and accelerator are monitored and affect the motion through the virtual world in real-time. Various hydraulics and motors provide realistic force feedback to driver controls to mimic actual driving.

The basic driving simulator software is a commercial product with a set of simulated sensors that, at the behavioral level, simulate a rich set of current and near future on-board sensors (www.drivesafety.com). This set consists of a radar for locating other traffic, a GPS system for position information, a camera system for lane positioning and lane marking detection, and a mapping data base for road names, directions, locations of points of interest, etc. There is also a complete car status system for determining the state of engine parameters (coolant temperature, oil pressure etc) and driving controls (transmission gear selection, steering angle, gas pedal, brake pedal, turn signal, window and seat belt status etc.). The simulator setup also has several video cameras, microphones, and infrared eye tracking sensors to record all driver actions during the drive in synchrony with all the sensor output and simulator tracking variables. The Seeing Machines eye tracking system is used to automatically acquire a driver's head and eye movements (www.seeingmachines.com). Because such eye tracking systems are not installed in current vehicles, head and eye movement variables do not enter into the machine learning algorithms as input. The 117 head and eye-tracker variables are recorded as two versions, real-time and filtered. Including both versions, there are altogether 476 variables describing an extensive scope of driving data — information about the auto, the driver, the environment, and associated conditions. An additional screen of video is digitally captured in MPEG4 format, consisting of a quad combiner providing four different views of the driver and environment. Combined, these produce around 400MB of data for each 10 minutes of drive time. Thus we are faced with processing massive data sets of mixed type; there are both numerical and categorical variables, and multimedia, if counting also the video and audio.

3 Database Creation

We describe now our approach to collecting and constructing a database of naturalistic driving data in the driving simulator. We concentrate on the machine learning aspect; making the database usable as the basis for learning driving/driver situation classifiers and detectors. Note that the same database can be (and will be) used in driver behavioral studies, too.

3.1 Experiment Design

Thirty-six participants took part in this study, with each participant completing about ten 1-hour driving sessions. In each session, after receiving practice drives to become accustomed to the simulated driving environment, participants were given a task for which they have to drive to a specific location. These drives were designed to be as natural and familiar for the participants as possible, and the simulated world replicated the local metropolitan area as much as possible, so that participants did not need navigation aids to drive to their destinations. The driving world was modeled on the local Phoenix, AZ topology. Signage corresponded to local street and Interstate names and numbers. The topography corresponded as closely as possible to local landmarks.

The tasks included driving to work, driving from work to pick up a friend at the airport, driving to lunch, and driving home from work. Participants were only instructed to drive as they normally would. Each drive varied in length from 10 to 25 minutes. As time allowed, participants did multiple drives per session.

This design highlights two crucial components promoting higher realism in driving and consequently in collected data: 1) familiarity of the driving environment, and 2) immersing participants in the tasks. The experiment produced a total of 132 hours of driving time with 315GB of collected data.

3.2 Annotation of the Database

We have created a semi-automatic data annotation tool to label the sensor data with 28 distinct driving maneuvers. This data annotation tool is unique in that we have made parts of the annotation process automatic, enabling the user just to verify automatically generated annotations, rather than annotating everything from scratch.

The purpose of the data annotation is to label the sensor data with meaningful classes. Supervised learning and modeling techniques then become available with labeled data. For example, one can train classifiers for maneuver detection [29] or for inattention detection [30]. Annotated data also provides a basis for research in characterizing driver behavior in different contexts.

The driver activity classes in this study were related to maneuvering the vehicle with varying degrees of required attention. An Alphabetical listing is presented in Table 1. Note that the classes are not mutually exclusive. An instant in time can be labeled simultaneously as "TurningRight" and "Starting", for example.

Table 1. Driving maneuvers used in the study. "Cruising" captures anything not included in the actual 28 classes.

ChangingLaneLeft	ChangingLaneRight
ComingToLeftTurnStop	ComingToRightTurnStop
Crash	CurvingLeft
CurvingRight	EnterFreeway
ExitFreeway	LaneChangePassLeft
LaneChangePassRight	LaneDepartureLeft
LaneDepartureRight	Merge
PanicStop	PanicSwerve
Parking	PassingLeft
PassingRight	ReversingFromPark
RoadDeparture	SlowMoving
Starting	StopAndGo
Stopping	TurningLeft
TurningRight	WaitingForGapInTurn
Cruising (other)	

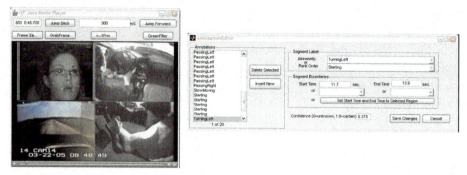
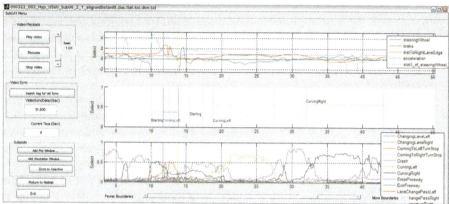

Fig. 2. The annotation tool. Top left - the video playback window. Top right - annotation label window. Bottom - sensor signal and classifier result display window. Note that colors have been re-used. See Fig 5 for a complete legend.

We developed a special purpose data annotation tool for the driving domain (Fig. 2). This was necessary because available video annotation tools do not provide a simultaneous view of the sensor data, and tools meant for signals, such as speech, do not allow simultaneous and synchronous playback of the video. The major properties of our annotation tool are listed below.

1. Ability to navigate through any portion of the driving sequence.
2. Ability to label (annotate) any portion of the driving sequence with proper time alignment.
3. Synchronization between video and other sensor data.
4. Ability to playback the video corresponding to the selected sensor signal segment.
5. Ability to visualize any number of sensor variables.
6. Provide persistent storage of the annotations.
7. Ability to modify existing annotations.

Since manual annotation is a tedious process, we automated parts of the process by taking advantage of automatic classifiers that are trained from data to detect the driving maneuvers. With these classifiers, annotation becomes an instance of active learning [7]. Only if a classifier is not very confident in its decision, its results are presented to the human to verify. The iterative annotation process is thus as follows:

1. Manually annotate a small portion of the driving data.
2. Train classifiers based on all annotated data.
3. Apply classifiers to a portion of database.
4. Present unsure classifications to the user to verify.
5. Add new verified and annotated data to the database.
6. Go to 2.

As the classifier improves due to increased size of training data, the decisions presented to the user improve, too, and the verification process takes less time [25]. The classifier is described in detail in the next section.

4 Driving Data Classification

We describe now a classifier that has turned out to be very appropriate for driving sensor data. Characteristics of this data are hundreds of variables (sensors), millions of observations, and mixed type data. Some variables have continuous values and some are categorical. The latter fact causes problems with conventional statistical classifiers that typically operate entirely with continuous valued variables. Categorical variables need to be converted first into binary indicator variables. If a categorical variable has a large number of levels (possible discrete values), each of them generates a new indicator variable, thus potentially multiplying the dimension of the variable vector. We attempted this approach using Support Vector Machines [24] as classifiers, but the results were inferior compared to ensembles of decision trees.

4.1 Decision Trees

Decision trees, such as CART [6], are an example of non-linear, fast, and flexible base learners that can easily handle massive data sets even with mixed variable types.

A decision tree partitions the input space into a set of disjoint regions, and assigns a response value to each corresponding region (see Fig. 3). It uses a greedy, top-down recursive partitioning strategy. At every step a decision tree uses exhaustive search by trying all combinations of variables and split points to achieve the maximum reduction in node impurity. In a classification problem, a node is "pure" if all the training data in the node has the same class label. Thus the tree growing algorithm tries to find a variable and a split point of the variable that best separates the data in the node into different classes. Training data is then divided among the resulting nodes according to the chosen decision test. The process is repeated for each resulting node until a certain maximum node depth is reached, or until the nodes become pure. The tree constructing process itself can be considered as a type of embedded variable selection, and

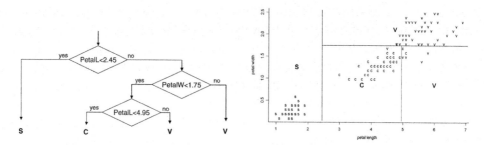

Fig. 3. An example of a decision tree for a three-class classification problem. The right side depicts the data with two variables. The final decision regions are also displayed.

the impurity reduction due to a split on a specific variable could indicate the relative importance of that variable to the tree model.

For a single decision tree, a measure of variable importance is proposed in [6]:

$$VI(x_i, T) = \sum_{t \in T} \Delta I(x_i, t) \quad (1)$$

where $\Delta I(x_i, t) = I(t) - p_L I(t_L) - p_R I(t_R)$ is the decrease in impurity due to an actual (or potential) split on variable x_i at a node t of the optimally pruned tree T. The sum in (1) is taken over all internal tree nodes where x_i is a primary splitter. Node impurity $I(t)$ for classification $I(t) = Gini(t)$ where $Gini(t)$ is the Gini index of node t:

$$Gini(t) = \sum_{i \neq j} p_i^t p_j^t \quad (2)$$

and p_i^t is the proportion of observations in t whose response label equals i ($y = i$) and i and j run through all response class numbers. The Gini index is in the same family of functions as *cross-entropy*, $-\sum_i p_i^t log(p_i^t)$, and measures node impurity. It is zero when t has observations only from one class, and reaches its maximum when the classes are perfectly mixed.

However, a single tree is inherently instable. The ability of a learner (a classifier in this case) to generalize to new unseen data is closely related to the stability of the learner. The stability of the solution could be loosely defined as a continuous dependence on the training data. A stable solution changes very little when the training data set changes a little. With decision trees, the node structure can change drastically even when one data point is added or removed from the training data set. A comprehensive treatment of the connection between stability and generalization ability can be found in [3].

Instability can be remedied by employing ensemble methods. Ensemble methods train multiple simple learners and then combine the outputs of the learners for the final decision. One well known ensemble method is bagging (bootstrap aggregation) [4]. Bagging decision trees is explained in detail in Sec. 4.2.

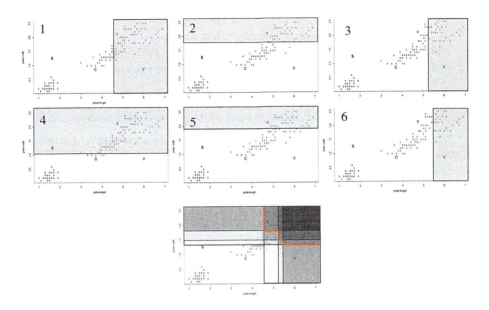

Fig. 4. A trivial example of Random Forest in action. The task is to separate class 'v' from 'c' and 's' (upper right corner) based on two variables only. Six decision trees are constructed sampling both examples and variables. Each tree becomes now a single node sampling one out of the two possible variables. Outputs (decision regions) are averaged and thresholded. The final nonlinear decision border is outlined as a thick line.

Bagging can dramatically reduce the variance of instable learners by providing a regularization effect. Each individual learner is trained using a different random sample set of the training data. Bagged ensembles do not overfit the training data. The keys to good ensemble performance are base learners that have a low bias, and a low correlation between their errors. Decision trees have a low bias, that is, they can approximate any nonlinear decision boundary between classes to a desirable accuracy given enough training data. Low correlation between base learners can be achieved by sampling the data, as described in the following section.

4.2 Random Forests

Random Forest (RF) is a representative of tree ensembles [5]. It grows a forest of decision trees on bagged samples (Fig. 4). The "randomness" originates from creating the training data for each individual tree by sampling both the data and the variables. This ensures that the errors made by individual trees are uncorrelated, which is a requirement for bagging to work properly.

Each tree in the Random Forest is grown according to the following parameters:

1. A number m is specified much smaller than the total number of total input variables M (typically m is proportional to $\sqrt{M}$).
2. Each tree of maximum depth (until pure nodes are reached) is grown using a bootstrap sample of the training set.
3. At each node, m out of the M variables are selected at random.
4. The split used is the best possible split on these m variables only.

Note that for each tree to be constructed, *bootstrap sampling* is applied. A different sample set of training data is drawn with replacement. The size of the sample set is the same as the size of the original dataset. This means that some individual samples will be duplicated, but typically 30% of the data is left out of this sample (out-of-bag). This data has a role in providing an unbiased estimate of the performance of the tree.

Also note that the sampled variable set does not remain constant while a tree is grown. Each new node in a tree is constructed based on a *different* random sample of m variables. The best split among these m variables is chosen for the current node, in contrast to typical decision tree construction, which selects the best split among all possible variables. This ensures that the errors made by each tree of the forest are not correlated. Once the forest is grown, a new sensor reading vector will be classified by every tree of the forest. Majority voting among the trees produces then the final classification decision.

We will be using RF throughout our experimentation because of it simplicity and excellent performance [1]. In general, RF is resistant to irrelevant variables, it can handle massive numbers of variables and observations, and it can handle mixed type data and missing data. Our data definitely is of mixed type, i.e., some variables are continuous, some variables are discrete, although we do not have missing data since the source is the simulator.

4.3 Random Forests for Driving Maneuver Detection

A characteristic of the driving domain and the chosen 29 driving maneuver classes is that the classes are not mutually exclusive. For example, an instance in time could be classified simultaneously as "SlowMoving" and "TurningRight". The problem cannot thus be solved by a typical multi-class classifier that assigns a single class label to a given sensor reading vector and excludes the rest. This dictates that the problem should be treated rather as a *detection* problem than a *classification* problem.

Furthermore, each maneuver is inherently a sequential operation. For example, "ComingToLeftTurnStop" consists of possibly using the turn signal, changing the lane, slowing down, braking, and coming to a full stop. Ideally, a model of a maneuver would thus describe this sequence of operations with variations

---

[1] We use Leo Breiman's Fortran version 5.1, dated June 15, 2004. An interface to Matlab was written to facilitate easy experimentation. The code is available at http://www.stat.berkeley.edu/users/breiman/.

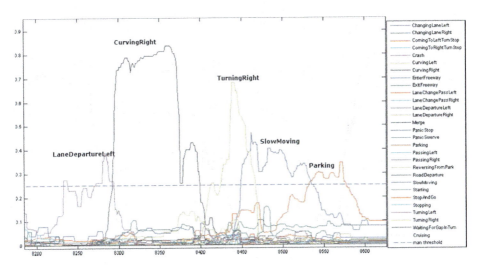

Fig. 5. A segment of driving with corresponding driving maneuver probabilities produced by one Random Forest trained for each maneuver class to be detected. Horizontal axis is the time in tenths of a second. Vertical axis is the probability of a particular class. These "probabilities" can be obtained by normalizing the random forest output voting results to sum to one.

that naturally occur in the data (as evidenced by collected naturalistic data). Earlier, we have experimented with Hidden Markov Models (HMM) for maneuver classification [31]. A HMM is able to construct a model of a sequence as a chain of hidden states, each of which has a probabilistic distribution (typically Gaussian) to match that particular portion of the sequence [22]. The sequence of sensor vectors corresponding to a maneuver would thus be detected as a whole.

The alternative to sequential modeling is instantaneous classification. In this approach, the whole duration of a maneuver is given just a single class label, and the classifier is trained to produce this same label for every time instant of the maneuver. Order, in which the sensor vectors are observed, is thus not made use of, and the classifier carries the burden of being able to capture all variations happening inside a maneuver under a single label. Despite these two facts, in our initial experiments the results obtained using Random Forests for instantaneous classification were superior to Hidden Markov Models.

Because the maneuver labels may be overlapping, we trained a separate Random Forest for each maneuver treating it as a binary classification problem – the data of a particular class against all the other data. This results in 29 trained "detection" forests.

New sensor data is then fed to all 29 forests for classification. Each forest produces something of a "probability" of the class it was trained for. An example plot of those probability "signals" is depicted in Fig. 5. The horizontal axis repesents the time in tenths of a second. About 45 seconds of driving is shown. None of the actual sensor signals are depicted here, instead, the "detector" signals from each of the forests are graphed. These show a sequence of driving maneuvers from "Cruising" through "LaneDepartureLeft", "CurvingRight", "TurningRight", and "SlowMoving" to "Parking".

The final task is to convert the detector signals into discrete and possibly overlapping labels, and to assign a confidence value to each label. In order to do this, we apply both median filtering and low-pass filtering to the signals. The signal at each time instant is replaced by the maximum of the two filtered signals. This has the effect of patching small discontinuities and smoothing the signal while still retaining fast transitions. Any signal exceeding a global threshold value for a minimum duration is then taken as a segment. Confidence of the segment is determined as the average of the detection signal (the probability) over the segment duration.

An example can be seen at the bottom window depicted in Fig 2. The top panel displays some of the original sensor signals, the bottom panel graphs the raw maneuver detection signals, and the middle panel shows the resulting labels.

We compared the results of the Random Forest maneuver detector to the annotations done by a human expert. On the average, the annotations agreed 85% of the time. This means that only 15% needed to be adjusted by the expert. Using this semi-automatic annotation tool, we can drastically reduce the time that is required for data processing.

5 Sensor Selection Using Random Forests

In this section we study which sensors are necessary for driving state classification. Sensor data is collected in our driving simulator; it is annotated with driving state classes, after which the problem reduces to that of feature selection [11]: "Which sensors contribute most to the correct classification of the driving state into various maneuvers?" Since we are working with a simulator, we have simulated sensors that would be expensive to arrange in a real vehicle. Furthermore, sensor behavioral models can be created in software. Goodness or noisiness of a sensor can be modified at will. The simulator-based approach makes it possible to study the problem without implementing the actual hardware in a real car.

Variable selection methods can be divided in three major categories [16, 11, 12]. These are

1. Filter methods, that evaluate some measure of relevance for all the variables and rank them based on the measure (but the measure may not necessarily be relevant to the task, and any interactions that variables may have will be ignored),
2. Wrapper methods, that using some learner, actually learn the solution to the problem evaluating all possible variable combinations (this is usually computationally too prohibitive for large variable sets), and
3. Embedded methods that use a learner with all variables, but infer the set of important variables from the structure of the trained learner.

Random Forests (Sec. 4.2) can act as an embedded variable selection system. As a by-product of the construction, a measure of variable importance can be derived from each tree, basically from how often different variables were used in the splits of the tree and from the quality of those splits [5]. For an ensemble of N trees the importance measure (1) is simply averaged over the ensemble.

$$M(x_i) = \frac{1}{N} \sum_{n=1}^{N} VI(x_i, T_n) \tag{3}$$

The regularization effect of averaging makes this measure much more reliable than a measure extracted from just a single tree.

One must note that in contrast to simple filter methods of feature selection, this measure considers multiple simultaneous variable interactions – not just two at a time. In addition, the tree is constructed for the exact task of interest. We apply now this importance measure to driving data classification.

5.1 Sensor Selection Results

As the variable importance measure, we use the tree node impurity reduction (1) summed over the forest (3). This measure does not require any extra computation in addition to the basic forest construction process. Since a forest was trained for each class separately, we can now list the variables in the order of importance for each class. These results are combined and visualized in Fig. 6.

This figure has the driving activity classes listed at the bottom, and variables on the left column. Head-and eye-tracking variables were excluded from the figure. Each column of the figure thus displays the importances of all listed variables, for the class named at the bottom of the column. White, through yellow, orange, red, and black, denote decreasing importance.

In an attempt to group together those driving activity classes that require a similar set of variables to be accurately detected, and to group together those variables that are necessary or helpful for a similar set of driving activity classes, we clustered first the variables in six clusters and then the driving activity classes in four clusters. Any clustering method can be used here, we used spectral clustering [19]. Rows and columns are then re-ordered according to the cluster identities, which are indicated in the names by alternating blocks of red and black font in Fig. 6.

Looking at the variable column on the left, the variables within each of the six clusters exhibit a similar behavior in that they are deemed important by approximately the same driving activity classes. The topmost cluster of variables (except "Gear") appears to be useless in distinguishing most of the classes. The next five variable clusters appear to be important but for different class clusters. Ordering of the clusters as well as variable ordering within the clusters is arbitrary. It can also be seen that there are variables (sensors) that are important for a large number of classes.

In the same fashion, clustering of the driving activity classes groups those classes together that need similar sets of variables in order to be successfully

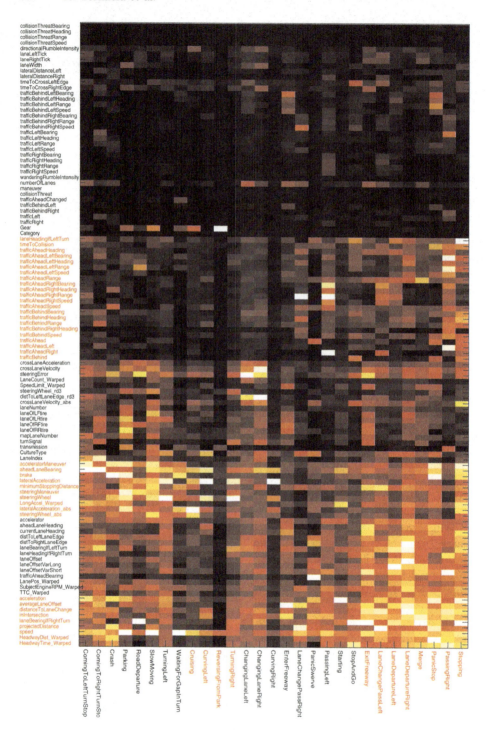

Fig. 6. Variable importances for each class. See text for explanation.

detected. The rightmost and the leftmost clusters are rather distinct, whereas the two middle clusters do not seem to be that clear. There are only a few classes that can be reliably detected using a handful of sensors. Most notable ones are "ReversingFromPark" that only needs "Gear", and "CurvingRight" that needs "lateralAcceleration" with the aid of "steeringWheel". This clustering shows that some classes need quite a wide array of sensors in order to be reliably detected. The rightmost cluster is an example of those classes.

5.2 Sensor Selection Discussion

We present first results of a large scale automotive sensor selection study aimed towards intelligent driver assistance systems. In order to include both traditional and advanced sensors, the experiment was done on a driving simulator. This study shows clusters of both sensors and driving state classes: what classes need similar sensor sets, and what sensors provide information for which sets of classes. It also provides a basis to study detecting an isolated driving state or a set of states of interest and the sensors required for it.

6 Driver Inattention Detection through Intelligent Analysis of Readily Available Sensors

Driver inattention is estimated to be a significant factor for 78% of all crashes [8]. A system that could accurately detect driver inattention could aid in reducing this number. In contrast to using specialized sensors or video cameras to monitor the driver we detect driver inattention by using only readily available sensors. A classifier was trained using Collision Avoidance Systems (CAS) sensors which was able to accurately identify 80% of driver inattention and could be added to a vehicle without incurring the cost of additional sensors.

Detection of driver inattention could be utilized in intelligent systems to control electronic devices [21] or redirect the driver's attention to critical driving tasks [23].

Modern automobiles contain many infotainment devices designed for driver interaction. Navigation modules, entertainment devices, real-time information systems (such as stock prices or sports scores), and communication equipment are increasingly available for use by drivers. In addition to interacting with onboard systems, drivers are also choosing to carry in mobile devices such as cell phones to increase productivity while driving. Because technology is increasingly available for allowing people to stay connected, informed, and entertained while in a vehicle many drivers feel compelled to use these devices and services in order to multitask while driving.

This increased use of electronic devices along with typical personal tasks such as eating, shaving, putting on makeup, reaching for objects on the floor or in the back seat can cause the driver to become inattentive to the driving task. The resulting driver inattention can increase risk of injury to the driver, passengers, surrounding traffic and nearby objects.

The prevailing method for detecting driver inattention involves using a camera to track the driver's head or eyes [9, 26]. Research has also been conducted on modeling driver behaviors through such methods as building control models [14, 15] measuring behavioral entropy [2] or discovering factors affecting driver intention [20, 10].

Our approach to detecting inattention is to use only sensors currently available on modern vehicles (possibly including Collision Avoidance Systems (CAS) sensors) without using head and eye tracking system. This avoids the additional cost and complication of video systems or dedicated driver monitoring systems. We derive several parameters from commonly available sensors and train an inattention classifier. This results in a sophisticated yet inexpensive system for detecting driver inattention.

6.1 Driver Inattention

What Is Driver Inattention?

Secondary activities of drivers during inattention are many, but mundane. The 2001 NETS survey in Table 2 found many activities that drivers perform in addition to driving. A study, by the American Automobile Association placed miniature cameras in 70 cars for a week and evaluated three random driving hours from each. Overall, drivers were inattentive 16.1 percent of the time they drove. About 97 percent of the drivers reached or leaned over for something and about 91 percent adjusted the radio. Thirty percent of the subjects used their cell phones while driving.

Causes of Driver Inattention

There are at least three factors affecting attention:

1. Workload. Balancing the optimal cognitive and physical workload between too much and boring is an everyday driving task. This dynamic varies from instant to instant and depends on many factors. If we chose the wrong fulcrum, we can be overwhelmed or unprepared.

Table 2. 2001 NETS Survey. Activities Drivers Engage in While Driving.

96% Talking to passengers
89% Adjusting vehicle climate/radio controls
74% Eating a meal/snack
51% Using a cell phone
41% Tending to children
34% Reading a map/publication
19% Grooming
11% Prepared for work

2. Distraction. Distractions might be physical (e.g., passengers, cars, signage) or cognitive (e.g., worry, anxiety, aggression). These can interact and create multiple levels of inattention to the main task of driving.
3. Perceived Experience. Given the overwhelming conceit that almost all drivers rate their driving ability as superior than others, it follows that they believe they have sufficient driving control to take part of their attention away from the driving task and give it to multi-tasking. This "skilled operator" over-confidence tends to underestimate the risk involved and reaction time required. This is especially true in the inexperienced younger driver and the physically challenged older driver.

Effects of Driver Inattention

Drivers involved in crashes often say that circumstances occurred suddenly and could not be avoided. However, due to laws of physics and visual perception, very few things occur suddenly on the road. Perhaps more realistically an inattentive driver will suddenly notice that something is going wrong. This inattention or lack of concentration can have catastrophic effects. For example, a car moving at a slow speed with a driver inserting a CD will have the same effect as an attentive driver going much faster. Simply obeying the speed limits may not be enough.

Measuring Driver Inattention

Many approaches to measuring driver inattention have been suggested or researched. Hankey et al suggested three parameters: average glance length, number of glances, and frequency of use [13]. The glance parameters require visual monitoring of the drivers face and eyes. Another approach is using the time and/or accuracy of a surrogate secondary task such as Peripheral Detection Task (PDT) [32]. These measures are yet not practical real time measures to use during everyday driving.

Boer [1] used a driver performance measure, steering error entropy, to measure workload, which unlike eye gaze and surrogate secondary-tasks, is unobtrusive, practical for everyday monitoring, and can be calculated in near real time.

We calculate it by first training a linear predictor from "normal" driving [1]. The predictor uses four previous steering angle time samples (200 ms apart) to predict the next time sample. The residual (prediction error) is computed for the data. A ten-bin discretizer is constructed from the residual, selecting the discretization levels such that all bins become equiprobable. The predictor and discretizer are then fixed, and applied to a new steering angle signal, producing a discretized steering error signal. We then compute the running entropy of the steering error signal over a window of 15 samples using the standard entropy definition $E = -\sum_{i=1}^{10} p_i \log_{10} p_i$, where p_i are the proportions of each discretization level observed in the window.

Our work indicates that steering error entropy is able to detect driver inattention while engaged in secondary tasks. Our current study expands and extends this approach and looks at other driver performance variables, as well as the steering error entropy, that may indicate driver inattention during a common driving task, such as looking in the "blind spot".

Experimental Setup

We designed the following procedure to elicit defined moments of normal driving inattention.

The simulator authoring tool, HyperDrive, was used to create the driving scenario for the experiment. The drive simulated a square with curved corners, six kilometers on a side, 3-lanes each way (separated by a grass median) beltway with on- and off-ramps, overpasses, and heavy traffic in each direction. All drives used daytime dry pavement driving conditions with good visibility.

For a realistic driving environment, high-density random "ambient" traffic was programmed. All "ambient" vehicles simulated alert, "good" driver behavior, staying at or near the posted speed limit, and reacted reasonably to any particular maneuver from the driver.

This arrangement allowed a variety of traffic conditions within a confined, but continuous driving space. Opportunities for passing and being passed, traffic congestion, and different levels of driving difficulty were thereby encountered during the drive.

After two orientation and practice drives, we collected data while drivers drove about 15 minutes in the simulated world. Drivers were instructed to follow all normal traffic laws, maintain the vehicle close to the speed limit (55 mph, 88.5 kph), and to drive in the middle lane without lane changes. At 21 "trigger" locations scattered randomly along the road, the driver received a short burst from a vibrator located in the seatback on either the left or right side of their backs. This was their alert to look in their corresponding "blind spot" and observe a randomly selected image of a vehicle projected there. The image was projected for 5 sec and the driver could look for any length of time he felt comfortable. They were instructed that they would receive "bonus" points for extra money for each correctly answered question about the images. Immediately after the image disappeared, the experimenter asked the driver questions designed to elicit specific characteristics of the image- i.e. What kind of vehicle was it?, Were there humans in the image?, What color was the vehicle?, etc.

Selecting Data for Inattention Detection

Though the simulator has a variety of vehicle, environment, cockpit, and driver parameters available for our use, our goal was to experiment with only readily extractable parameters that are available on modern vehicles. We experimented with two subsets of these parameter streams: one which used only traditional driver controls (steering wheel position and accelerator pedal position), and a second subset which included the first subset but also added variables available from CAS systems (lane boundaries, and upcoming road curvature). A list of variables used and a brief description of each is displayed in Table 3.

Table 3. Variables Used to Detect Inattention

Variable	Description
steeringWheel	Steering wheel angle
accelerator	Position of accelerator pedal
distToLeftLaneEdge	Perpendicular distance of left front wheel from left lane edge
crossLaneVelocity	Rate of change of distToLeftLaneEdge
crossLaneAcceleration	Rate of change of crossLaneVelocity
steeringError	Difference between steering wheel position and ideal position for vehicle to travel exactly parallel to lane edges
aheadLaneBearing	Angle of road 60m in front of current vehicle position

Eye/Head Tracker

In order to avoid having to manually label when the driver was looking away from the simulated road, an eye/head tracker was used (Figure 7).

When the driver looked over their shoulder at an image in their blind spot this action caused the eye tracker to lose eye tracking ability (Figure 8). This loss sent the eye tracking confidence to a low level. These periods of low confidence were used as the periods of inattention. This method avoided the need for hand labeling.

Fig. 7. Eye/head tracking during attentive driving

Fig. 8. Loss of eye/head tracking during inattentive driving

6.2 Inattention Data Processing

Data was collected from six different drivers as described above. This data was later synchronized and re-sampled at a constant sampling rate of 10Hz. resulting in 40,700 sample vectors. In order to provide more relevant information to the task at hand, further parameters were derived from the original sensors. These parameters are as follows

1. ra9: Running average of the signal over nine previous samples (smoothed version of the signal).
2. rd5: Running difference 5 samples apart (trend).
3. rv9: Running variance of 9 previous samples according to the standard definition of sample variance.

4. ent15: Entropy of the error that a linear predictor makes in trying to predict the signal as described in [1]. This can be thought of as a measure of randomness or unpredictability of the signal.
5. stat3: Multivariate stationarity of a number of variables simultaneously three samples apart as described in [27]. Stationarity gives an overall rate of change for a group of signals. Stationarity is one if there are no changes over the time window and approaches zero for drastic transitions in all signals of the group.

The operations can be combined. For example, "_rd5_ra9" denotes first computing a running difference five samples apart and then computing the running average over nine samples.

Two different experiments were conducted.

1. The first experiment used only two parameters: steeringWheel and accelerator, and derived seven other parameters: steeringWheel_rd5_ra9, accelerator_rd5_ra9, stat3_of_steeringWheel_accel, steeringWheel_ent15_ra9, accelerator_ent15_ra9, steeringWheel_rv9, and accelerator_rv9.
2. The second experiment used all seven parameters in Table I and derived 13 others as follows: steeringWheel_rd5_ra9, steeringError_rd5_ra9, distToLeftLaneEdge_rd5_ra9, accelerator_rd5_ra9, aheadLaneBearing_rd5_ra9, stat3_of_steeringWheel_accel, stat3_of_steeringError_crossLaneVelocity_distToLeftLaneEdge_aheadLaneBearing, steeringWheel_ent15_ra9, accelerator_ent15_ra9, steeringWheel_rv9, accelerator_rv9, distToLeftLaneEdge_rv9, and crossLaneVelocity_rv9.

Variable Selection for Inattention

In variable selection experiments we are attempting to determine the relative importance of each variable to the task of inattention detection. First a Random Forest classifier is trained for inattention detection (see also Sec. 6.2). Variable importances can then be extracted from the trained forest using the approach that was outlined in Sec 5.

We present the results in Table 4 a nd Table 5. These tables provide answers to the question "Which sensors are most important in detecting driver's inattention?" When just the two basic "driver control" sensors were used, some new derived variables may provide as much new information as the original signals, namely the running variance and entropy of steering. When CAS sensors are combined, the situation changes: lane position (distToLeftLaneEdge) becomes the most important variable together with the accelerator pedal. Steering wheel variance becomes the most important variable related to steering.

Inattention Detectors

Detection tasks always have a tradeoff between desired recall and precision. Recall denotes the percentage of total events of interest detected. Precision denotes

Table 4. Important sensor signals for inattention detection derived from steering wheel and accelerator pedal

Variable	Importance
steeringWheel	100.00
accelerator	87.34
steeringWheel_rv9	68.89
steeringWheel_ent15_ra9	58.44
stat3_of_steeringWheel_accelerator	41.38
accelerator_ent15_ra9	40.43
accelerator_rv9	35.86
steeringWheel_rd5_ra9	32.59
accelerator_rd5_ra9	29.31

Table 5. Important sensor signals for inattention detection derived from steering wheel, accelerator pedal and CAS sensors

Variable	Importance
distToLeftLaneEdge	100.00
accelerator	87.99
steeringWheel_rv9	73.76
distToLeftLaneEdge_rv9	65.44
distToLeftLaneEdge_rd5_ra9	65.23
steeringWheel	64.54
Stat3_of_steeringWheel_accel	60.00
steeringWheel_ent15_ra9	57.39
steeringError	57.32
aheadLaneBearing_rd5_ra9	55.33
aheadLeneBearing	51.85
crossLaneVelocity	50.55
Stat3_of_steeringError_crossLaneVelocity_distToLeftLaneEdge_aheadLaneBearing	38.07
crossLaneVelocity_rv9	36.50
steeringError_rd5_ra9	33.52
Accelerator_ent15_ra9	29.83
Accelerator_rv9	28.69
steeringWheel_rd5_ra9	27.76
Accelerator_rd5_ra9	20.69
crossLaneAcceleration	20.64

the percentage of detected events that are true events of interest and not false detections. A trivial classifier that classifies every instant as a true event would have 100% recall (since none were missed), but its precision would be poor. On the other hand, if the classifier is so tuned that only events having high certainty are classified as true events, the recall would be low, missing most of the events, but its precision would be high, since among those that were classified as true events, only a few would be false detections. Usually any classifier has some means of tuning the threshold of detection. Where that threshold will be set depends on the demands of the application. It is also noteworthy to mention

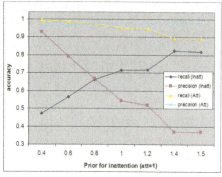 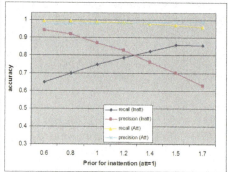

Fig. 9. Precision/recall figures for detection of inattention using only driver controls (left) and driver controls combined with CAS sensors (right). Random Forest is used as the classifier. Horizontal axis, prior for inattention, denotes a classifier parameter which can be tuned to produce different precision/recall operating points. This can be thought of as a weight or cost given to missing inattention events. Typically, if it is desirable not to miss events (high recall - in this case a high parameter value), the precision may be low - many false detections will be made. Conversely, if it is desirable not to make false detections (high precision), the recall will be low - not all events will be detected. Thus, as a figure of merit we use equal accuracy, the operating point where precision equals recall.

that in tasks involving detection of rare events, overall classification accuracy is not a meaningful measure. In our case only 7.3% of the database was inattention so a trivial classifier classifying everything as attention would thus have an accuracy of 92.7%. Therefore we will report our results using the recall and precision statistics for each class.

First, we constructed a Random Forests (RF) classifier of 75 trees using either the driver controls or the driver controls combined with the CAS sensors. Figure 9 depicts the resulting recall/precision graphs.

One simple figure of merit that allows comparison of two detectors is equal error rate, or equal accuracy, which denotes the intersection of recall and precision curves. By that figure, basing the inattention detector only on driver control sensors results in an equal accuracy of 67% for inattention and 97% for attention, whereas adding CAS sensors raises the accuracies up to 80% and 98%, respectively. For comparison, we used the same data to train a quadratic classifier [29]. Compared to the RF classifier, the quadratic classifier performs poorly in this task. We present the results in Table 6 for two different operating points of the quadratic classifier. The first one (middle rows) is tuned not to make false alarms, but its recall rate remains low. The second one is compensated for the less frequent occurrences of inattention, but it makes false alarms about 28% of the time. Random Forest clearly outperforms the quadratic classifier with almost no false alarms and good recall.

Table 6. Comparison of Random Forest (RF) to Quadratic classifier using equal accuracy as the figure of merit (operating point where recall of the desired events equals the precision of the detector)

Detector	Sensors	Inattention	Attention
RF	Driver control sensors only	67%	97%
RF	Additional CAS sensors	80%	98%
Quadratic	Driver control sensors only	29%	91%
Quadratic	Additional CAS sensors	33%	93%
Quadratic (prior compensated)	Driver control sensors only	58%	59%
Quadratic (prior compensated)	Additional CAS sensors	72%	72%

Future Driver Inattention Work

The experiments described in this section are only our first steps in investigating how driver attention can be detected. We have several ideas of how to improve the accuracy of detectors based on modifications to our described approach. The first technique will be to treat inattentive periods as longer time segments that actually have a weighting mechanism that prohibits rapid toggling between states. Also, an edge detector could be trained to detect the transitions between attention/inattention states instead of states for individual time samples. Even with improvements we will end up with a less than perfect inattention detector and we will have to study user experiences in order to define levels of accuracy required before inattention detectors could be used as an acceptable driver assistant tool.

Future work should also include modeling how drivers "recover" from inattentive periods. Even with perfect eye tracking it is unlikely that when a driver's eyes return to the road that the driver is instantly attentive and aware of his environment. This is a general attention modeling problem and is not specific to our technique.

Once driver inattention is detected there still needs to be experimentation on how to best assist the driver. Good driving habits will include periods that we have defined as inattentive, such as a "blind spot" glance before changing lanes. The system must understand the appropriate frequency and duration of these "blind spot" glances and not annoy the driver by offering counter-productive or unreasonable advice.

7 Conclusion

In this chapter, we have demonstrated three instances of using computational intelligence techniques such as the random forest method in developing intelligent driver assistance systems. Random Forest appears to be a very well suited tool for massive heterogeneous data sets that are generated from the driving domain. A driver activity classifier based on Random Forests is also fast enough to run in real time in a vehicle.

First, the random forest method is used to create a semi-automatic data annotation tool for driving database creation to support data-driven approaches in the driving domain such as driving state classification. The tool significantly reduces the manual annotation effort and thus enables the user to verify automatically generated annotations, rather than annotating from scratch. In experiments going through the whole database annotation cycle, we have observed six-fold reductions in the annotation time by a human annotator [25].

Second, the random forest method is employed to identify the sensor variables that are important for determining the driving maneuvers. Different combinations of sensor variables are identified for different driving maneuvers. For example, steering wheel angle is important for determining the turning maneuver, and brake status, acceleration, and headway distance are important for determining the panic brake maneuver.

Third, our random forest technique enabled us to detect driver inattention through the use of sensors that are available in modern vehicles. We compared both traditional sensors (detecting only steering wheel angle and accelerator pedal position) and CAS sensors against the performance of a state-of-the-art eye/head tracker. As expected, the addition of CAS sensors greatly improve the ability for a system to detect inattention. Though not as accurate as eye tracking, a significant percentage of inattentive time samples could be detected by monitoring readily available sensors (including CAS sensors) and it is believed that a driver assistant system could be built to use this information to improve driver attention. The primary advantage of our system is that it requires only a small amount of code to be added to existing vehicles and avoids the cost and complexity of adding driver monitors such as eye/head trackers.

A driving simulator is an excellent tool to investigate driver inattention since this allows us to design experiments and collect data on driver behaviors that may impose a safety risk if these experiments were performed in a real vehicle. There is still much to be learned about the causes and effects of driver inattention, but even small steps toward detecting inattention can be helpful in increasing driver performance.

References

[1] Boer, E.R.: Behavioral entropy as an index of workload. In: Proceedings of the IEA/HFES 2000 Congress, pp. 125–128 (2000)
[2] Boer, E.R.: Behavioral entropy as a measure of driving performance. In: Driver Assessment, pp. 225–229 (2001)
[3] Bousquet, O., Elisseeff, A.: Algorithmic stability and generalization performance. In: Proc. NIPS, pp. 196–202 (2000)
[4] Breiman, L.: Bagging predictors. Machine Learning 24(2), 123–140 (1996)
[5] Breiman, L.: Random forests. Machine Learning 45(1), 5–32 (2001)
[6] Breiman, L., Friedman, J.H., Olshen, R.A., Stone, C.J.: Classification and Regression Trees. CRC Press, Boca Raton (1984)
[7] Cohn, D., Atlas, L., Ladner, R.: Improving generalization with active learning. Machine Learning 15(2), 201–221 (1994)

8. Dingus, T.A., Klauer, S.G., Neale, V.L., Petersen, A., Lee, S.E., Sudweeks, J., Perez, M.A., Hankey, J., Ramsey, D., Gupta, S., Bucher, C., Doerzaph, Z.R., Jermeland, J., Knipling, R.R.: The 100 car naturalistic driving study: Results of the 100-car field experiment performed by virginia tech transportation institute. Report DOT HS 810 593, National Highway Traffic Safety Administration, Washington D.C. (April 2006)
9. Fletcher, L., Apostoloff, N., Petersson, L., Zelinsky, A.: Driver assistance systems based on vision in and out of vehicles. In: Proceedings of IEEE Intelligent Vehicles Symposium, pp. 322–327. IEEE, Los Alamitos (2003)
10. Forbes, J., Huang, T., Kanazawa, K., Russell, S.: The BATmobile: Towards a Bayesian automated taxi. In: Proc. Fourteenth International Joint Conference on Artificial Intelligence, Montreal, Canada (1995)
11. Guyon, I., Elisseeff, A.: An introduction to feature selection. Journal of Machine Learning Research 3, 1157–1182 (2003)
12. Guyon, I., Gunn, S., Nikravesh, M., Zadeh, L.: Feature Extraction, Foundations and Applications. Springer, Heidelberg (2006)
13. Hankey, J., Dingus, T.A., Hanowski, R.J., Wierwille, W.W., Monk, C.A., Moyer, M.J.: The development of a design evaluation tool and model of attention demand. Report 5/18/00, National Highway Traffic Safety Administration, Washington D.C. (May 18, 2000)
14. Hess, R.A., Modjtahedzadeh, A.: A preview control model of driver steering behavior. In: Proceedings of IEEE International Conference on Systems, Man and Cybernetics, pp. 504–509 (November 1989)
15. Hess, R.A., Modjtahedzadeh, A.: A control theoretic model of driver steering behavior. IEEE Control Systems Magazine 10(5), 3–8 (1990)
16. Liu, H., Motoda, H.: Feature Selection for Knowledge Discovery and Data Mining. Kluwer Academic Publishers, Dordrecht (1998)
17. McCall, J.C., Trivedi, M.M.: Driver behavior and situation aware brake assistance for intelligent vehicles. Proceedings of the IEEE 95(2), 374–387 (2007)
18. Neale, V.L., Klauer, S.G., Knipling, R.R., Dingus, T.A., Holbrook, G.T., Petersen, A.: The 100 car naturalistic driving study: Phase 1-experimental design. Interim Report DOT HS 809 536, Department of Transportation, Washington D.C, Contract No: DTNH22-00-C-07007 by Virginia Tech Transportation Institute (November 2002)
19. Ng, A., Jordan, M., Weiss, Y.: On spectral clustering: Analysis and an algorithm. In: Advances in Neural Information Processing Systems 14: Proceedings of the NIPS 2001 (2001)
20. Oza, N.: Probabilistic models of driver behavior. In: Proceedings of Spatial Cognition Conference, Berkeley, CA (1999)
21. Pompei, F.J., Sharon, T., Buckley, S.J., Kemp, J.: An automobile-integrated system for assessing and reacting to driver cognitive load. In: Proceedings of Convergence 2002, pp. 411–416. IEEE SAE, Los Alamitos (2002)
22. Rabiner, L.R.: A tutorial on Hidden Markov Models and selected applications in speech recognition. Proceedings of the IEEE 77(2), 257–286 (1989)
23. Remboski, D., Gardner, J., Wheatley, D., Hurwitz, J., MacTavish, T., Gardner, R.M.: Driver performance improvement through the driver advocate: A research initiative toward automotive safety. In: Proceedings of the 2000 International Congress on Transportation Electronics, SAE P-360, pp. 509–518 (2000)
24. Schölkopf, B., Smola, A.: Learning with Kernels. MIT Press, Cambridge (2002)

[25] Schreiner, C., Torkkola, K., Gardner, M., Zhang, K.: Using machine learning techniques to reduce data annotation time. In: Proceedings of the 50th Annual Meeting of the Human Factors and Ergonomics Society, San Francisco, CA, October 16-20 (2006)
[26] Smith, P., Shah, M., da Vitoria Lobo, N.: Adetermining driver visual attention with one camera. IEEE Transactions on Intelligent Transportation Systems 4(4), 205 (2003)
[27] Torkkola, K.: Automatic alignment of speech with phonetic transcriptions in real time. In: Proceedings of the IEEE International Conference on Acoustics, Speech and Signal Processing (ICASSP 1988), New York City, USA, April 11-14, pp. 611–614 (1988)
[28] Torkkola, K., Gardner, M., Schreiner, C., Zhang, K., Leivian, B., Summers, J.: Sensor selection for driving state recognition. In: Proceedings of the World Congress on Computational Intelligence (WCCI), IJCNN, Vancouver, Canada, June 16-21, pp. 9484–9489 (2006)
[29] Torkkola, K., Massey, N., Leivian, B., Wood, C., Summers, J., Kundalkar, S.: Classification of critical driving events. In: Proceedings of the International Conference on Machine Learning and Applications (ICMLA), Los Angeles, CA, USA, June 23-24, pp. 81–85 (2003)
[30] Torkkola, K., Massey, N., Wood, C.: Driver inattention detection through intelligent analysis of readily available sensors. In: Proceedings of the 7th Annual IEEE Conference on Intelligent Transportation Systems (ITSC 2004), Washington, D.C., USA, October 3-6, pp. 326–331 (2004)
[31] Torkkola, K., Venkatesan, S., Liu, H.: Sensor sequence modeling for driving. In: Proceedings of the 18th International FLAIRS Conference, Clearwater Beach, FL, USA, May 15-17. AAAI Press, Menlo Park (2005)
[32] Van Winsum, W., Martens, M., Herland, L.: The effects of speech versus tactile driver support messages on workload, driver behaviour and user acceptance. TNO-report TM-00-C003, TNO, Soesterberg, The Netherlands (1999)
[33] Wood, C., Leivian, B., Massey, N., Bieker, J., Summers, J.: Driver advocate tool. In: Driver Assessment (2001)

Computer Vision and Machine Learning for Enhancing Pedestrian Safety

Tarak Gandhi and Mohan Manubhai Trivedi

Laboratory for Safe and Intelligent Vehicles (LISA), University of California
San Diego, La Jolla, CA 92093
tgandhi@ucsd.edu, mtrivedi@ucsd.edu

Summary. Accidents involving pedestrians is one of the leading causes of death and injury around the world. Intelligent driver support systems hold a promise to minimize accidents and save many lives. Such a system would detect the pedestrian, predict the possibility of collision, and then warn the driver or engage automatic braking or other safety devices. This chapter describes the framework and issues involved in developing a pedestrian protection system. It is emphasized that the knowledge of the state of the environment, vehicle, and driver are important for enhancing safety. Classification, clustering, and machine learning techniques for effectively detecting pedestrians are discussed, including the application of algorithms such as SVM, Neural Networks, and AdaBoost for the purpose of distinguishing pedestrians from background. Pedestrians unlike vehicles are capable of sharp turns and speed changes, therefore their future paths are difficult to predict. In order to estimate the possibility of collision, a probabilistic framework for pedestrian path prediction is described along with related research. It is noted that sensors in vehicle are not always sufficient to detect all the pedestrians and other obstacles. Interaction with infrastructure based systems as well as systems from other vehicles can provide a wide area situational awareness of the scene. Furthermore, in infrastructure based systems, clustering and learning techniques can be applied to identify typical vehicle and pedestrian paths and to detect anomalies and potentially dangerous situations. In order to effectively integrate information from infrastructure and vehicle sources, the importance of developing and standardizing vehicle-vehicle and vehicle-infrastructure communication systems is also emphasized.

1 Introduction

Intelligent Transportation Systems (ITS) show promise of making road travel safer and comfortable. Automobile companies have recently taken considerable interest in developing Intelligent Driver Support Systems (IDSS) for high-end vehicles. These include active cruise control, lane departure warning, blind spot monitoring, and pedestrian detection systems based on sensors such as visible light and thermal infrared cameras, RADARs, or LASER scanners. However, for an effective driver support system, it is desirable to take a holistic approach, using all available data from the environment, vehicle dynamics, and the driver that can be obtained using various sensors incorporated in vehicle and infrastructure [35].

Infrastructure based sensors can complement the vehicle sensors by filling gaps and providing more complete information about the surroundings. Looking in the vehicle at driver's state is as important as looking out in surroundings in order to convey warnings to the driver in the most effective and least distracting manner. Furthermore, due to the highly competitive nature of automobile manufacturing, it is necessary to make such systems cost effective. This makes multi-functional sensors that are used by several of these systems highly desirable.

Accidents involving pedestrians and other vulnerable road users such as bicyclists are one of the leading causes of death and injury around the world. In order to reduce these accidents, pedestrian protection systems need to detect pedestrians, track them over time, and predict the possibility of collision based on the paths that the pedestrian and the vehicle are likely to take. The system should relay the information to the driver in efficient and non-distracting manner or to the control system of the vehicle in order to take preventive actions. Considerable efforts have been made on enhancing pedestrian safety by programs in United states [3; 4], Europe [2; 5] and Japan [1]. Conferences such as Intelligent Vehicles Symposium [6] and Intelligent Transportation Systems Conference [7] have a number of publications related to pedestrian detection every year. The recent survey on pedestrian protection [19] has covered the current research on pedestrian detection, tracking, and collision prediction. It is observed that detecting pedestrians in cluttered scenes from a moving vehicle is a challenging problem that involves a number of computational intelligence techniques spanning image processing, computer vision, pattern recognition, and machine learning. This paper focuses on specific computational intelligence techniques used in stages of sensor based pedestrian protection for detecting and classifying pedestrians, and predicting their trajectories to assess the possibility of collision.

2 Framework for Pedestrian Protection System

Fig. 1 shows the components of a general pedestrian protection system. The data from one or more types of sensors can be processed using computer vision algorithms to detect pedestrians and determine their trajectories. The trajectories can then be sent to collision prediction module that would predict the probability of collision between the host vehicle and pedestrians. In the case of high probability of collision, the driver is given appropriate warning that enables corrective action. If the collision is imminent, the automatic safety systems could also be triggered to decelerate the vehicle and reduce the impact of collision. In the following sections we illustrate these components using examples focusing on approaches used for these tasks.

3 Techniques in Pedestrian Detection

Pedestrian detection is usually divided into two stages. The candidate generation stage processes raw data using simple cues and fast algorithms to identify potential pedestrian candidates. The classification and verification stage then applies

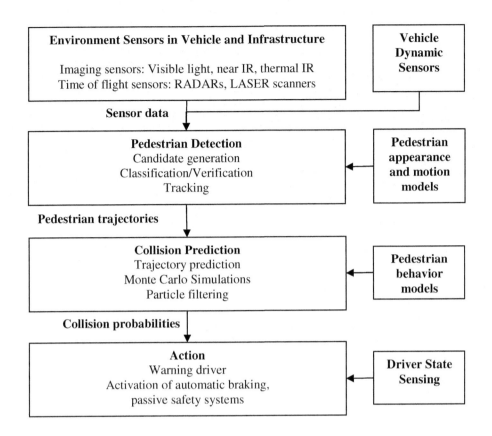

Fig. 1. Data flow diagram for pedestrian protection systems

more complex algorithms to the candidates from the attention focusing stage in order to separate genuine pedestrians from false alarms. However, the line between these stages is often blurred and some approaches combine the stages into one. Table 1 shows the approaches used by researchers for stages in pedestrian detection.

3.1 Candidate Generation

Cues such as shape, appearance, motion, and distance can be used to generate potential pedestrian candidates. Here, we describe selected techniques used for generating pedestrian candidates using these cues.

Chamfer Matching

Chamfer matching is a generic technique to recognize objects based on their shape and appearance using hierarchical matching with a set of object templates from training images. The original image is converted to a binary image using

Table 1. Approaches used in stages of pedestrian protection

Publication	Candidate generation	Feature extraction	Classification
Gavrila ECCV00 [20], IJCV07 [21], Munder PAMI06 [27]	Chamfer matching	Image ROI pixels	LRF Neural Network
Gandhi MM04 [15], MVA05 [16]	Omni camera based planar motion estimation		
Gandhi ICIP05 [17], ITS06 [18]	Stereo based U disparity analysis		
Krotosky IV06 [25]	Stereo based U and V disparity analysis	Histogram of oriented gradients	Support Vector Machine
Papageorgiou IJCV00 [28]		Haar wavelets	Support Vector Machine
Dalal CVPR05 [13]		Histogram of oriented gradients	Support Vector Machine
Viola, IJCV05 [37]		Haar-like features in spatial and temporal domain	AdaBoost
Park ISI07 [29]	Background subtraction, homography projection	Shape and size of object	

an edge detector. A distance transform is then applied to the edge image. Every pixel $r = (x, y)$ in the distance transformed image has a value $d_I(t)$ equal to the distance to the nearest edge pixel:

$$d_I(r) = \min_{r' \in edges(I)} \|r' - r\| \tag{1}$$

The distance transformed image is matched to the binary templates generated from examples. For this purpose, the template is slid over the image and at every displacement, the chamfer distance between the image and template is obtained by taking the mean of all the pixels in the distance transform image that have an 'on' pixel in template image:

$$D(T, I) = \frac{1}{|T|} \sum_{r \in T} d_I(r) \tag{2}$$

Positions in the image where the chamfer distance is less than a threshold are considered as successful matches.

In order to account for the variations between individual objects, the image is matched with number of templates. For efficient matching, a template hierarchy is generated using bottom up clustering. All the training templates are grouped into K clusters, each represented by a prototype template p_k and the set of templates S_k in the cluster. Clustering is performed using using an iterative optimization algorithm that minimizes an objective function:

$$E = \sum_{k=1}^{K} \max_{t_i \in S_k} D_{min}(t_i, p_k) \tag{3}$$

where $D_{min}(t_i, p_k)$ denotes the minimum chamfer distance between the template t_i and the prototype p_k for all relative displacements between them. The process is repeated recursively by treating the prototypes as templates and re-clustering them to form a tree as shown in Fig. 2.

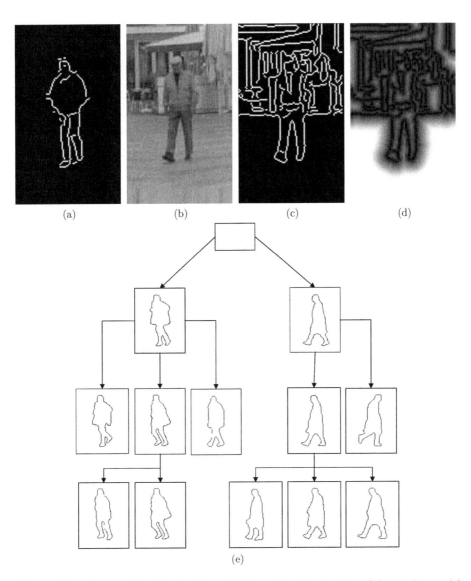

Fig. 2. Chamfer matching illustration: (a) Template of pedestrian (b) Test Image (c) Edge image (d) Distance Transform image (e) Template Hierarchy (partial) used for matching with distance transform. (Figure based on [27]).

For recognition, the given image is recursively matched with the nodes of the template tree, starting from the root. At any level, the branches where the minimum chamfer distance is greater than a threshold are pruned to reduce the search time. For remaining nodes, matching is repeated for all children nodes. Candidates are generated at image positions where the chamfer distance with any of the template nodes is below threshold.

Motion-based Detection

Motion is an important cue in detecting pedestrians. In the case of moving platforms, the background undergoes ego-motion that depends on camera motion as well as the scene structure, which needs to be accounted for. In [15; 16], a parametric planar motion model is used to describe the ego-motion of the ground in an omnidirectional camera. The perspective coordinates of a point P on the ground in two consecutive camera positions is governed by a homography matrix H as:

$$\begin{pmatrix} X_b \\ Y_b \\ Z_b \end{pmatrix} = \lambda \begin{pmatrix} h_{11} & h_{12} & h_{13} \\ h_{21} & h_{32} & h_{33} \\ h_{31} & h_{32} & 1 \end{pmatrix} \begin{pmatrix} X_a \\ Y_a \\ Z_a \end{pmatrix} \quad (4)$$

The perspective camera coordinates can be mapped to pixel coordinates (u_a, v_a) and (u_b, v_b) using the internal calibration of the camera:

$$\begin{aligned} (u_a, v_a) &= F_{int}([X_a, Y_a, Z_a]^T), \\ (u_b, v_b) &= F_{int}([X_b, Y_b, Z_b]^T) = F_{int}(HF_{int}^{-1}(u_a, v_a)) \end{aligned} \quad (5)$$

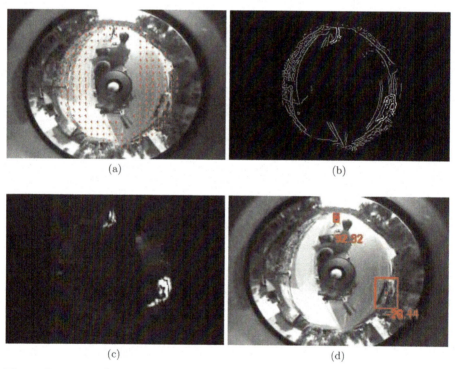

Fig. 3. Detection of people and vehicles using an omnidirectional camera on a moving platform [15] (a) Estimated motion based on parametric motion of ground plane (b) Image pixels used for estimation. Gray pixels are inliers, and white pixels are outliers. (c) Motion-compensated image difference that captures independently moving objects (d) Detected vehicle and person.

The image motion of the point satisfies the optical flow constraint:

$$g_u(u_b - u_a) + g_v(v_b - v_a) = -g_t + \nu \tag{6}$$

where g_u, g_v, and g_t are the spatial and temporal image gradients and ν is the noise term.

Based on these relations, the parameters of the homography matrix can be estimated using the spatio-temporal image gradients at every point (u_a, v_a) in the first image using non-linear least squares. Based on these parameters, every point (u_a, v_a) on the ground plane in first frame corresponds to a point (u_b, v_b) in the second frame. Using this transformation, the second image can be transformed to first frame, compensating the image motion of the ground plane. The objects that have independent motion or height above ground do not obey the motion model and their motion is not completely compensated. Taking the motion compensated frame difference between adjacent video frames highlights these areas that are likely to contain pedestrians, vehicles, or other obstacles. These regions of interest can then be classified using the classification stage. Fig. 3 shows detection of a pedestrian and vehicle from a moving platform. The details of the approach are described in [15].

Depth Segmentation using Binocular Stereo

Binocular stereo imaging can provide useful information about the depth of the objects from the cameras. This information has been used for disambiguating pedestrians and other objects from features on ground plane [18; 25; 26], segmenting images based on layers with different depths [14], handling occlusion between pedestrians [25], and using the size-depth relation to eliminate extraneous objects [32]. For a pair of stereo cameras with focal length f and baseline distance of B between the cameras situated at the height of H above the road as shown in Fig. 4 (a). An object at distance D will have disparity of $d = Bf/D$ between the two cameras, that is inversely proportional to object distance. The ground in the same line of sight is farther away and has a smaller disparity of $d_{bg} = Bf/D_{bg}$. Based on this difference, objects having height above the ground can be separated.

Stereo disparity computation can be performed using software packages such as SRI Stereo Engine [24]. Such software produces a disparity map that gives disparities of individual pixels in the image. In [17], the concept of U-disparity proposed by Labayrade et al. [26] is used to identify potential obstacles in the scene using images from a stereo pair of omnidirectional cameras as shown in Fig. 4 (b). The disparity image $disp(u,v)$ generated by a stereo engine separates the pedestrian in a layer of nearly constant depth. U-disparity $udisp(u,d)$ image counts occurrences of every disparity d for each column u in the image. In order to suppress the ground plane pixels, only the pixels with disparity significantly greater than ground plane disparity are used.

$$udisp(u,d) = \#\{v | disp(u,v) = d, \ d > d_{bg}(u,v) + d_{thresh}\} \tag{7}$$

where $\#$ stands for number of elements in the set.

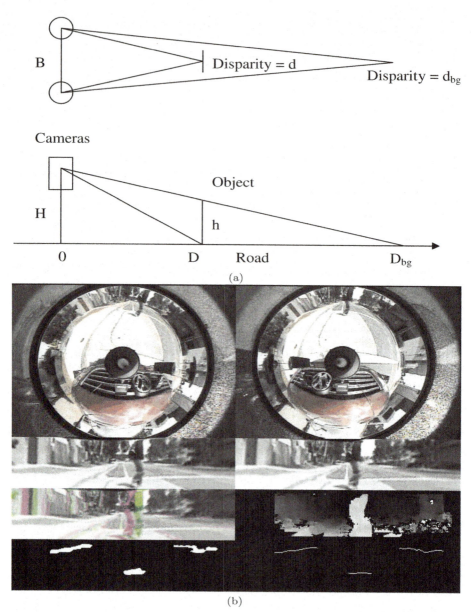

Fig. 4. (a) Stereo geometry (top and side views). The disparity between image positions in the two cameras decrease with object distance. (b) Stereo based pedestrian candidate generation [17]: Row 1: Color images from a stereo pair of omni camera containing pedestrian. Row 2: Virtual front view images generated from omni images. Row 3: Superimposed front view images and disparity image with lighter shades showing nearer objects with larger disparity. Row 4: U-disparity image taking histogram of disparities for each column. The lower middle segment in U-disparity image corresponds to the pedestrian. Other segments corresponds to more distant structures above the ground. Figure based on [17].

Pixels in an object at a particular distance would have nearly same disparity and therefore form a horizontal ridge in the disparity histogram image. Even if disparities of individual object pixels are inaccurate, the histogram image clusters the disparities and makes it easier to isolate the objects. Based on the position of the line segments, the regions containing obstacles can be identified. The nearest (lowest) region with largest disparity corresponds to the pedestrian. The parts of the virtual view image corresponding to the U-disparity segments can then be sent to classifier for distinguishing between pedestrians and other objects.

3.2 Candidate Validation

The candidate generation stage generates regions of interest (ROI) that are likely to contain a pedestrian. Characteristic features are extracted from these ROIs and a trained classifier is used to separate pedestrian from the background and other objects. The input to the classifier is a vector of raw pixel values or characteristic features extracted from them, and the output is the decision showing whether a pedestrian is detected or not. In many cases, the probability or a confidence value of the match is also returned. Fig. 5 shows the flow diagram of validation stage.

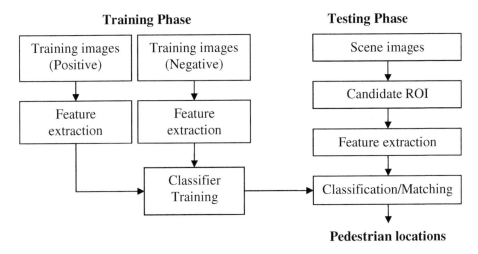

Fig. 5. Validation stage for pedestrian detection. Training phase uses positive and negative images to extract features and train a classifier. Testing phase applies feature extractor and classifier to candidate regions of interest in the images.

Feature Extraction

The features used for classification should be insensitive to noise and individual variations in appearance and at the same time able to discriminate pedestrians from other objects and background clutter. For pedestrian detection features such as Haar wavelets [28], histogram of oriented gradients [13], and Gabor filter outputs [12], are used.

Haar Wavelets

An object detection system needs to have a representation that has high inter-class variability and low intra-class variability [28]. For this purpose, features must be identified at resolutions where there will be some consistency throughout the object class, while at the same time ignoring noise. Haar wavelets extract local intensity gradient features at multiple resolution scales in horizontal, vertical, and diagonal directions and are particularly useful in efficiently representing the discriminative structure of the object. This is achieved by sliding the wavelet functions in Fig. 6 over the image and taking inner products as:

$$w_k(m,n) = \sum_{m=0}^{2^k-1} \sum_{n=0}^{2^k-1} \psi_k(m',n') f(2^{k-j}m + m', 2^{k-j}n + n') \quad (8)$$

where f is the original image, ψ_k is any of the wavelet functions at scale k with support of length 2^k, and 2^j is the over-sampling rate. In the case of standard wavelet transforms, $k = 0$ and the wavelet is translated at each sample by the length of the support as shown in Fig. 6. However, in over-complete representations, $k > 0$ and the wavelet function is translated only by a fraction of the length of support. In [28] the over-complete representation with quarter length sampling is used in order to robustly capture image features. The wavelet transform can be concatenated to form a feature vector that is sent to a classifier. However, it is observed that some components of the transform have more discriminative information than others. Hence, it is possible to select such components to form a truncated feature vector as in [28] to reduce complexity and speed up computations.

Histograms of Oriented Gradients

Histograms of oriented gradients (HOG) have been proposed by Dalal and Triggs [13] to classify objects such as people and vehicles. For computing HOG, the region of interest is subdivided into rectangular blocks and histogram of gradient orientations is computed in each block. For this purpose, sub-images corresponding to the regions suspected to contain pedestrian are extracted from the original image. The gradients of the sub-image are computed using Sobel operator [22]. The gradient orientations are quantized into K bins each spanning an interval of $2\pi/K$ radians, and the sub-image is divided into $M \times N$ blocks. For each block (m, n) in the subimage, the histogram of gradient orientations is computed by counting the number of pixels in the block having the gradient direction of each bin k. This way, an $M \times N \times K$ array consisting of $M \times N$ local histograms is formed. The histogram is smoothed by convolving with averaging kernels in position and orientation directions to reduce sensitivity to discretization. Normalization is performed in order to reduce sensitivity to illumination changes and spurious edges. The resulting array is then stacked into a $B = MNK$ dimensional feature vector **x**. Fig. 7 shows examples with pedestrian snapshots along

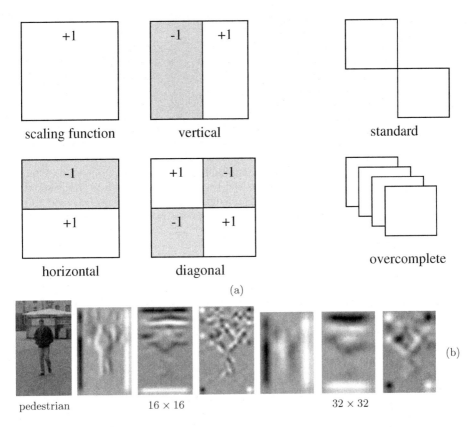

Fig. 6. Haar wavelet transform framework. Left: Scaling and wavelet functions at a particular scale. Right: Standard and overcomplete wavelet transforms. (Figure based on [28]).

with the HOG representation shown by red lines. The value of a histogram bin for a particular position and orientation is proportional to the length of the respective line.

Classification

The classifiers employed to distinguish pedestrians from non-pedestrian objects are usually trained using feature vectors extracted from a number of positive and negative examples to determine the decision boundary between them. After training, the classifier processes unknown samples and decides the presence or absence of the object based on which side of the decision boundary the feature vector lies. The classifiers used for pedestrian detection include Support Vector Machines (SVM), Neural Networks, and AdaBoost, which are described here.

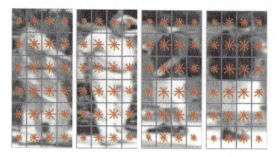

Fig. 7. Pedestrian subimages with computed Histograms of Oriented Gradients (HOG). The image is divided into blocks and the histogram of gradient orientations is individually computed for each block. The lengths of the red lines correspond to the frequencies of image gradients in the respective directions.

Support Vector Machines

The Support Vector Machine (SVM) forms a decision boundary between two classes by maximizing the 'margin' i.e. the separation between nearest examples on either side of the boundary [11]. SVM in conjunction with various image features are widely used for pedestrian recognition. For example, Papageorgiou and Poggio [28] have designed a general object detection system that they have applied to detect pedestrians for a driver assistance. The system uses SVM classifier on Haar wavelet representation of images. A support vector machine is trained using a large number of positive and negative examples from which the image features are extracted. Let $\mathbf{x}_i$ denote the feature vector of sample i and y_i denote one of the two class labels in $\{0, 1\}$. The feature vector $\mathbf{x}_i$ is projected into a higher dimensional kernel space using a mapping function Φ which allows complex non-linear decision boundaries. The classification can be formulated as an optimization problem to find a hyperplane boundary in the kernel space:

$$\mathbf{w}^T \Phi(\mathbf{x}_i) + b = 0 \tag{9}$$

using

$$\min_{\mathbf{w},b,\boldsymbol{\xi},\rho} \mathbf{w}^T \mathbf{w} - \nu\rho + \frac{1}{L}\sum_{i=1}^{L}\xi_i \tag{10}$$

subject to

$$\mathbf{w}^T \Phi(\mathbf{x}_i) + b \geq \rho - \xi_i \ , \ \xi_i \geq 0, i = 1\ldots L, \rho \geq 0$$

where ν is the parameter to accommodate training errors and $\boldsymbol{\xi}$ is used to account for some samples that are not separated by the boundary. Figure 8 illustrates the principle of SVM for classification of samples. The problem is converted into the dual form which is solved using quadratic programming [11]:

$$\min_{\boldsymbol{\alpha}} \sum_{i=1}^{L}\sum_{j=1}^{L} \alpha_i y_i K(\mathbf{x}_i, \mathbf{x}_j) y_j \alpha_j \tag{11}$$

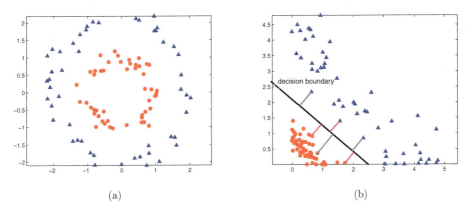

Fig. 8. Illustration of Support Vector Machine principle. (a) Two classes that cannot be separated by a single straight line. (b) Mapping into Kernel space. SVM finds a line separating two classes to minimize the 'margin', i.e. the distance to the closest samples called 'Support Vectors'.

subject to

$$0 \leq \alpha_i \leq 1/L, \sum_{i=1}^{L} \alpha_i \geq \nu, \sum_{i=1}^{L} \alpha_i y_i = 0 \quad (12)$$

where $K(\mathbf{x}_i, \mathbf{x}_j) = \Phi(\mathbf{x}_i)^T \Phi(\mathbf{x}_j)$ is the kernel function derived from the mapping function Φ, and represents the distance in the high-dimensional space. It should be noted that the kernel function is usually much easier to compute than the mapping function Φ. The classification is then given by the decision function:

$$D(\mathbf{x}) = \sum_{i=1}^{L} \alpha_i y_i K(\mathbf{x}_i, \mathbf{x}) + b \quad (13)$$

Neural Networks

Neural networks have been used to address problems in vehicle diagnostics and control [31]. They are particularly useful when the phenomenon to be modeled is highly complex but one has large amount of training data to enable learning of patterns from them. Neural networks can obtain highly non-linear boundaries between classes based on the training samples, and therefore can account for large shape variations. Zhao and Thorpe [41] have applied neural networks on gradient images of regions of interest to identify pedestrians. However, unconstrained neural networks require training of a large number of parameters necessitating very large training sets. In [21; 27], Gavrila and Munder use Local receptive fields (LRF) proposed by Wöhler and Anlauf [39] (Fig. 9) to reduce the number of weights by connecting each hidden layer neuron only to a local region of input image. Furthermore, the hidden layer is divided into a number of branches, each encoding a local feature, with all neurons within a branch sharing the same set of weights. Each hidden layer can be represented by the equation:

$$G_k(r) = f\left[\sum_i W_{ki} F(T(r) + \Delta r_i)\right] \quad (14)$$

where $F(p)$ denotes the input image as a function of pixel coordinates $p = (x, y)$, $G_k(r)$ denotes the output of the neuron with coordinate $r = (r_x, r_y)$ in the branch k of the hidden layer, W_{ki} are the shared weights for branch k, and $f(\cdot)$ is the activation function of the neuron. Each neuron with coordinates of r is associated with a region in the image around the transformed pixel $t = T(r)$, and Δr_i denote the displacements for pixels in the region. The output layer is a standard fully connected layer given by:

$$H_m = f\left[\sum_i w_{mk}(x, y) G_k(x, y)\right] \quad (15)$$

where H_m is the output of neuron m in output layer, w_{mk} is the weight for connection between output neuron m and hidden layer neuron in branch k with coordinate (x, y).

LeCun et al. [40] describe similar weight-shared and grouped networks for application in document analysis.

Adaboost Classifier

Adaboost is a scheme for forming a strong classifier using a linear combination of a number of weak classifiers based on individual features [36; 37]. Every weak classifier is individually trained on a single feature. For boosting the weak classifier, the training examples are iteratively re-weighted so that the samples which are incorrectly classified by the weak classifier are assigned larger weights. The final strong classifier is a weighted combination of weak classifiers followed by a thresholding step. The boosting algorithm is described as follows [36; 8]:

- Let $\mathbf{x}_i$ denote the feature vector and y_i denote one of the two class labels in $\{0, 1\}$ for negative and positive examples, respectively.
- Initialize weights w_i to $1/2M$ for each of the M negative samples and $1/2L$ for each of the L positive samples
- Iterate for $t = 1 \ldots T$
  - Normalize weights: $w_{t,i} \leftarrow w_{t,i} / \sum_k w_{t,k}$
  - For each feature j, train classifier h_j that uses only that feature. Evaluate weighted error for all samples as: $\epsilon_j = \sum_i w_{t,i} |h_j(\mathbf{x}_i) - y_i|$
  - Choose classifier h_t with lowest error ϵ_t
  - Update weights: $w_{t+1,i} \leftarrow w_{t,i} \left(\frac{\epsilon_t}{1-\epsilon_t}\right)^{1-|h_j(\mathbf{x}_i)-y_i|}$
  - The final strong classifier decision is given by the linear combination of weak classifiers and thresholding the result: $\sum_t \alpha_t h_t(\mathbf{x}) \geq \sum_t \alpha_t / 2$ where $\alpha_t = \log\left(\frac{1-\epsilon_t}{\epsilon_t}\right)$

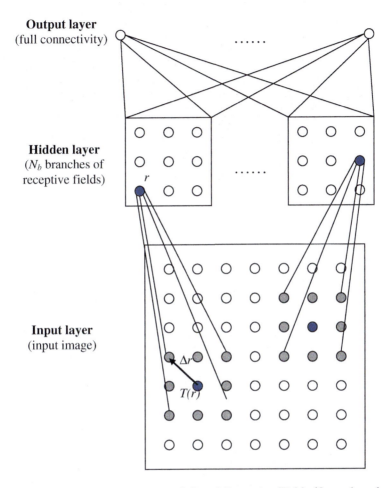

Fig. 9. Neural network architecture with Local Receptive Fields (figure based on [27])

4 Infrastructure Based Systems

Sensors mounted on vehicles are very useful for detecting pedestrians and other vehicles around the host vehicle. However, these sensors often cannot see objects that are occluded by other vehicles or stationary structures. For example, in the case of the intersection shown in Fig. 10, the host vehicle X cannot see the pedestrian P occluded by a vehicle Y as well as the vehicle Z occluded by buildings. Sensor C mounted on infrastructure would be able to see all these objects and help to fill the 'holes' in the fields of view of the vehicles. Furthermore, if vehicles can communicate with each other and the infrastructure, they can exchange information about objects that are seen by one but not seen by others. In the future, infrastructure based scene analysis as well as infrastructure-vehicle

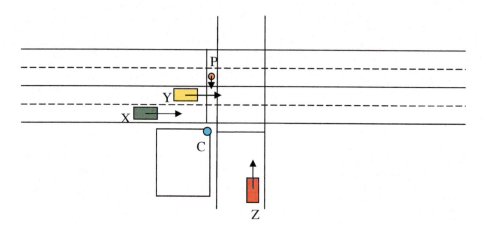

Fig. 10. Contribution of sensors mounted in infrastructure. Vehicle X cannot see pedestrian P or vehicle Z, but the infrastructure mounted camera C can see all of them.

and vehicle-vehicle communication will contribute towards robust and effective working of Intelligent Transportation Systems.

Cameras mounted in infrastructure have been extensively applied to video surveillance as well as traffic analysis [34]. Detection and tracking of objects from these cameras is easier and more reliable due to absence of camera motion. Background subtraction which is one of the standard methods to extract moving objects from stationary background is often employed, followed by classification of objects and activities.

4.1 Background Subtraction and Shadow Suppression

In order to separate moving objects from background, a model of the background is generated from multiple frames. The pixels not satisfying the background model are identified and grouped to form regions of interest that can contain moving objects. A simple approach for modeling the background is to obtain the statistics of each pixel described by color vector $\mathbf{x} = (R, G, B)$ over time in terms of mean and variance. The mean and variance are updated at every time frame using:

$$\boldsymbol{\mu} \leftarrow (1-\alpha)\boldsymbol{\mu} + \alpha \mathbf{x}$$
$$\sigma^2 \leftarrow (1-\alpha)\sigma^2 + \alpha(\mathbf{x}-\boldsymbol{\mu})^T(\mathbf{x}-\boldsymbol{\mu}) \qquad (16)$$

If for a pixel at any given time, $\|\mathbf{x}-\boldsymbol{\mu}\|/\sigma$ is greater than a threshold (typically 2.5), the pixel is classified as foreground. Schemes have been designed that adjust the background update according to the pixel currently being in foreground or background. More elaborate models such as Gaussian Mixture Models [33] and codebook model [23] are used to provide robustness against fluctuating motion such as tree branches, shadows, and highlights.

An important problem in object-background segmentation is the presence of shadows and highlights of the moving objects, which need to be suppressed in order to get meaningful object boundaries. Prati et al. [30] have conducted a survey of approaches used for shadow suppression. An important cue for distinguishing shadows from background is that the shadow reduces the luminance value of a background pixel, with little effect on the chrominance. Highlights similarly increase the value of luminance. On the other hand, objects are more likely to have different color from the background and brighter than the shadows. Based on these cues, bright objects can often be separated from shadows and highlights.

4.2 Robust Multi-camera Detection and Tracking

Multiple cameras offer superior scene coverage from all sides, provide rich 3D information, and enable robust handling of occlusions and background clutter. In particular, they can help to obtain the representation of the object that is independent of viewing direction. In [29], multiple cameras with overlapping fields of view are used to track persons and vehicles. Points on the ground plane can be projected from one view to another using a planar homography mapping. If (u_1, v_1) and (u_2, v_2) are image coordinates of a point on ground plane in two views, they are related by the following equations:

$$u_2 = \frac{h_{11}u_1 + h_{12}v_1 + h_{13}}{h_{31}u_1 + h_{32}v_1 + h_{33}}, \; v_2 = \frac{h_{21}u_1 + h_{22}v_1 + h_{23}}{h_{31}u_1 + h_{32}v_1 + h_{33}} \qquad (17)$$

The matrix H formed from elements h_{ij} is the Homography matrix. Multiple views of the same object are transformed by planar homography which assumes that pixels lie on ground plane. Pixels that violate this assumption result in mapping to a skewed location. Hence, the common footage region of the object on ground can be obtained by intersecting multiple projections of the same object on the ground plane. The footage area on the ground plane gives an estimate of the size and the trajectory of the object, independent of the viewing directions of the cameras. Fig. 11 depicts the process of estimating the footage area using homography. The locations of the footage areas are then tracked using Kalman filter in order to obtain object trajectories.

4.3 Analysis of Object Actions and Interactions

The objects are classified into persons and vehicles based on their footage area. The interaction among persons and vehicles can then be analyzed at semantic level as described in [29]. Each object is associated with spatio-temporal interaction potential that probabilistically describes the region in which the object can be subsequent time. The shape of the potential region depends on the type of object (vehicle/pedestrian) and speed (larger region for higher speed), and is modeled as a circular region around the current position. The intersection of interaction potentials of two objects represents the possibility of interaction

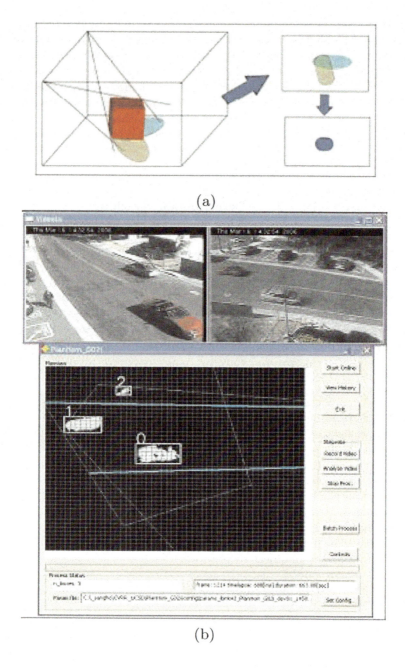

Fig. 11. (a) Homography projection from two camera views to virtual top views. The footage region is obtained by the intersection of the projections on ground plane. (b) Detection and mapping of vehicles and a person in virtual top view showing correct sizes of objects. [29]

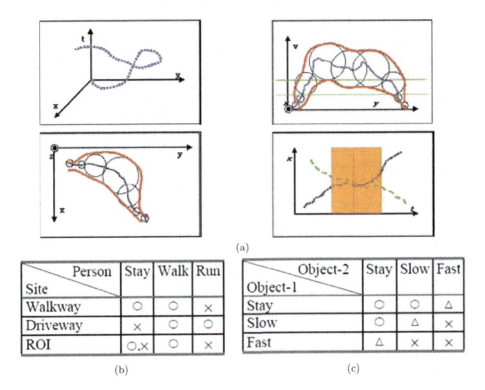

Fig. 12. (a) Schematic diagrams for trajectory analysis in spatio-temporal space. Circles represent interaction potential boundaries at a given space/time. Red curves represent the envelopes of the interaction boundary along tracks. (b) Spatial context dependency of human activity (c) Temporal context dependency of interactivity between two objects. Track patterns are classified into normal (○), cautious (△) and abnormal (×). [29].

between them as shown in Fig. 12 (a). They are categorized as safe or unsafe depending on the site context such as walkway or driveway, as well as motion context in terms of trajectories. For example, as shown in Fig. 12 (b), a person standing on walkway is normal scenario, whereas the person standing on driveway or road represents a potentially dangerous situation. Also, when two objects are moving fast, the possibility of collision is higher than when they are traveling slowly. This domain knowledge can be fed into the system in order to predict the severity of the situation.

5 Pedestrian Path Prediction

In addition to detection of pedestrians and vehicles, it is important to predict what path they are likely to take in order to estimate the possibility of collision. Pedestrians are capable of making sudden maneuvers in terms of the speed and direction of motion. Hence, probabilistic methods are most suitable for predicting the pedestrian's future path and potential collisions with vehicles. In fact, even for vehicles whose paths are easier to predict due to simpler dynamics, predictions beyond 1 or 2 seconds is still very challenging, making probabilistic methods valuable even for vehicles.

For probabilistic prediction, Monte-Carlo simulations can be used to generate a number of possible trajectories based on the dynamic model. The collision probability is then predicted based on the fraction of trajectories that eventually collide with the vehicle. Particle filtering [10] gives a unified framework for integrating the detection and tracking of objects with risk assessment as in [8]. Such a framework is shown in Figure 13 (a) with following steps:

1. Every tracked object can be modeled using a state vector consisting of properties such as 3-D position, velocity, dimensions, shape, orientation, and other appropriate attributes. The probability distribution of the state can then be modeled using a number of weighted samples randomly chosen according to the probability distribution.
2. The samples from the current state are projected to the sensor fields of view. The detection module would then produce hypotheses about the presence of vehicles. The hypotheses can then be associated with the samples to produce likelihood values used to update the sample weights.
3. The object state samples can be updated at every time instance using the dynamic models of pedestrians and vehicles. These models put constraints on how the pedestrian and vehicle can move over short and long term.
4. In order to predict collision probability, the object state samples are extrapolated over a longer period of time. The number of samples that are on collision course divided by the total number of samples gives the probability of collision.

Various dynamic models can be used for predicting the positions of the pedestrians at subsequent time. For example, in [38], Wakim et al. model the pedestrian dynamics using Hidden Markov Model with four states corresponding to standing still, walking, jogging, and running as shown in Figure 13 (b). For each state, the probability distributions of absolute speed as well as the change of direction is modeled by truncated Gaussians. Monte Carlo simulations are then used to generate a number of feasible trajectories and the ratio of the trajectories on collision course to total number of trajectories give the collision probability. The European project CAMELLIA [5] has conducted research in pedestrian detection and impact prediction based in part on [8; 38]. Similar to [38], they use a model for pedestrian dynamics using HMM. They use the position of pedestrian (sidewalk or road) to determine the transition probabilities between different gaits and orientations. Also, the change in orientation is modeled according to the side of the road that the pedestrian is walking.

In [9], Antonini et al. another approach called "Discrete Choice Model" which a pedestrian makes a choice at every step about the speed and direction of the next step. Discrete choice models associate a utility value to every such choice and select the alternative with the highest utility. The utility of each

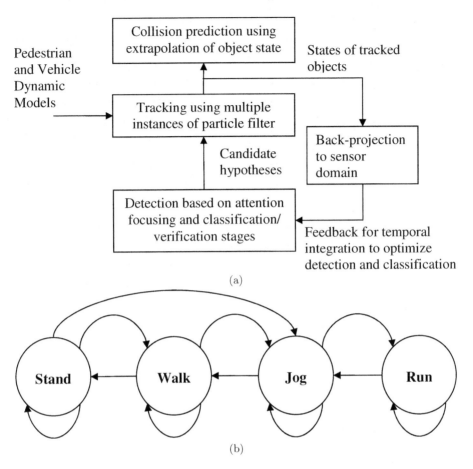

Fig. 13. (a) Integration of detection, tracking, and risk assessment of pedestrians and other objects based on particle filter [10] framework. (b) Transition diagram between states of pedestrians in [38]. The arrows between two states are associated with non-zero probabilities of transition from one state to another. Arrows on the same state corresponds to the pedestrian remaining in the same state in the next time step.

alternative is a latent variable depending on the attributes of the alternative and the characteristics of the decision-maker. This model is integrated with person detection and tracking from static cameras in order to improve performance. Instead of making hard decisions about target presence on every frame, it integrates evidence from a number of frames before making a decision.

6 Conclusion and Future Directions

Pedestrian detection, tracking, and analysis of behavior and interactions between pedestrians and vehicles are active research areas having important application in protection of pedestrians on road. Pattern classification approaches are particularly useful in detecting pedestrians and separating them from background. It is seen that pedestrian detection can be performed using sensors on vehicle itself or in the infrastructure. Vehicle based sensing gives continuous awareness of the scene around the vehicle and systems are being designed to detect pedestrians from vehicles. However, it is also seen that relying on vehicle sensors is not always sufficient to give full situational awareness of the scene. Infrastructure based sensors can play a complementary role of providing wide area scene coverage. For seamless integration of information from vehicle and infrastructure, efficient and reliable communication is needed. Communication can be performed at image level or object level. However, transmitting full images over the network is likely to be very expensive in terms of bandwidth. Hence, it would be desirable to perform initial detection locally and transmit candidate positions and trajectories along with sub-images of candidate bounding boxes as needed. Future research will be directed towards developing and standardizing these communications between vehicle and infrastructure to efficiently convey all the information needed to get complete situational awareness of the scene.

Acknowledgment

The authors thank the UC Discovery Grant with several automobile companies, Department of Defense Technical Support Working Group, National Science Foundation for the sponsorship of the research. The authors also thank the members in the Computer Vision and Robotics Research Laboratory including Dr. Stephen Krotosky, Brendan Morris, Erik Murphy-Chutorian, and Dr. Sangho Park for their contributions.

References

[1] Advanced Highway Systems Program, Japanese Ministry of Land, Infrastructure and Transport, Road Bureau, http://www.mlit.go.jp/road/ITS/index.html
[2] http://prevent.ertico.webhouse.net/en/home.htm, http://prevent.ertico.webhouse.net/en/prevent_subprojects/vulnerable_road_users_collision_mitigation/apalaci/.
[3] http://www.path.berkeley.edu/
[4] http://www.walkinginfo.org/pedsmart
[5] Deliverable 3.3b report on initial algorithms 2. Technical Report IST-2001-34410, CAMELLIA: Core for Ambient and Mobile intELLigent Imaging Applications (December 2003)
[6] IEEE Intelligent Vehicle Symposium, Istanbul, Turkey (June 2007)
[7] IEEE International Transportation Systems Conference, Seattle, WA (September 2007)
[8] Abramson, Y., Steux, B.: Hardware-friendly pedestrian detection and impact prediction. In: IEEE Intelligent Vehicle Symposium, pp. 590–595 (June 2004)
[9] Antonini, G., Venegas, S., Thiran, J.P., Bierlaire, M.: A discrete choice pedestrian behavior model for pedestrian detection in visual tracking systems. In: Proc. Advanced Concepts for Intelligent Vision Systems (September 2004)

[10] Arulampalam, S., Maskell, S., Gordon, N., Clapp, T.: A tutorial on particle filters for on-line non-linear/non-gaussian bayesian tracking. IEEE Transactions on Signal Processing 50(2), 174–188 (2002)
[11] Chang, C.-C., Lin, C.-J.: LIBSVM: A Library for Support Vector Machines (Last updated, June 2007)
[12] Cheng, H., Zheng, N., Qin, J.: Pedestrian detection using sparse gabor filters and support vector machine. In: IEEE Intelligent Vehicle Symposium, pp. 583–587 (June 2005)
[13] Dalal, N., Triggs, B.: Histograms of oriented gradients for human detection. In: Proc. IEEE Conference on Computer Vision and Pattern Recognition (June 2005)
[14] Franke, U.: Real-time stereo vision for urban traffic scene understanding. In: IEEE Intelligent Vehicle Symposium, pp. 273–278 (2000)
[15] Gandhi, T., Trivedi, M.M.: Motion analysis for event detection and tracking with a mobile omni-directional camera. Multimedia Systems Journal, Special Issue on Video Surveillance 10(2), 131–143 (2004)
[16] Gandhi, T., Trivedi, M.M.: Parametric ego-motion estimation for vehicle surround analysis using an omnidirectional camera. Machine Vision and Applications 16(2), 85–95 (2005)
[17] Gandhi, T., Trivedi, M.M.: Vehicle mounted wide fov stereo for traffic and pedestrian detection. In: Proc. International Conference on Image Processing, pp. 2:121–124 (September 2005)
[18] Gandhi, T., Trivedi, M.M.: Vehicle surround capture: Survey of techniques and a novel omni video based approach for dynamic panoramic surround maps. IEEE Transactions on Intelligent Transportation Systems 7(3), 293–308 (2006)
[19] Gandhi, T., Trivedi, M.M.: Pedestrian protection systems: Issues, survey, and challenges. IEEE Transactions on Intelligent Transportation Systems 8(3) (2007)
[20] Gavrila, D.M.: Pedestrian detection from a moving vehicle. In: Proc. European Conference on Computer Vision, pp. 37–49 (2000)
[21] Gavrila, D.M., Munder, S.: Multi-cue pedestrian detection and tracking from a moving vehicle. International Journal of Computer Vision 73(1), 41–59 (2007)
[22] Gonzalez, R.C., Woods, R.E.: Digital Image Processing, 3rd edn. Prentice Hall, Upper Saddle River (2008)
[23] Kim, K., Chalidabhongse, T.H., Harwood, D., Davis, L.S.: Real-time foreground-background segmentation using codebook model. Real-Time Imaging 11(3), 172–185 (2005)
[24] Konolige, K.: Small vision system: Hardware and implementation. In: Eighth International Symposium on Robotics Research, pp. 111–116 (October 1997), http://www.ai.sri.com/~konolige/papers
[25] Krotosky, S.J., Trivedi, M.M.: A comparison of color and infrared stereo approaches to pedestrian detection. In: IEEE Intelligent Vehicles Symposium (June 2007)
[26] Labayrade, R., Aubert, D., Tarel, J.-P.: Real time obstacle detection in stereovision on non flat road geometry through V-disparity representation. In: IEEE Intelligent Vehicles Symposium, vol. II, pp. 646–651 (2002)
[27] Munder, S., Gavrila, D.M.: An experimental study on pedestrian classification. IEEE Transactions on Pattern Analysis and Machine Intelligence 28(11), 1863–1868 (2006)
[28] Papageorgiou, C., Poggio, T.: A trainable system for object detection. International Journal of Computer Vision 38(1), 15–33 (2000)

[29] Park, S., Trivedi, M.M.: Video Analysis of Vehicles and Persons for Surveillance. In: Chen, H., Yang, C. (eds.) Intelligent and Security Informatics: Techniques and Applications. Springer, Heidelberg (2007)
[30] Prati, A., Mikic, I., Trivedi, M.M., Cucchiara, R.: Detecting moving shadows: Algorithms and evaluation. IEEE Trans. on Pattern Analysis and Machine Intelligence, pp. 918–923 (July 2003)
[31] Prokhorov, D.V.: Neural Networks in Automotive Applications. In: Computational Intelligence in Automotive Applications. Studies in Computational Intelligence. Springer, Heidelberg (2008)
[32] Soga, M., Kato, T., Ohta, M., Ninomiya, Y.: Pedestrian detection with stereo vision. In: International Conference on Data Engineering (April 2005)
[33] Stauffer, C., Grimson, W.E.L.: Adaptive background mixture model for real-time tracking. In: Proc. IEEE Intl. Conf. on Computer Vision and Pattern Recognition, pp. 246–252 (1999)
[34] Trivedi, M.M., Gandhi, T., Huang, K.S.: Distributed interactive video arrays for event capture and enhanced situational awareness. IEEE Intelligent Systems, Special Issue on AI in Homeland Security 20(5), 58–66 (2005)
[35] Trivedi, M.M., Gandhi, T., McCall, J.: Looking-in and looking-out of a vehicle: Computer vision based enhanced vehicle safety. IEEE Transactions on Intelligent Transportation Systems 8(1), 108–120 (2007)
[36] Viola, P., Jones, M.J.: Rapid object detection using a boosted cascade of simple features. In: Proc. IEEE Conference on Computer Vision and Pattern Recognition, pp. I: 511–518 (June 2001)
[37] Viola, P., Jones, M.J., Snow, D.: Detecting pedestrians using patterns of motion and appearance. International Journal of Computer Vision 63(2), 153–161 (2005)
[38] Wakim, C., Capperon, S., Oksman, J.: A markovian model of pedestrian behavior. In: Proc. IEEE Int. Conf. on Systems, Man, and Cybernetics, pp. 4028–4033 (October 2004)
[39] Wöhler, C., Anlauf, J.: An adaptable time-delay neural-network algorithm for image sequence analysis. IEEE Trans. on Neural Networks 10(6), 1531–1536 (1999)
[40] Bengio, Y., LeCun, Y., Bottou, L.: Gradient-based learning applied to document recognition. Proceedings of the IEEE 86(11), 2278–2324 (1998)
[41] Zhao, L., Thorpe, C.: Stereo and neural network-based pedestrian detection. IEEE Trans. Intelligent Transportation 1(3), 148–154 (2000)

Application of Graphical Models in the Automotive Industry

Matthias Steinbrecher, Frank Rügheimer, and Rudolf Kruse

Department of Knowledge Processing and Language Engineering
Otto-von-Guericke University of Magdeburg
Universitätsplatz 2, 39106 Magdeburg, Germany
{msteinbr,ruegheim,kruse}@iws.cs.uni-magdeburg.de

1 Introduction

The production pipeline of present day's automobile manufacturers consists of a highly heterogeneous and intricate assembly workflow that is driven by a considerable degree of interdependencies between the participating instances as there are suppliers, manufacturing engineers, marketing analysts and development researchers. Therefore, it is of paramount importance to enable all production experts to quickly respond to potential on-time delivery failures, ordering peaks or other disturbances that may interfere with the ideal assembly process. Moreover, the fast moving evolvement of new vehicle models require well-designed investigations regarding the collection and analysis of vehicle maintenance data. It is crucial to track down complicated interactions between car components or external failure causes in the shortest time possible to meet customer-requested quality claims.

To summarize these requirements, let us turn to an example which reveals some of the dependencies mentioned in this chapter. As we will later see, a normal car model can be described by hundreds of variables each of which representing a feature or technical property. Since only a small number of combinations (compared to all possible ones) will represent a valid car configuration, we will present a means of reducing the model space by imposing restrictions. These restrictions enter the mathematical treatment in the form of dependencies since a restriction may cancel out some options, thus rendering two attributes (more) dependent. This early step produces qualitative dependencies like "engine type and transmission type are dependent". To quantify these dependencies some uncertainty calculus is necessary to establish the dependence strengths. In our cases probability theory is used to augment the model, e. g. "whenever engine type 1 is ordered, the probability is 56% of having transmission type 2 ordered as well". There is a multitude of sources to estimate or extract this information from. When ordering peaks occur like an increased demand of convertibles during the Spring, or some supply shortages arise due to a strike in the transport industry, the model is used to predict vehicle configurations that may run into delivery delays in order to forestall such a scenario by e. g. acquiring alternative supply chains or temporarily shifting production load. Another part of the model may contain similar information for the aftercare, e. g. "whenever a warranty claim contained battery type 3, there is a 30% chance of having radio type 1 in the car". In this case dependencies are contained in the quality assessment data and are not known beforehand but are extracted to reveal possible hidden design flaws.

These examples — both in the realm of planning and subsequent maintenance measures — call for treatment methods that exploit the dependence structures embedded inside the application domains. Furthermore, these methods need to be equipped with dedicated updating, revision and refinement techniques in order to cope with the above-mentioned possible supply and demand irregularities. Since every production and planning stage involves highly specialized domain experts, it is necessary to offer intuitive system interfaces that are less prone to inter-domain misunderstandings.

The next section will sketch the underlying theoretical frameworks, after which we will present and discuss successfully applied planning and analysis methods that have been rolled out to production sites of two large automobile manufacturers. Section 3 deals with the handling of production planning at Volkswagen. The underlying data is sketched in section 3.1 which also covers the description of the model structure. Section 3.2 introduces three operations that serve the purpose of modifying the model and answering user queries. Finally, section 3.3 concludes the application report at Volkswagen. The Daimler AG application is introduced in section 4 which itself is divided to explain the data and model structure (section 4.1, to propose the visualization technique for data exploration (section 4.2 and finally to present empirical evidence of the usability in section 4.3.

2 Graphical Models

As motivated in the introduction, there are a lot of dependencies and independencies that have to be taken into account when to approach the task of planning and reasoning in complex domains. Graphical models are appealing since they provide a framework of modeling independencies between attributes and influence variables. The term "graphical model" is derived from an analogy between stochastic independence and node separation in graphs. Let $V = \{A_1, \ldots, A_n\}$ be a set of random variables. If the underlying probability distribution $P(V)$ satisfies some criteria (see e. g. [5; 13]), then it is possible to capture some of the independence relations between the variables in V using a graph $G = (V, E)$, where E denotes the set of edges. The underlying idea is to decompose the joint distribution $P(V)$ into lower-dimensional marginal oder conditional distributions from which the original distribution can be reconstructed with no or at least as few errors as possible [12; 14]. The named independence relations allow for a simplification of these factor distributions. We claim, that every independence that can be read from a graph also holds in the corresponding joint distribution. The graph is then called an independence map (see e. g. [4]).

2.1 Bayesian Networks

If we are dealing with an acyclic and directed graph structure G, the network is referred to as a Bayesian network. The decomposition described by the graph consists of a set of conditional distributions assigned to each node given its direct predecessors (parents). For each value of the attribute domains (dom), the original distribution can be reconstructed as follows:

$$\forall a_1 \in \text{dom}(A_1) : \cdots \forall a_n \in \text{dom}(A_n) :$$
$$P(A_1 = a_1, \ldots, A_n = a_n) = \prod_{A_i \in V} P\left(A_i = a_i \mid \bigwedge_{(A_j, A_i) \in E} A_j = a_j\right)$$

2.2 Markov Networks

Markov networks rely on undirected graphs where the lower-dimensional factor distributions are defined as marginal distributions on the cliques $C = \{C_1, \ldots, C_m\}$ of the graph G. The original joint distribution $P(V)$ can then be recombined as follows:

$$\forall a_1 \in \text{dom}(A_1) : \cdots \forall a_n \in \text{dom}(A_n) :$$
$$P(A_1 = a_1, \ldots, A_n = a_n) = \prod_{C_i \in C} \phi_{C_i}\left(\bigwedge_{A_j \in C_i} A_j = a_j\right)$$

For a detailed discussion on how to choose the functions ϕ_{C_i}, see e.g. [4].

3 Production Planning at Volkswagen Group

One goal of the project described here was to develop a system which plans parts demand for the production sites of the Volkswagen Group [8]. The market strategy is strongly customer-focused—based on adaptable designs and special emphasis on variety. Consequently, when ordering an automobile, the customer is offered several options of how each feature should be realized. The result is a very large number of possible car variants. Since the particular parts required for an automobile depend on the variant of the car, the overall parts demand can not be successfully estimated from total production numbers alone. The modeling of domains with such a large number of possible states is very complex. Therefore, decomposition techniques were applied and augmented by a set of operations on these subspaces that allow for a flexible parts demand planning and also provide a useful tool to simulate capacity usage in projected market development scenarios.

3.1 Data Description and Model Induction

The first step towards a feasible planning system consists of the identification of valid vehicle variants. If cars contain components that only work when combined with specific versions of other parts, changes in the predicted rates for one

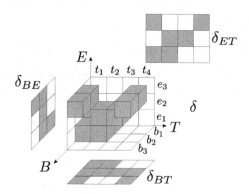

Fig. 1. The 3-dimensional space $\text{dom}(E) \times \text{dom}(T) \times \text{dom}(B)$ is thinned out by a rule set, sparing only the depicted value combinations. Further, one can reconstruct the 3-dimensional relation δ from the two projections δ_{ET} and δ_{BT}.

component may have an influence on the demand for other components. Such relations should be reflected in the design of the planning system.

A typical model of car is described by approximately 200 attributes, each consisting of at least 2, but up to 50 values. This scaffolds a space of possible car variants with a cardinality of over 10^{60}. Of course, not every combination corresponds to a valid specification. To ensure only valid combinations, restrictions are introduced in form of a rule system. Let us assume we are dealing with three variables E, T and B representing engine type, transmission type and brake type with the following respective domains:

$$\text{dom}(E) = \{e_1, e_2, e_3\}, \quad \text{dom}(T) = \{t_1, t_2, t_3, t_4\}, \quad \text{dom}(B) = \{b_1, b_2, b_3\}$$

A set of rules could for example contain statements like

$$\text{If } T = t_3 \text{ then } B = b_2$$
$$\text{or}$$
$$\text{If } E = e_2 \text{ then } T \in \{t_2, t_3\}$$

A comprehensive set of rules cancels out invalid combinations and may result in our example in a relation as depicted in figure 1.

It was decided to employ a probabilistic Markov network to represent the distribution of the value combinations. Probabilities are thus interpreted in terms of estimated relative frequencies. Therefore, an appropriate decomposition has to be found. Starting from a given rule base R and a production history to estimate relative frequencies from, the graphical component is generated as follows: We start out with an undirected graph $G = (V, E)$ where two variables F_i and F_j are connected by an edge $(F_i, F_j) \in E$ if there is a rule in R that contains both variables. To make reasoning efficient, it is desirable that the graph has hypertree structure. This includes the triangulation of G, as well as the identification of its cliques. This process is depicted in figure 2. To complete the model, for every clique a joint distribution for the variables of that clique has to be estimated from the production history.

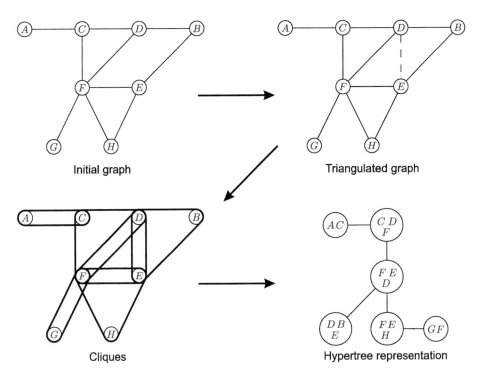

Fig. 2. Transformation of the model into hypertree structure. The initial graph is derived from the rule base. For reasoning, the hypertree cliques have to have the running intersection property which basically allows for a composition of the original distribution from the clique distributions. See [5] for details. This property can be asserted by requiring the initial graph to be triangulated.

3.2 Operations on the Model

A planning model that was generated using the above method, usually does not reflect the whole potential of available knowledge. For instance, experts are often aware of differences between the production history and the particular planning interval the model is meant to be used with. Thus, a mechanism to modify the represented distribution is required. Planning operators have been developed [10] to efficiently handle this kind of problem, so modification of the distribution and restoration of a consistent state can be supported.

Updating

Consider a situation where previously forbidden item combinations become valid. This can result for example from changes in the rule base. The relation in figure 1 does not allow engine type 2 to be combined with transmission type 1 because $(e_2, t_1) \notin E \times T$. If this option becomes valid probability mass, it has to be transferred to the respective distribution. Another scenario would be the advent of a new engine type, i. e. a change in the domain itself. Then, a multitude of new probabilities have to be assessed. Another related problem arises when subsets of cliques are altered while the information of the remaining network is retained. Both scenarios are addressed with the updating operation.

This operation marks these combinations as valid by assigning a positive near-zero probability to their respective marginals. Due to this small value, the quality of the estimation is not affected by this alteration. Now instead of using the same initialization for all new combinations, the proportion of the values is chosen in accordance to an existing combination, i. e. the probabilistic interaction structure is copied from reference item combinations.

Since updating only provides the qualitative aspect of the dependence structure, it is usually followed by the subsequent application of the revision operation, which is used to reassign probability mass to the new item combinations.

Revision

The revision operation, while preserving the network structure, serves to modify quantitative knowledge in such a way that the revised distribution becomes consistent with the new specialized information. There is usually no unique solution to this task. However, it is desirable to retain as much of the original distribution as possible so that the principle of minimal change [7] should be applied. Given that, a successful revision holds a unique result [9]. As an example for a specification, experts might predict a rise of the popularity of a recently introduced navigation system and set the relative frequency of this respective item from 20% to 30%.

Focusing

While revision and updating are essential operations for building and maintaining a distribution model, it is much more common activity to apply the model for the exploration of the represented knowledge and its implications with respect to user decisions. Typically users would want to concentrate on those aspects of the represented knowledge that fall into their domain of expertise. Moreover, when predicting parts demand from the model, one is only interested in estimated rates for particular item combinations. Such activities require a focusing operation. It is implemented by performing evidence-driven conditioning on a subset of variables and distributing the information through the network. Apart from predicting parts demand, focusing is often employed for market analyses and simulation. By analyzing which items are frequently combined by customers, experts can tailor special offers for different customer groups. To support planning of buffer capacities, it is necessary to deal with the eventuality of temporal logistic restrictions. Such events would entail changes in short-term production planning so that consumption of the concerned parts is reduced.

3.3 Application

The development of the planning system explained was initiated in 2001 by the Volkswagen Group. System design and most of the implementation is currently done by Corporate IT. The mathematical modeling, theoretical problem solving, and the development of efficient algorithms have been entirely provided by Intelligent Systems Consulting (ISC) Gebhardt. Since 2004 the system is being rolled out to all trademarks of the Volkswagen Group. With this software, the increasing planning quality, based on the many innovative features and the appropriateness of the chosen model of knowledge representation, as well as a considerable reduction of calculation time turned out to be essential prerequisites for advanced item planning and calculation of parts demand in the presence of structured products with an extreme number of possible variants.

4 Vehicle Data Mining at Daimler AG

While the previous section presented techniques that were applied ahead-of-time, i. e., prior and during the manufacturing process, we will now turn to the area of assessing the quality of cars after they left the assembly plant. For every car that is sold, a variety of data is collected and stored in corporate-wide databases. After every repair or checkup the respective records are updated to reflect the technical treatment. The analysis scenario discussed here is the interest of the automobile manufacturer to investigate car failures by identifying common properties that are exposed by specific subsets of cars that have a higher failure rate.

4.1 Data Description and Model Induction

As stated above, the source of information consists of a database that contains for every car a set of up to 300 attributes that describe the configuration of every car that has been sold.

The decision was made to use Bayesian networks to model the dependence structure between these attributes to be able to reveal possible interactions of vehicle components that cause higher failure rates. The induction of a Bayesian network consists of identifying a good candidate graph that encodes the independencies in the database. The goodness of fit is estimated by an evaluation measure. Therefore, usual learning algorithms consist of two parts: a search method and the mentioned evaluation measure which may guide the search. Examples for both parts are studied in [6; 11; 2].

Given a network structure, an expert user will gain first insights into the corresponding application domain. In figure 3 one could identify the road surface conditions to have a major (stochastic) impact on the failure rate and type. Of course, arriving at such a model is not always a straightforward task since the available database may lack some entries requiring the treatment of missing values. In this case possibilistic networks [3] may be used. However, with full information it might still be problematic to extract significant statistics since there may be value combinations that occur too scarcely. Even if we are in the favorable position to have sufficient amounts of complete data, the bare network structure does not reveal information about which *which* road conditions have *what kind* of impact on *which* type of failure. Fortunately, this information can be

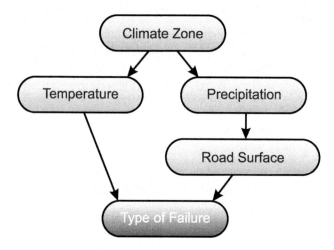

Fig. 3. An example of a Bayesian network illustrating qualitative linkage of components

retrieved easily in form of conditional probabilities from the underlying dataset, given the network structure. This becomes clear if the sentence above is restated: Given a specific road surface condition, what is the failure probability of a randomly picked vehicle?

4.2 Model Visualization

Every attribute with its direct parent attributes encodes a set of conditional probability distributions. For example, given a database D, the sub-network consisting of Failure, RoadSurface and Temperature in figure 3 defines the following set of distributions:

$$P_D(\text{Failure} \mid \text{Temperature}, \text{RoadSurface})$$

For every distinct combination of values of the attributes RoadSurface and Temperature, the conditional probability of the attribute Failure is estimated (counted) from the database D. We will argue in the next section, that it is this information that enables the user to gain better insight into the data under consideration [15].

Given an attribute of interest (in most cases the class variable like Failure in the example setting) and its conditioning parents, every probability statement like

$$P(\text{Failure} = \text{Suspension} \mid \text{RoadSurface} = \text{rough}, \text{Temperature} = \text{low}) \quad = \quad p^*$$

can be considered an association rule:[1]

> If RoadSurface = rough ∧ Temperature = low, then there will be a suspension failure in $100 \cdot p^*\%$ of all cases.

---
[1] See [1] for details.

The value p^* is then the confidence of the corresponding association rule. Of course, all known evaluation measures can be applied to assess the rules. With the help of such measures one can create an intuitive visual representation according to the following steps:

- For every probabilistic entry (i.e., for every rule) of the considered conditional distribution $P(C \mid A_1, \ldots, A_m)$ a circle is generated to be placed inside a two-dimensional chart.
- The graylevel (or color in the real application) of the circle corresponds the the value of attribute C.
- The circle's area corresponds to the value of some rule evaluation measure selected before displaying. For the remainder of this chapter, we choose this measure to be the support, i.e., the relative number of vehicles (or whatever instances) specified by the values of C and $A_1, \ldots, A_m$. Therefore, the area of the circle corresponds to the number of vehicles.
- In the last step these circles are positioned. Again, the value of the x- and y-coordinate are determined by two evaluation measures selected in advance. We suggest these measures to be recall[2] and lift.[3] Circles above the darker horizontal line in every chart mark subsets with a lift greater than 1 and thus indicate that the failure probability is larger given the instantiation of $A_1, \ldots, A_n$ in contrast to the marginal failure probability $P(C = c)$.

With these prerequisites we can recommend to the user the following heuristic in order to identify suspicious subsets:

Sets of instances in the upper right hand side of the chart may be good candidates for a closer inspection.

The greater the y-coordinate (i.e., the lift value) of a rule, the stronger is the impact of the conditioning attributes' values on the class variable. Larger x-coordinates correspond to higher recall values.

4.3 Application

This section illustrates the proposed visualization method by means of three real-world datasets that were analyzed during a cooperate research project with a automobile manufacturer. We used the K2 algorithm[4] [6] to induce the network structure and visualized the class variable according to the given procedure.

---

[2] The recall is defnided as $P(A_1 = a_1, \ldots, A_k = a_k \mid C = c)$.
[3] The lift of a rule indicates the ratio between confidence and the marginal failure rate: $\frac{P(C=c|A_1=a_1,\ldots,A_k=a_k)}{P(C=c)}$.
[4] It is a greedy approach that starts with a single attribute (here: the class attribute) and tries to add parent attributes greedily. If no addition of an attribute yields a better result, the process continues at the just inserted parent attributes. The quality of a given network is measured with the K2 metric (a Bayesian model averaging metric).

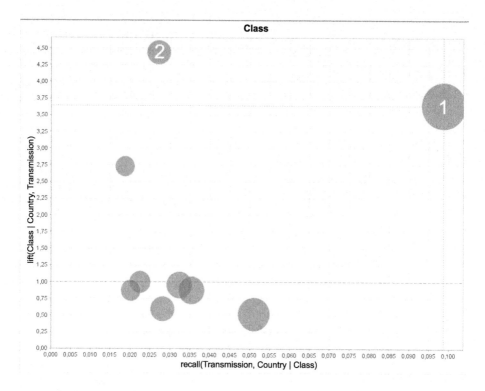

Fig. 4. The subset marked 1 corresponds to approx. 1000 vehicles whose attributes values of Country and Transmission yielded a causal relationship with the class variable. Unfortunately, there was not found a causal description of subset 2. The cluster of circles below the lift-1 line corresponds to sets of cars that fail less often, if their instantiantions of attributes become known.

Example 1

Figure 4 shows the analysis result of 60000 vehicles. The chart only depicts failed cars. Attributes Transmission and Country had most (stochastic) impact on the Class variable. The subset labeled 1 was re-identified by experts as a problem already known. Set 2 could not be given a causal explanation.

Example 2

The second dataset consisted of 300000 cars that exposed a many-valued class variable, hence the different gray levels of the circles in figure 5. Although there was no explanation for the sets 3, the subset 4 represented 900 cars the increased failure rate of which could be tracked down to the respective values of the attributes Mileage and RoadSurface.

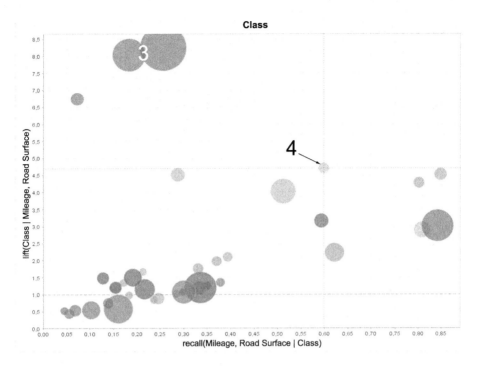

Fig. 5. Although it was not possible to find a reasonable description of the vehicles contained in subsets 3, the attribute values specifying subset 4 were identified to have a causal impact on the class variable.

Example 3

As a last example, the same dataset as in example 2 yielded the result as shown in figure 6. Here, an expert user changed the conditioning attributes manually and identified the set 5 which represented a subset of cars whose failure type and rate were affected by the respective attribute values.

User Acceptance

The proposed visualization technique has proven to be a valuable tool that facilitates the identification of subsets of cars that may expose a critical dependence between configuration and failure type. Generally, it represents an intuitive way of displaying high-dimensional, nominal data. A pure association rule analysis needs heavy postprocessing of the rules since a lot of rules are generated due to the commonly small failure rate. The presented approach can be considered a visual exploration aid for association rules. However, one has to admit that the rules represented by the circles share the same attributes in the antecedence, hence the sets of cars covered by these rules are mutually disjoint, which is a considerable difference to general rule sets.

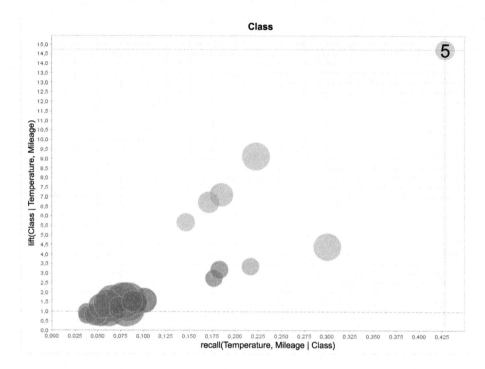

Fig. 6. In this setting the user selected the parent attributes manually and was able to identify the subset 5, which could be given a causal interpretation in terms of the conditioning attributes Temperature and Mileage

5 Conclusion

This paper presented an empirical evidence that graphical models can provide a powerful framework for data- and knowledge-driven applications with massive amounts of information. Even though the underlying data structures can grow highly complex, both presented projects implemented at two automotive companies result in effective complexity reduction of the methods suitable for intuitive user interaction.

References

[1] Agrawal, R., Imielinski, T., Swami, A.N.: Mining Association Rules between Sets of Items in Large Databases. In: Buneman, P., Jajodia, S. (eds.) Proceedings of the 1993 ACM SIGMOD International Conference on Management of Data, May 26-28, pp. 207–216. ACM Press, Washington (1993)

[2] Borgelt, C., Kruse, R.: Some experimental results on learning probabilistic and possibilistic networks with different evaluation measures. In: 1st International Joint Conference on Qualitative and Quantitative Practical Reasoning (ECSQARU/FAPR 1997), Bad Honnef, Germany, pp. 71–85 (1997)

[3] Borgelt, C., Kruse, R.: Probabilistic and possibilistic networks and how to learn them from data. In: Kaynak, O., Zadeh, L., Turksen, B., Rudas, I. (eds.) Computational Intelligence: Soft Computing and Fuzzy-Neuro Integration with Applications. NATO ASI Series F, pp. 403–426. Springer, New York (1998)

[4] Borgelt, C., Kruse, R.: Graphical Models — Methods for Data Analysis and Mining. John Wiley & Sons, United Kingdom (2002)

[5] Castillo, E., Gutiérrez, J.M., Hadi, A.S.: Expert Systems and Probabilistic Network Models. Springer, Heidelberg (1997)

[6] Cooper, G.F., Herskovits, E.: A Bayesian Method for the Induction of Probabilistic Networks from Data. Machine Learning 9, 309–347 (1992)

[7] Gärdenfors, P.: Knowledge in the Flux—Modeling the Dynamics of Epistemic States. MIT Press, Cambridge (1988)

[8] Gebhardt, J., Detmer, H., Madsen, A.L.: Predicting Parts Demand in the Automotive Industry — An Application of Probabilistic Graphical Models. In: Proc. International Joint Conference on Uncertainty in Artificial Intelligence (UAI 2003), Bayesian Modelling Applications Workshop, Acapulco, Mexico, August 4-7 (2003)

[9] Gebhardt, J., Borgelt, C., Kruse, R., Detmer, H.: Knowledge Revision in Markov Networks. Journal on Mathware and Soft Computing, Special Issue From Modelling to Knowledge Extraction XI(2–3), 93–107 (2004)

[10] Gebhardt, J., Kruse, R.: Knowledge-Based Operations for Graphical Models in Planning. In: Godo, L. (ed.) ECSQARU 2005. LNCS (LNAI), vol. 3571, pp. 3–14. Springer, Heidelberg (2005)

[11] Heckerman, D., Geiger, D., Chickering, D.M.: Learning Bayesian Networks: The Combination of Knowledge and Statistical Data. Technical Report MSR-TR-94-09, Microsoft Research, Advanced Technology Division, Redmond, WA, 1994 (Revised February 1995)

[12] Lauritzen, S.L., Spiegelhalter, D.J.: Local Computations with Probabilities on Graphical Structures and Their Application to Expert Systems. Journal of the Royal Statistical Society, Series B 2(50), 157–224 (1988)

[13] Pearl, J.: Aspects of Graphical Models Connected with Causality. In: 49th Session of the International Statistics Institute (1993)

[14] Pearl, J.: Probabilistic Reasoning in Intelligent Systems: Networks of Plausible Inference. Morgan Kaufmann, San Mateo (1988)

[15] Steinbrecher, M., Kruse, R.: Visualization of Possibilistic Potentials. In: Melin, P., Castillo, O., Aguilar, L.T., Kacprzyk, J., Pedrycz, W. (eds.) IFSA 2007. LNCS (LNAI), vol. 4529, pp. 295–303. Springer, Heidelberg (2007)

Extraction of Maximum Support Rules for the Root Cause Analysis

Tomas Hrycej[1] and Christian Manuel Strobel[2]

[1] Formerly with DaimlerChrysler Research, Ulm, Germany
  tomas_hrycej@yahoo.de
[2] University of Karlsruhe (TH), Germany
  mstrobel@statistik.uni-karlsruhe.de

Summary. Rule extraction for root cause analysis in manufacturing process optimization is an alternative to traditional approaches to root cause analysis based on process capability indices and variance analysis. Process capability indices alone do not allow to identify the process parameters having the major impact on quality since these indices are only based on measurement results and do not consider the explaining process parameters. Variance analysis is subject to serious constraints concerning the data sample subject to the analysis. In this work a rule search approach using Branch and Bound principles is presented, considering both the numerical measurement results and the nominal process factors. This combined analysis allows to associate the process parameters with the measurement results and therefore to identify the main drivers for quality deterioration of a manufacturing process.

1 Introduction

An important group of intelligent methods is concerned with discovering interesting information in large data sets. This discipline is generally referred to as Knowledge Discovery or Data Mining.

In the automotive domain, large data sets may arise through on-board measurements in cars. However, more typical sources of huge data amounts are in the vehicle, aggregate or component manufacturing process. One of the most prominent applications is the manufacturing quality control, which is the topic of this chapter.

Knowledge discovery subsumes a broad variety of methods. A rough classification may be into

- Machine learning methods
- Neural net methods
- Statistics

This partitioning is neither complete nor exclusive. The methodical frameworks of machine learning methods and neural nets have been extended by aspects covered by classical statistics, resulting in a successful symbiosis of these methods.

An important stream within the machine learning methods is committed to a specific (but very general) representation of discovered knowledge: to the rule based representation. A rule has the form $x \to y$, x and y being, respectively the *antecedent* and the *consequent*. The meaning of the rule is: if the antecedent (which has the form of a logical expression) is satisfied, the consequent is sure or probable to be true.

The discovery of rules of data can be simply defined as a search for highly informative (i.e., interesting from the application point of view) rules. So the most important subtasks are:

1. Formulating the criterion to decide to which extent a rule is interesting
2. Using an appropriate search algorithm to find those rules that are the most interesting according to this criterion

The research of the last decades has resulted in the formulation of various systems of interestingness criteria (well known of which are, e.g., support, confidence or lift), and the corresponding search algorithms.

However, general algorithms may miss the goal of a particular application. In such cases, dedicated algorithms are useful. This is the case in the application domain reported here: the root cause analysis for process optimization.

The indices for quality measurement and the application example involved are briefly presented in Section 2. The goal of the application is to find manufacturing parameters to which the quality level can be attributed. To this aim, rules expressing this relationships are searched for. This is what the rule extraction search algorithm based on Branch and Bound principles of Section 3 performs. Section 5 shows the computing results documenting the efficiency of the algorithm.

2 Root Cause Analysis for Process Optimization

The quality of a manufacturing process can be seen as the ability to manufacture a certain product within its specification limits U, L and as close as possible to its target value T, describing the point where its quality is optimal. A deviation from T generally results in minor quality and to minimize this deviation is crucial to a company to be competitive in the marketplace. In literature, numerous *process capability indices* (PCIs) have been proposed in order to provide a unitless quality measures to determine the performance of a manufacturing process, relating the preset specification limits to the actual behavior (6).

The behavior of a manufacturing process can be described by the process variation and process location. Therefore, to assign a quality measure to a process, the produced goods are continuously tested and the performance of the process is determined by calculating its PCI using the measurement results. In some cases it is not feasible to test and to measure all goods of a manufacturing process, as it might be too time consuming, too costly or the inspection might be destructive. Then, only a sample is drawn and the quality is determine upon this sample set. In order to state future quality of a manufacturing process based on the past performance, the process is supposed to be stable or in control. This means that both process mean and process variation has to be, in the long run, in between pre-defined limits. A common technique to monitor this are control charts, which are an essential part of the Statistical Process Control.

The basic idea for the most common indices is to assume the considered manufacturing process follows a normal distribution and the distance between the upper and lower specification limit U and L equals 12σ. This requirement implies a lot fraction defective of the manufacturing process of no more than $0.00197\text{ppm} \cong 0\%$ and reflects the widespread *Six-Sigma* principle (see (7)). The commonly recognized *basic* PCIs C_p, C_{pm}, C_{pk} and C_{pmk} can be summarized by a superstructure first introduced by Vännman (9) and referred to in literature as $C_p(u,v)$:

$$C_p(u,v) = \frac{d - u|\mu - M|}{3\sqrt{\sigma^2 + v(\mu - T)^2}} \quad (1)$$

where σ is the process standard deviation, μ the process mean, $d = (U-L)/2$ tolerance width, $m = (U+L)/2$ the mid-point between the two specification limits and T the target value. The *basic* PCIs can be obtained by choosing u and v according to

$$\begin{array}{ll} C_p \equiv C_p(0,0); & C_{pk} \equiv C_p(1,0) \\ C_{pm} \equiv C_p(0,1); & C_{pmk} \equiv C_p(1,1) \end{array} \quad (2)$$

The estimators for these indices are obtained by substituting μ by the sample mean $\bar{X} = \sum_{i=1}^{n} X_i/n$ and σ by the sample variance $S^2 = \sum_{i=1}^{n}(X_i - \bar{X})^2/(n-1)$. They provide stable and reliable point estimators for processes following a normal distribution. However, in practice, normality is hardly encountered. Consequently the basic PCIs as defined in (1) are not appropriate for processes with non-normal distributions. What is really needed are indices which do not make assumptions about the distribution, in order to be useful for measuring quality of a process.

$$C'_p(u,v) = \frac{d - u|m - M|}{3\sqrt{[\frac{F_{99.865} - F_{0.135}}{6}]^2 + v(m-T)^2}} \quad (3)$$

In 1997, Pearn and Chen introduced in their paper (8) a non-parametric generalization of the PCIs superstructure (1) in order to cover those cases, where the underlying data does not follow a Gaussian distribution. The authors replaced the process standard deviation σ by the 99.865 and 0.135 quantiles of the empiric distribution function and μ by the median of the process. The rationale behind that is that the difference between the $F_{99.865}$ and $F_{0.135}$ quantiles equals again 6σ or $C'_p(u,v) = 1$, under the standard normal distribution with $m = M = T$. As an analogy to the parametric superstructure (1), the special non-parametric PCIs C'_p, C'_{pm}, C'_{pk} and C'_{pk} can be obtained by applying u and v as in (2).

Assuming that the following assumptions hold, a class of non-parametric process indices and a particular specimen thereof can be introduced: Let $\mathbf{Y}: \Omega \to \mathbb{R}$ be a random variable with $\mathbf{Y}(\omega) = (Y^1, \ldots, Y^m) \in \mathcal{S} = \{S^1 \times \cdots \times S^m\}$, $S^i \in \{s_1^i, \ldots, s_{m_i}^i\}$ where $s_j^i \in \mathbb{N}$ describe the possible influence variables or process parameters. Furthermore, let $X: \Omega \to \mathbb{R}$ be the corresponding measurements results with $X(\omega) \in \mathbb{R}$. Then the pair $\mathcal{X} = (X, \mathbf{Y})$ denotes a manufacturing process and a class of *process indices* can be defined as:

Definition 1. *Let $\mathcal{X} = (X, \mathbf{Y})$ describe a manufacturing process as described above. Furthermore, let $f(x, y)$ be the density function of the underlying process and $w : \mathbb{R} \to \mathbb{R}$ an arbitrary measurable function. Then*

$$Q_{w,\mathcal{X}} = E\left(w(x)|\mathbf{Y} \in \mathcal{S}\right) = \frac{E\left(w(x)\mathbb{1}_{\{\mathbf{Y}\in\mathcal{S}\}}\right)}{P(\mathbf{Y} \in \mathcal{S})} \quad (4)$$

defines a class of process indices.

Obviously, if $w(x) = x$ or $w(x) = x^2$ we obtain, respectively, the first and the second moment of the process, as $P(\mathbf{Y} \in \mathcal{S}) = 1$. However, to determine the quality of a process, we are interested in the relationship between the designed specification limits U, L and the process behavior described by its variation and location. A possibility is to chose the function $w(x)$ in such way, that it becomes a function of the designed limits U and L. Given a particular manufacturing process $\mathcal{X}$ with $(x_i, \mathbf{y}_i), i = 1, \ldots, n$ we can define:

Definition 2. *Let $\mathcal{X} = (X, Y)$ be a particular manufacturing process with realizations $(x_i, \mathbf{y}_i), i = 1, \ldots, n$ and U, L be specification limits. Then, the* Empirical Capability Index *(E_{ci}) is defined as:*

$$\hat{E}_{ci} = \frac{\sum_{i=1}^{n} \mathbb{1}_{\{L \leq x_i \leq U\}} \mathbb{1}_{\{\mathbf{y}_i \in \mathcal{S}\}}}{\sum_{i=1}^{n} \mathbb{1}_{\{\mathbf{y}_i \in \mathcal{S}\}}} \quad (5)$$

By choosing the function $w(x)$ as the identity function $\mathbb{1}_{(L \leq x \leq U)}$, the E_{ci} measures the percentage of data points which are within the specification limits U and L. A disadvantage is, that for processes with a relatively good quality, it may happen that all sampled data points are within the Six-Sigma specification limits (i.e. $C_p' > 1$), and so the sample E_{ci} being one. To avoid this, the specification limits U and L have to be relaxed to values realistic for a given sample size, in order to get "further into the sample", by linking them to the behavior of the process. One possibility is to choose empirical quantiles:

$$[\bar{L}, \bar{U}] = [F_\alpha, F_{1-\alpha}]$$

The drawback of using empirical quantiles as specification limits is, that $\bar{L}$ and $\bar{U}$ do not depend anymore on the actual specification limits U and L. But exactly the relation of process behavior and the designed limits is essential to determine the quality of a manufacturing process. A combined solution, which on the one hand depends on the actual behavior and on the other hand incorporates the designed specification limit U and L can be obtained by:

$$[\bar{L}, \bar{U}] = \left[\hat{\mu}_{0,5} - \frac{\hat{\mu}_{0,5} - LSL}{t}, \hat{\mu}_{0,5} + \frac{USL - \hat{\mu}_{0,5}}{t}\right]$$

with $t \in \mathbb{R}$ being a adjustment factor. When setting $t = 4$ the new specification limits incorporate the *Six-Sigma* principle, assuming the special case of a centralized normally distributed process.

Extraction of Maximum Support Rules for the Root Cause Analysis

As stated above, the described PCIs only provide a quality measure but do not identify the major influence variables responsible for minor or superior quality. But knowing these factors is necessary to sustainable improve a manufacturing process in order to produce high quality on the long run. In practice it is desirable to know, whether there are subsets of influence variables values, such that the quality of a process becomes better, if constraining the process only to this parameters. In the following section a non-parametric, numerical approach for identifying those parameters is derived and an algorithm, which efficiently solves this problem is presented.

Application Example

To illustrate the basic ideas of the used methods and algorithms, an example is used throughout this paper, including an evaluation in the last section. This example is a simplified and anonymized version of a manufacturing process optimization task of a foundry of a premium automotive manufacturer.

Table 1. Measurement results and process parameters for the optimization of a foundry of an automotive manufacturer

Result	Tool	Shaft	Location
6.0092	1	1	right
6.008	4	2	right
6.0061	4	2	right
6.0067	1	2	left
...	...	...	...
6.0076	4	1	right
6.0082	2	2	left
6.0075	3	1	right
6.0077	3	2	right
6.0061	2	1	left
6.0063	1	1	right
6.0063	1	2	right

In Table 1 an extract of the documentation of such a manufacturing process is shown which will be used for further explanations. There are some typical influence variables (i.e. process parameters, relevant for the quality of the considered product) as the used tools, locations and used shafts, each with their specific values, for each manufacture specimen . Additionally, the corresponding quality measurement (column Result) - a geometric property or the size of a drilled hole - is a part of a data record.

Manufacturing Process Optimization - The Traditional Approach

A common technique to identify significant discrete parameters having an impact on numeric variables like measurement results, is the Analysis of Variance

(ANOVA). Unfortunately, ANOVA techniques are only useful if the problem is of a relatively low dimension. Additionally, the considered variables ought to have a simple structure and should be well balanced. Another constraint is the assumption that the analyzed data follows a multivariate Gaussian distribution. In most real world applications these requirements are hardly complied with. The distribution of the parameters describing the measure variable is in general non-parametric and of higher dimension. Also the combinations of the cross product of the parameters are non-uniformly and sparely populated or have a simple dependence structure. Therefore, the method of Variance Analysis is only applicable in some special cases. What is really needed, is a more general, non-parametric approach to determine a set of influence variables, responsible for minor or superior quality of a manufacturing process.

3 Rule Extraction Approach to Manufacturing Process Optimization

A manufacturing process X is defined as a pair $(X, \mathbf{Y})$ where $\mathbf{Y}(\omega)$ describe the influence variables (i.e. process parameters) and $X(\omega)$ the corresponding goal variables (measurement results). As we will see later, it is sometimes useful to constrain the manufacturing process to a particular subset of influence variables.

Definition 3. *Let $\mathcal{X}$ describe a manufacturing process as stated in definition 1 and $\mathbf{Y}_0 : \Omega \to \mathbb{R}$ be a random variable with $\mathbf{Y}_0(\omega) \in \mathcal{S}_0 \subset \mathcal{S}$. Then a sub-process of $\mathcal{X}$ is defined by the pair $\mathcal{X}_0 = (X, \mathbf{Y}_0)$.*

This subprocess constitutes the antecedent (i.e., precondition) of a rule to be discovered. The consequent of the rule is defined by the quality level (as measured by some process capability index) implied by the antecedent. To remain consistent with the terminology of application domain, we will speak about sub-processes and process capability indices, rather than about rule antecedents and consequents.

Given a manufacturing process $\mathcal{X}$ with a particular realization $(x_i, \mathbf{y}_i), i = 1\ldots, n$ the support of a sub-process $\mathcal{X}_0$ can be written as

$$N_{\mathcal{X}_0} = \sum_{i=1}^{n} \mathbb{1}_{\{\mathbf{y}_i \in \mathcal{S}_0\}} \tag{6}$$

and consequently, a conditional PCI is defined as $Q_{\mathcal{X}_0}$. Any of the indices defined in the previous section can be used, whereby the value of the respective index is calculated on the conditional subset $X_0 = \{x_i : \mathbf{y}_i \in \mathcal{S}_0, i = 1, \ldots, n\}$. In the following we use the notation $\tilde{\mathcal{X}} \subseteq \mathcal{X}$ to denote possible sub-processes of a given manufacturing process $\mathcal{X}$. An extraction of possible sub-process of the introduced example with their support and conditional E_{ci} is given in Table 2.

To determine those parameters having the greatest impact on quality, an optimal sub-process, consisting of optimal influence combinations, has to be identified. A first approach could be, to maximize $Q_{\tilde{\mathcal{X}}}$ over all sub-processes $\tilde{\mathcal{X}}$ of $\mathcal{X}$. In

Table 2. Possible sub-processes with support and conditional E_{ci} for the foundry's example

$N_{\mathcal{X}_0}$	$Q_{\mathcal{X}_0}$	Sub-process $\mathcal{X}_0$
123	0.85	Tool in (2,4) and Location in (left)
126	0.86	Shaft in (2) and Location in (right)
127	0.83	Tool in (2,3) and Shaft in (2)
130	0.83	Tool in (1,4) and Location in (right)
133	0.83	Tool in (4)
182	0.81	Tool not in (4) and Shaft in (2)
183	0.81	Tool not in (1) and Location in (right)
210	0.84	Tool in (1,2)
236	0.85	Tool in (2,4)
240	0.81	Tool in (1,4)
244	0.81	Location in (right)
249	0.83	Shaft in (2)
343	0.83	Tool not in (3)

general, this approach would yield a "optimal" sub-process $\tilde{\mathcal{X}}^*$, which has only a limited support ($N_{\tilde{\mathcal{X}}^*} \ll n$) (the fraction of cases that meet the constraints defining this subprocess). Such a formal optimum is usually of limited practical value since it is no possible to constrain arbitrary parameters to arbitrary values. For example, constraining the parameter "working shift" to the value "morning shift" would not be economically acceptable even if a quality increase were attained.

A better approach is to think in economic terms and to weight the factors responsible for minor quality, which we want to eliminate, by the costs of removing them. In practise this is not feasible, as tracking the actual costs is too expensive. But it is likely that infrequent influence factors, which are responsible for lower quality are *cheaper* to remove than frequent influences. In other words, sub-processes with high support are preferable over those sub-processes yielding a high quality measure but having a low support.

In most applications, the available sample set for process optimization is small, having numerous influence variables but only few measurement results. By limiting ourselves only to combinations of variables, we might get too small sub-process (having low support). Therefore, we extend the possible solutions to combinations of variables and their values - the search space for optimal sub-processes is spanned by the powerset of the influence parameters $\mathcal{P}(\mathbf{Y})$. The two sided problem, to find the parameter set combining on the one hand an optimal quality measure and on the other hand a maximal support, can be summarized, according to the above notation, by the the following optimization problem:

Definition 4

$$(P_{\mathcal{X}}) = \begin{cases} N_{\tilde{\mathcal{X}}} \to max \\ Q_{\tilde{\mathcal{X}}} \geq q_{min} \\ \tilde{\mathcal{X}} \subseteq \mathcal{X} \end{cases}$$

The solution $\tilde{\mathcal{X}}^*$ of the optimization problem is the subset of process parameters with maximal support among those processes, having a better quality than the given threshold q_{min}. Often, q_{min} is set to the common values for process capability of 1.33 or 1.67. In those cases, where the quality is poor, it is preferable to set q_{min} to the unconditional PCIs, to identify whether there is any process optimization potential.

Due to the nature of the application domain, the investigated parameters are discrete which inhibits an analytical solution but allows the use of *Branch and Bound* techniques. In the following section a root cause algorithm (RCA) which efficiently solves the optimization problem (4) is presented. To avoid the the exponential amount of possible combinations, spanned by the cross product of the influence parameters, several efficient cutting rules for the presented algorithm are derived and proven in the next subsection.

4 Manufacturing Process Optimization

Root Cause Analysis Algorithm

In order to access and efficiently store the necessary information and to apply Branch and Bound techniques, a multi-tree was chosen as representing data structure. Each node of the tree represents a possible combination of the influence parameters (sub-process) and is build out of the combination of the parents influence set and a new influence variable and it's value(s). Figure 1 depicts the data structure, whereby each node represents the set of sub-processes generated by the powerset of the considered variable(s). More precisely: Let I, J be to index sets with $I = \{1, \ldots, m\}$ and $J \subseteq I$. Then $\tilde{\mathcal{X}}_J$ denotes the set of sub-processes constraint by the powerset of $Y^j, j \in J$ and arbitrary other variables $(Y^i, i \in I \setminus J)$.

To find the optimal solution to the optimization problem (4), a combination of depth-first and breadth-first search is applied to traverse the multitree (compare Algorithm 1) using two Branch and Bound principles. The first, an generally applicable principle is based on the following relationship: by descending a branch of the tree, the number of constraints is increasing, as new influence variables

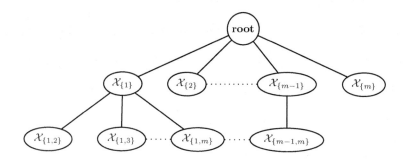

Fig. 1. Data structure for root cause analysis algorithm

Algorithm 1. Branch & Bound algorithm for process optimization

1: **procedure** TRAVERSETREE($\tilde{\mathcal{X}}$)
2:     **X** = GenerateSubProcesses($\tilde{\mathcal{X}}$)
3:     **for all** $\tilde{x} \in$ **X** **do**
4:         TraverseTree($\tilde{x}$)
5:     **end for**
6: **end procedure**

are added and therefore the sub-process support decreases (compare Figure 1). Considering Table 2, two variables (sub-processes), i.e. $\mathcal{X}_1$ = Shaft in (2) and $\mathcal{X}_2$ = Location in (right) have supports of $N_{\mathcal{X}_1} = 249$ and $N_{\mathcal{X}_2} = 244$, respectively. The joint condition of both has a lower (or equal) support than any of them ($N_{\mathcal{X}_1,\mathcal{X}_2} = 126$).

Thus, if a node has a support lower than a actual minimum support, there is no possibility to find a node (sub-process) with a higher support in the branch below. This reduces the time to find the optimal solution significantly, as a good portion of the tree to traverse, can be omitted. This first principle is realized in the function GENERATESUBPROCESSES as listed in Algorithm 2 and can be seen as the breadth-first-search of the RCA. This function takes as argument a sub-process and generates all the sub-processes, having a a support higher than the actual n_{max}.

Algorithm 2. Branch & Bound algorithm for process optimization

1: **procedure** GENERATESUBPROCESSES($\mathcal{X}$)
2:     **for all** $\tilde{\mathcal{X}} \subseteq \mathcal{X}$ **do**
3:         **if** $N_{\tilde{\mathcal{X}}} > n_{max}$ and $Q_{\tilde{\mathcal{X}}} \geq q_{min}$ **then**
4:             $n_{max} = N_{\tilde{\mathcal{X}}}$
5:         **end if**
6:         **if** $N_{\tilde{\mathcal{X}}} > n_{max}$ and $Q_{\tilde{\mathcal{X}}} < q_{min}$ **then**
7:             **X** = {**X** $\cup \tilde{\mathcal{X}}$}
8:         **end if**
9:     **end for**
10:     **return X**
11: **end procedure**

The second principle is considering disjoint value sets. For the support of a sub-process the following holds: Let $\mathcal{X}_1, \mathcal{X}_2$ be two sub-sets with $Y_1(\omega) \in \mathcal{S}_1 \subseteq \mathcal{S}$, $Y_2(\omega) \in \mathcal{S}_2 \subseteq \mathcal{S}$ with $\mathcal{S}_1 \cap \mathcal{S}_2 = \varnothing$ and $\mathcal{X}_1 \cup \mathcal{X}_2$ denote the unification of two sub-processes. It is obvious, that $N_{\mathcal{X}_1 \cup \mathcal{X}_2} = N_{\mathcal{X}_1} + N_{\mathcal{X}_2}$, what implies, that by extending the codomain of the influence variables, the support $N_{\mathcal{X}_1 \cup \mathcal{X}_2}$ can only increase. For the a class of convex process indices, as defined in (1), the second Branch and Bound principle can be derived, based on the next theorem:

Theorem 1. *Given two sub-processes* $\mathcal{X}_1 = (X, \mathbf{Y}_1)$, $\mathcal{X}_2 = (X, \mathbf{Y}_2)$ *of a manufacturing process* $\mathcal{X} = (X, \mathbf{Y})$ *with* $\mathbf{Y}_1(\omega) \in \mathcal{S}_1 \subseteq \mathcal{S}$, $\mathbf{Y}_2(\omega) \in \mathcal{S}_2 \subseteq \mathcal{S}$ *and*

$S_1 \cap S_2 = \varnothing$. Then for the class of process indices as defined in (4), the following inequality holds:

$$\min_{\mathcal{Z} \in \{\mathcal{X}_1, \mathcal{X}_2\}} Q_{w,\mathcal{Z}} \leq Q_{w, \mathcal{X}_1 \cup \mathcal{X}_2} \leq \max_{\mathcal{Z} \in \{\mathcal{X}_1, \mathcal{X}_2\}} Q_{w,\mathcal{Z}}$$

Proof. With $p = \frac{P(\mathbf{Y} \in S_1)}{P(\mathbf{Y} \in S_1 \cup S_2)}$ the following convex property holds:

$$\begin{aligned}
Q_{w,\mathcal{X}_1 \cup \mathcal{X}_2} &= E\left(w(x) | \mathbf{Y}(\omega) \in S_1 \cup S_2\right) \\
&= \frac{E\left(w(x) \mathbb{1}_{\{\mathbf{Y}(\omega) \in S_1 \cup S_2\}}\right)}{P(\mathbf{Y}(\omega) \in S_1 \cup S_2)} \\
&= \frac{E\left(w(x) \mathbb{1}_{\{\mathbf{Y}(\omega) \in S_1\}}\right) + E\left(w(x) \mathbb{1}_{\{\mathbf{Y}(\omega) \in S_2\}}\right)}{P\left(\mathbf{Y}(\omega) \in S_1 \cup S_2\right)} \\
&= p \frac{E\left(w(x) \mathbb{1}_{\{\mathbf{Y}(\omega) \in S_1\}}\right)}{P\left(\mathbf{Y}(\omega) \in S_1\right)} + (1-p) \frac{E\left(w(x) \mathbb{1}_{\{\mathbf{Y}(\omega) \in S_2\}}\right)}{P\left(\mathbf{Y}(\omega) \in S_2\right)}
\end{aligned}$$

Therefore, by combining two disjoint combination sets, the E_{ci} of the union of these two sets lies in between the maximum and minimum E_{ci} of these sets. This can be illustrated considering once more Table 2. The two disjoint sub-processes $\mathcal{X}_1$ = Tool in (1,2) and $\mathcal{X}_2$ = Tool in (4) yield a conditional E_{ci} of $Q_{\mathcal{X}_1} = 0.84$ and $Q_{\mathcal{X}_2} = 0.82$. The union of both sub-processes yields an E_{ci} value of $Q_{\mathcal{X}_1 \cup \mathcal{X}_2} = Q_{\text{Tool not in (3)}} = 0.82$. This value is within the interval $< 0.82, 0.84 >$, as stated by the theorem. This convex property reduces the number of times the E_{ci} actually has to be calculated, as in some special cases we can estimate the value of E_{ci} by the upper and lower limit and compare it to q_{min}.

In root cause analysis for process optimization, we are in general not interested in one global optimal solution but in a list of processes, having a quality better than the defined threshold q_{min} and maximal support. An expert might choose out of the n-best processes the one, he wishes to use as benchmark. To get the n-best sub-processes is realized by traversing also those branches, which already exhibit a (local) optimal solution. The rationales are, that a (local) optimum $\tilde{\mathcal{X}}^*$ with $N_{\tilde{\mathcal{X}}^*} > n_{max}$ might have a child node in its branch, which might yield the second best solution. Therefore, line 4 in algorithm (2) has to be adapted by *postpone* the found solution $\tilde{\mathcal{X}}$ to the set of sub-nodes $\mathbf{X}$. Hence, the actual maximal support is no longer defined by the (actual) best solution, but by the (actual) n-th best solution.

In many real world applications, the influence domain is mixed, consisting of discrete data and numerical variables. To enable a joint evaluation of both influence types, the numerical data is transformed into nominal data by mapping the continuous data onto pre-set quantiles. In most our applications, the 10%, 20%, 80% and 90% quantile have performed best. Additionally, only those influence sets have to be accounted for, which are successional.

Verification

As in practice the samples to analyze are small and the used PCIs are point estimators, the optimum of the problem (4) can only be defined in statistical terms. To get a more valid statement of the true value of the considered PCI, confidence intervals have to be used. In the special case, where the underlying data follows a known distribution, it is straight forward to construct a confidence interval. For example, if a normal distribution can be assumed, the distribution of $\frac{C_p}{\hat{C}_p}$ ($\hat{C}_p$ denotes the estimator of C_p) is known, and a $(1-\alpha)\%$ confidence interval for C_p is given by

$$C(X) = \left[\hat{C}_p \sqrt{\frac{\chi^2_{n-1;\frac{\alpha}{2}}}{n-1}}, \hat{C}_p \sqrt{\frac{\chi^2_{n-1;1-\frac{\alpha}{2}}}{n-1}} \right] \qquad (7)$$

For the other parametric basic indices, in general there exits no analytical solution as they all have a non-centralized χ^2 distribution. In (10), (2) or (4), for example, the authors derive different numerical approximation for the *basic* PCIS, assuming a normal distribution.

If there is no possibility to make an assumption about the distribution of the data, computer based, statistical methods as the well known Bootstrap method (5) are used to determine confidence intervals for process capability indices. In (1), three different methods for calculating confidence intervals are derived and a simulation study is performed for these intervals. As result of this study, the bias-corrected-method (BC) outperformed the other two methods (standard-bootstrap and percentile-bootstrap-method). In our applications, an extension to the BC-Method the Bias-corrected-accelerated-method (BCa) as described in (3) was used for determining confidence intervals for the non-parametric basic PCIs, as described in (3). For the Empirical Capability Index E_{ci} a simulation study showed that the standard-bootstrap-method, as used in (1), performed the best. A $(1-\alpha)\%$ confidence interval for the E_{ci} can be obtained using

$$C(X) = \left[\hat{E}_{ci} - \Phi^{-1}(1-\alpha)\sigma_B, \hat{E}_{ci} + \Phi^{-1}(1-\alpha)\sigma_B \right] \qquad (8)$$

where $\hat{E}_{ci}$ denotes an estimator for E_{ci}, σ_B the Bootstrap standard deviation and Φ^{-1} the inverse standard normal.

As all statements that are made using the RCA algorithm are based on sample sets, it is important to verify the soundness of the results. Therefore, the sample set to analyze is to be randomly divided into two disjoint sets: training and test set. A list of the n best sub-processes is generated, by first applying the describe RCA algorithm and second the referenced Bootstrap-methods to calculate confidence intervals. In a second step, the root cause analysis algorithm is applied to the test set. The final output is a list of sub-processes, having the same influence sets and a comparable level for the used PCI.

5 Experiments

An evaluation of the concept was performed on data of a foundry plant and engine manufacturing in the premium automotive industry (see Section 2). Three

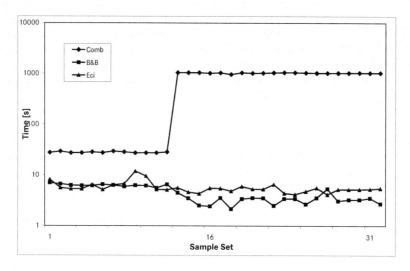

Fig. 2. Computational time for combinatorial search vs. Branch and Bound using the C'_{pk} and E_{ci}

different groups of data sets where used with a total of 33 different data sets of samples to evaluate the computational performance of the used algorithms. Each of the analyzed data sets comprises measurement results describing geometric characteristics like the position of drill holes or surface texture of the produced products and the corresponding influence sets like a particular machine number or a workers name. The first group of analyzed data, consists of 12 different measurement variables with 4 different influence variables, each with 2 to 9 different values. The second group of data sets comprises 20 different sample sets is made up of 14 variables with up to 7 values each. An additional data set, recording the results of a cylinder twist measurement having 76 influence variables, was used to evaluated the algorithm for numerical parameter sets. The output for each sample set was a list of the 20ies best sub-processes in order to cross check with the quality expert of the foundry plant. q_{min} was chosen to the unconditional PCI value. The analyzed data sets had at least 500 and at most 1000 measurement results.

The first computing series was performed using the empirical capability index E_{ci} and the non-parametric C'_{pk}. To demonstrate the efficiency of the first Branch and Bound principle, an additional combinatorial search was conducted. The reduction of computational time, using the first Branch and Bound principle, amounted to two orders of magnitude in comparison to the combinatorial search as can be seen in Figure 2. Obviously, the computational time for finding the n best sub-processes increases with the number of influence variables. This fact explains the jump of the combinatorial computing time in Figure 2 (the first 12 data sets correspond to the first group introduced in the section above). In average, the algorithm using the first Branch and Bound principle outperformed

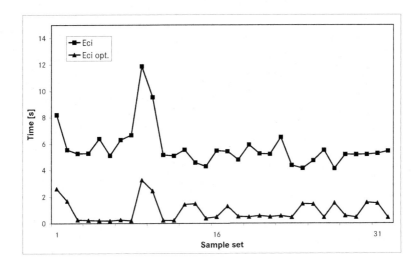

Fig. 3. Computational time first Branch and Bound vs. second Branch and Bound Principle using E_{ci}

the combinatorial search by a factor of 160. Using the combinatorial search it took in average 18 minutes to evaluate the available data sets. However, using the first Branch and Bound principle decreased the computing time to only 4.4 seconds for C'_{pk} and to 5.7 seconds using the E_{ci}. The evaluation was performed to a search up to a depth of 4, which means, that all sub-process have no more than 4 different influence variables. A higher depth level did not yield any other results, as the support of the sub-processes diminishes with increasing number of influence variables, used as constraints.

Applying the second Branch and Bound principle reduced the computational time even further. As the graph of Figure 2 depicts, the identification of the 20 optimal sub-processes using the E_{ci} was in average reduced by a factor of 5 in comparison to the first Branch and Bound principle and resulted in an average computational time of only 0.92 seconds vs. 5.71. Over all analyzed sample sets, the second principle reduced the computing time by 80%.

As the number of influence parameters of the numerical data set are, compared to the other data sets, significantly larger, it took, even using the E_{ci} and the second Branch and Bound principle, 20 seconds and for the non parametric calculation using the first Branch and Bound principle approximately 2 minutes. In this special case, the combinatorial search was omitted, as the evaluation of 76 influence variables with 4 values each would have taken too long.

Optimum Solution

Applying the sub-processes found to the original data set, the original, unconditional PCI improves. More precisely, considering for example the sub-process

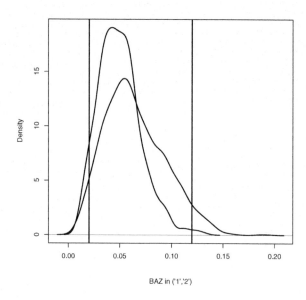

Fig. 4. Density plot for optimal sub-process *(narrower plot)* and its original process *(broader plot)* using E_{ci}

$\mathcal{X}$ = Tool in (1,2) and using the E_{ci} the index improves from originally 0.49 to 0.70. As Figure 4 shows, the quality of the sub-process clearly outperforms the original process, having less variance and a better process location.

On the test set, the performance of the optimum solution, characterized by Q_{Test} is over its lower bound determined by the bootstrap procedure on the training set, as shown in Table 3.

Table 3. Results for the process optimization for one data set

Index	N_{Test}	Q_{Test}	N_{Train}	C_B^l
E_{ci}	244	0.85	210	0.84

6 Conclusion

We have introduced an algorithm for efficient rule extraction in the domain of root cause analysis. The application goal is the manufacturing process optimization, with the intention to detect those process parameters having a major impact on the quality of a manufacturing process. The basic idea was to transform the search for those quality drivers into a optimization problem and to identify a set of optimal parameter subsets using two different Branch and Bound principles. These two methods allow for a considerable reduction of the computational time for identifying optimal solutions, as the computational results show.

A new class of convex process capability indices, E_{ci}, was introduced and its superiority, with regard to computing time, to common PCIs is shown. As the identification of major quality drivers is crucial to industrial practice and quality management, the presented solution may be useful and applicable for a broad scope of quality and reliability problems.

References

[1] Kalyanasundaram, M., Balamurali, S.: Bootstrap lower confidence limits for the process capability indices cp, cpk and cpm. International Journal of Quality & Reliability Management 19, 1088–1097 (2002)

[2] Bissel, A.F.: How reliable is your capability index. Applied Statistics 39(3), 331–340 (1990)

[3] Tibshirani, R.J., Efron, B.: An introduction to the bootstrap. Chapman and Hall, Boca Raton (1993)

[4] Wasserman, G.S., Franklin, L.A.: Bootstrap lower confidence limits for capability indices. Journal of quality technology 24(4), 196–210 (1992)

[5] Urban Hjorth, J.S.: Computer intensive statistical methods, 1st edn. Chapman and Hall, Boca Raton (1994)

[6] Johanson, N.L., Kotz, S.: Process capability indices - a review, 1992 - 2000. Journal of Quality Technology 34 (January 2002)

[7] Douglas, C.: Montgomery. In: Introduction to statistical quality control, 2nd edn., Wiley, Chichester (1991)

[8] Pearn, W., Chen, K.: Capability indices for non-normal distributions with an application in electrolytic capacitor manufacturing. Microelectronics Reliability 37, 1853–1858 (1997)

[9] Vännman, K.: A unified approach to capability indices. Statistica Sina 5, 805–820 (1995)

[10] Stenback, G.A., Wardrop, D.M., Zhang, N.F.: Interval estimation of process capability index cpk. Communications in statistics. Theory and methods (Commun. stat., Theory methods) 19(12), 4455–4470 (1990)

Neural Networks in Automotive Applications

Danil Prokhorov

Toyota Technical Center - a division of Toyota Motor Engineering and Manufacturing (TEMA), Ann Arbor, MI 48105, USA

Neural networks are making their ways into various commercial products across many industries. As in aerospace, in automotive industry they are not the main technology. Automotive engineers and researchers are certainly familiar with the buzzword, and some have even tried neural networks for their specific applications as models, virtual sensors, or controllers (see, e.g., [1] for a collection of relevant papers). In fact, a quick search reveals scores of recent papers on automotive applications of NN, fuzzy, evolutionary and other technologies of computational intelligence (CI); see, e.g., [2], [3], [4]. However, such technologies are mostly at the stage of research and not in the mainstream of product development yet. One of the reasons is "black-box" nature of neural networks. Other, perhaps more compelling reasons are business conservatism and existing/legacy applications (trying something new costs money and might be too risky) [5], [6].

NN technology which complements, rather than replace, the existing non-CI technology in applications will have better chances of wide acceptance (see, e.g., [8]). For example, NN is usually better at learning from data, while systems based on first principles may be better at modeling underlying physics. NN technology can also have greater chances of acceptance if it either has no alternative solution, or any other alternative is much worse in terms of the cost-benefit analysis. A successful experience with CI technologies at the Dow Chemical Company described in [7] is noteworthy.

Ford Motor Company is one of the pioneers in automotive NN research and development [9],[10]. Relevant Ford papers are referenced below and throughout this volume.

Growing emphasis on model based development is expected to help pushing mature elements of the NN technology into the mainstream. For example, a very accurate hardware-in-the-loop (HIL) system is developed by Toyota to facilitate development of advanced control algorithms for its HEV platforms [11]. As discussed in this chapter, some NN architectures and their training methods make possible an effective development process on high fidelity simulators for subsequent on-board (in-vehicle) deployment. While NN can be used both on-board and outside the vehicle, e.g., in a vehicle manufacturing process, only on-board applications usually impose stringent constraints on the NN system, especially in terms of available computational resources.

Here we provide a brief overview of NN technology suitable for automotive applications and discuss a selection of NN training methods. Other surveys are also available, targeting broader application base and other non-NN methods in general; see, e.g., [12].

Three main roles of neural network in automotive applications are distinguished and discussed: models (Section 1), virtual sensors (Section 2) and controllers (Section 3). Training of NN is discussed in Section 4, followed by a simple example illustrating importance of recurrent NN (Section 5). The issue of verification and validation is then briefly discussed in Section 6, concluding this chapter.

1 Models

Arguably the most popular way of using neural networks is shown in Figure 1. NN receives inputs and produces outputs which are compared with target values of the outputs from the system/process to be modeled or identified. This arrangement is known as supervised training because the targets for NN training are always provided by the system ("supervisor") to be modeled by NN. Figure 1 pertains to not only supervised modeling but also decision making, e.g., when it is required to train a NN classifier.

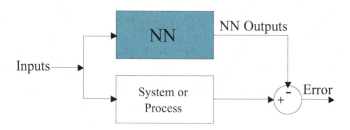

Fig. 1. A very popular arrangement for training NN to model another system or process including decision making is termed supervised training. The inputs to the NN and the system are not necessarily identical. The error between the NN outputs and the corresponding outputs of the system may be used to train the NN.

A general architecture of discrete-time NN is shown in Figure 2. The neurons or nodes of the NN are labeled as 1 through N. The links or connections may have adjustable or fixed parameters, or NN weights. Some nodes in the NN serve as inputs of signals external to the NN, others serve as outputs from the NN to the external world. Each node can sum or multiply all the links feeding it. Then the node transforms the result through any of a variety of functions such as soft (sigmoidal) and hard thresholds, linear, quadratic, or trigonometric functions, Gaussians, etc.

The blocks Z^{-1} indicates one time step delay for the NN signals. A NN without delays is called feedforward NN. If the NN has delays but no feedback connections, it is called time delay NN. A NN with feedback is called recurrent NN (RNN).

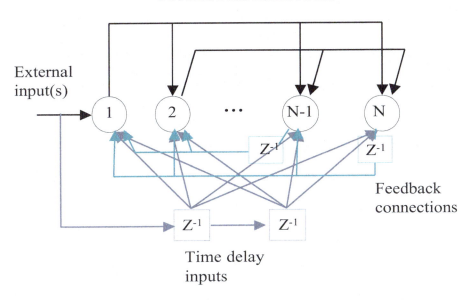

Fig. 2. Selected nodes in this network may be declared outputs. Any connectivity pattern, e.g., a popular layered architecture such as in multilayer perceptron (MLP), can be created by specifying the NN connectivity table. The order in which the nodes "fire", or get activated, also needs to be specified to preserve causality. Furthermore, explicit delays longer than one time step can also be included.

A large variety of NN exists. The reader is referred to [13] for a comprehensive discussion about many of the NN architectures and their training algorithms.

Clearly, many problem specific issues must be addressed to achieve successful NN training. They include pre- and (sometimes) post-processing of the data, the use of training data sufficiently representative of the system to be modeled, architectural choices, the optimal accuracy achievable with the given NN architecture, etc.

For a NN model predicting next values of its inputs it is useful to verify whether iterative predictions of the model are meaningful. A model trained to predict its input for the next time step might have a very large error predicting the input two steps into the future. This is usually the sign of overfitting. A single-step prediction might be too simple a task, especially for a slowly changing time series. The model might quickly learn that predicting the next value to be the same as its current value is good enough; the iterative prediction test should quickly reveal this problem with the model.

The trick above is just one of many useful tricks in the area of NN technology. The reader is referred to [14] and others for more information [15], [13].

Automotive engine calibration is a good example for relatively simple application of NN models. Traditionally, look-up tables have been used within the engine control system. For instance, a table linking engine torque production (output) with engine controls (inputs), such as spark angle (advance or retard), intake/exhaust valve timing, etc. Usually the table is created by running many experiments with the engine on a test stand. In experiments the space of engine controls is explored (in some fashion), and steady state engine torque values are recorded. Clearly, the higher the dimensionality of the look-up table, and the finer the required resolution, the more time it takes to complete the look-up table.

The least efficient way is full factorial experimental design (see, e.g., [16]), where the number of necessary measurement increases exponentially with the number of the table inputs. A modern alternative is to use model-based optimization with design of experiment [17], [18], [19], [20]. This methodology uses optimal experimental design plans (e.g., D- or V-optimal) to measure only a few predetermined points. A model is then fitted to the points, which enables the mapping interpolation in between the measurements. Such a model can then be used to optimize control strategy, often with significant computational savings (see, e.g., [21]).

In terms of models, a radial basis function (RBF) network [22], [23], a probabilistic NN which is implementationally simpler form of the RBF network [24], or MLP ([25], [26]) can all be used. So-called cluster weighted models (CWM) may be advantageous over RBF even in low-dimensional spaces [27], [28], [29]. CWM is capable of essentially perfect approximation of linear mappings because each cluster in CWM is paired with a linear output model. In contrast, RBF needs many more clusters to approximate even linear mappings to high accuracy (see Figure 3).

More complex illustrations of NN models are available (see, e.g., [30], [31]). For example, [32] discusses how to use a NN model for an HEV battery diagnostics. The NN is trained to predict an HEV battery state-of-charge (SOC) for a healthy battery. The NN monitors the SOC evolution and signals about abnormally rapid discharges of the battery if the NN predictions deviate significantly from the observed SOC dynamics.

2 Virtual Sensors

A modern automobile has a large number of electronic and mechanical devices. Some of them are actuators (e.g., brakes), while others are sensors (e.g., speed gauge). For example, transmission oil temperature is measured using a dedicated sensor. A virtual or soft sensor for oil temperature would use existing signals from other available sensors (e.g., air temperature, transmission gear, engine speed) and an appropriate model to create a virtual signal, an estimate of oil temperature in the transmission. Accuracy of this virtual signal will naturally depend on both accuracy of the model parameters and accuracies of existing signals feeding the model. In addition, existing signals will need to be chosen with care, as the presence of irrelevant signals may complicate the virtual sensor design. The modern term "virtual sensor" appears to be used sometimes interchangeably with its older counterpart "observer". Virtual sensor assumes less knowledge of the physical process, whereas observer assumes more of such knowledge. In other

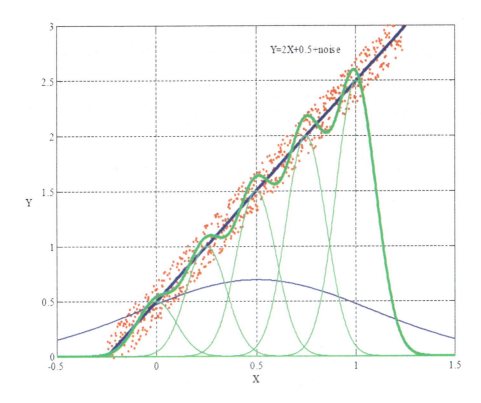

Fig. 3. A simple linear mapping $Y = 2X + 0.5$ subjected to a uniform noise (red) is to be approximated by CWM and RBF network. CMW needs only one cluster in the input space X and the associated mean-value linear model (the cluster and its model are shown in blue). In contrast, even with five clusters (green) the RBF network approximation (thick green line) is still not as good as the CWM approximation.

words, the observer model is often based on physical principles, and it is more transparent than that of virtual sensor. Virtual sensors are often "black boxes" such as neural networks, and they are especially valuable when the underlying physics is too complex or uncertain while there is plenty of data to develop/train a virtual sensor.

In the automotive industry, significant resources are devoted to the development of continuous monitoring of on-board systems and components affecting tailpipe emissions of vehicles. Ideally, on-board sensors specialized to measuring the regulated constituents of the exhaust gases in the tailpipe (mainly hydrocarbons, nitrogen oxides (NOx) and carbon monoxide) would check whether the vehicle is in compliance with government laws on pollution control. Given that such sensors are either unavailable or impractical, the on-board diagnostic system must rely on limited observations of system behavior and inferences based on those observations to determine whether the vehicle emissions are in compliance with the law. Our example is the diagnostics of engine combustion failures, known as misfire detection. This task must be performed with very high accuracy and under virtually all operating conditions (often the error rate of far less than 1% is necessary). Furthermore, the task requires the identification of the misfiring cylinder(s) of the engine quickly (on the order of seconds) to prevent any significant deterioration of the emission control system (catalytic converter). False alarm immunity becomes an important concern since on the order of one billion events must be monitored in the vehicle's lifetime.

The signals available to analyze combustion behavior are derived primarily from crankshaft position sensors. The typical position sensor is an encoder (toothed) wheel placed on the crankshaft of the engine prior to torque converter, transmission or any other engine load component. This wheel, together with its electronic equipment, provides a stream of accurately measured time intervals, with each interval being the time it takes for the wheel to rotate by one tooth. One can infer speed or acceleration of the crankshaft rotation by performing simple numerical manipulations with the time intervals. Each normal combustion event produces a slight acceleration of the crankshaft, whereas misfires exhibit acceleration deficits following a power stroke with little or no useful work (Figure 4).

The accurate detection of misfires is complicated by two main factors:

1) the engine exhibits normal accelerations and decelerations, in response to the driver input and from changing road conditions, and
2) the crankshaft is a torsional oscillator with finite stiffness. Thus, the crankshaft is subject to torsional oscillations which may turn the signature of normal events into that of misfires, and vice versa.

Analyzing complex time series of acceleration patterns requires a powerful signal processing algorithm to infer the quality of the combustion events. It turns out that the best virtual sensor of misfires can be developed on the basis of a recurrent neural network (RNN) [33]; see Section 4 as well as [34] and [35] for training method details and examples. The RNN is trained on a very large data set (on the order of million of events) consisting of many recordings of driving sessions. It uses engine context variables (such as crankshaft speed and engine load) and crankshaft acceleration as its inputs, and it produces estimates of the binary signal (normal or misfire) for each combustion event. During each engine cycle, the network is run as many times as the number of cylinders in the engine. The reader is referred to [36] for illustrative misfire data sets used in a competition organized at the International Joint Conference on Neural Networks (IJCNN) in 2001. The misfire detection NN is currently in production.

The underlying principle of misfire detection (dependence of crankshaft torsional vibrations on engine operation modes) is also useful for other virtual sensing opportunities, e.g., engine torque estimation.

Concluding this section, we list a few representative applications of NN in the virtual sensor category:

- NN can be trained to estimate emissions from engines based on a number of easily measured engine variables, such as load, RPM, etc. [37], [38], [39], or in-cylinder pressure [40] (note that using a structure identification by genetic

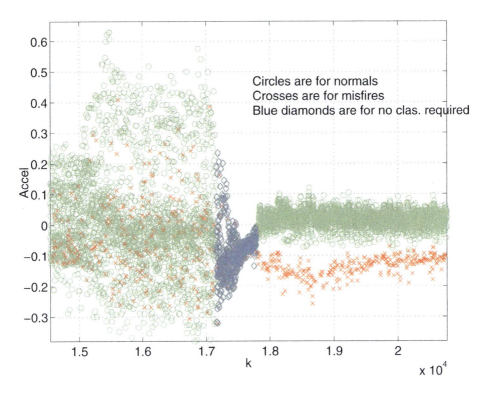

Fig. 4. A representative segment of the engine misfire data. Each cylinder firing is either normal (circle) or abnormal (misfire; cross). The region where misfire detection is not required is shown in the middle by diamonds. In terms of the crankshaft acceleration (y axis), sometimes misfires are easily separable from normal (the region between $k = 18,000$ and $k \sim 20,000$), and othertimes they are not (the region around $k = 16,000$).

algorithms for NOx estimation can result in performance better than that of a NN estimator; see [41] for details).
- air flow rate estimating NN is described in [19], and air-fuel ratio (AFR) estimation with NN is developed in [42], as well as in [43] and [44].
- a special processor Vindax is developed by Axeon, Ltd. (http://www.axeon.com/), to support a variety of virtual sensing applications [45], e.g., a mass airflow virtual sensor [46].

3 Controllers

NN as controllers have been known for years; see, e.g., [47], [48], [49], [50], [51]. We discuss only a few popular schemes in this section, referring the reader to a useful overview in [52], as well as [42] and [8], for additional information.

Regardless of the specific schemes employed to adapt or train NN controllers, there is a common issue of linkage between the cause and its effect. Aside of the causes which we mostly do not control such as disturbances applied to *plant*

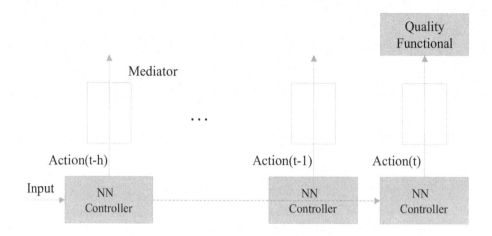

Fig. 5. The NN controller affects the quality functional through a mediator such as a plant. Older values of controls or actions (prior to t) may still have an effect on the plant as dynamic system with feedback, which necessitates the use of dynamic or temporal derivatives discussed in Section 4. For NN adaptation or training, the mediator may need to be complemented by its model and/or an adaptive critic.

(an object or system to be controlled), the NN controller outputs or *actions* in reinforcement learning literature [53] also affect the plant and influence a quality or performance functional. This is illustrated in Figure 5.

To enable NN training/adaptation, the linkage between NN actions and the quality functional can be achieved through a model of the plant (and a model of the quality functional if necessary), or without a model. Model-free adaptive control is implemented sometimes with the help of a reinforcement learning module called *critic* [54], [55]. Applications of reinforcement learning and approximate dynamic programming to automotive control have been attempted and not without success (see, e.g., [56] and [57]).

Figure 6 shows a popular scheme known as model reference adaptive control. The goal of adaptive controller which can be implemented as a NN is to make the plant behave as if it were the reference model which specifies the desired behavior for the plant. Often the plant is subject to various disturbances such as plant parameter drift, measurement and actuator noise, etc. (not shown in the figure).

Shown by dashed lines in Figure 6 is the plant model which may also be implemented as a NN (see Section 1). The control system is called *indirect* if the plant model is included, otherwise it is called *direct* adaptive control system.

We consider a process of indirect training NN controllers by an iterative method. Our goal is to improve the (ideal) performance measure I through training weights $\mathbf{W}$ of the controller:

$$I(\mathbf{W}(i)) = E_{\mathbf{x}_0 \in \mathbf{x}} \{ \sum_{t=0}^{\infty} U(\mathbf{W}(i), \mathbf{x}(t), \mathbf{e}(t)) \} \qquad (1)$$

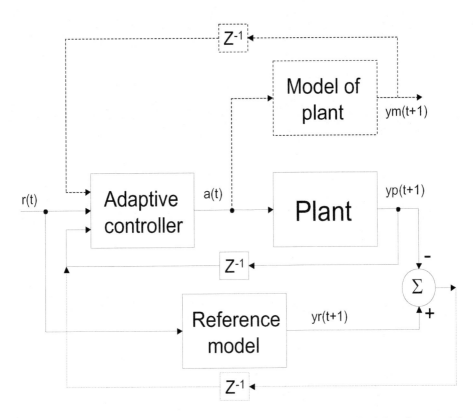

Fig. 6. The closed-loop system in model reference adaptive control. If the plant model is used (dashed lines), then the system is called indirect adaptive system; otherwise the system is called direct. The state vector **x** (not shown) includes state vectors of not only the plant but also the controller, the model of the plant and the reference model. The output vector **y** includes **a**, **yp**, **ym** and **yr**. The error between **yp** and **yr** may be used as an input to the controller and for controller adaptation (dotted lines).

where $\mathbf{W}(i)$ is the controller weight vector at the ith training iteration (or epoch), $E_{\mathbf{X}}$ is a suitable expectation operator (e.g., average) in the domain of permissible initial state vectors $\mathbf{x}_0 \equiv \mathbf{x}(0)$, and $U(\cdot)$ is a non-negative definite function with second-order bounded derivatives often called the instantaneous utility (or cost) function. We assume that the goal is to increase $I(\mathbf{W}(i))$ with i. The state vector **x** evolves according to the closed-loop system

$$\mathbf{x}(t+1) = \mathbf{f}(\mathbf{x}(t), \mathbf{e}(t), \mathbf{W}(i)) + \boldsymbol{\varepsilon}(t) \qquad (2)$$
$$\mathbf{y}(t) = \mathbf{h}(\mathbf{x}(t)) + \boldsymbol{\mu}(t) \qquad (3)$$

where $\mathbf{e}(t)$ is a vector of external variables, e.g., reference signals **r**, $\boldsymbol{\varepsilon}$ and $\boldsymbol{\mu}$ are noise vectors adding stochasticity to otherwise deterministic dynamic system, and $\mathbf{y}(t)$ is a vector of relevant outputs. Our closed-loop system includes not just the plant and its controller, which are usual components of the closed-loop system, but also the plant model and the reference model.

In reality, both E and ∞ in (1) must be approximated. Assuming that all initial states $\mathbf{x}_0(k)$ are equiprobable, the average operator can be approximated as

$$R(\mathbf{W}(i)) = \frac{1}{N} \sum_{\mathbf{x}_0(k) \in \mathbf{X}, k=1,2,\ldots,N} \sum_{t=0}^{T} U(i,t) \qquad (4)$$

where N is the total number of trajectories of length T along which the closed-loop system performance is evaluated at the iteration i. The first evaluation trajectory begins at time $t = 0$ in $\mathbf{x}_0(1)$, i.e., $\mathbf{x}(0) = \mathbf{x}_0(1)$, the second trajectory starts at $t = 0$ in $\mathbf{x}(0) = \mathbf{x}_0(2)$, etc. The coverage of the domain $\mathbf{X}$ should be as broad as practically possible for a reasonably accurate approximation of I.

Training the NN controller may impose computational constraints on our ability to compute (4) many times during our iterative training process. It may be necessary to contend with this approximation of R:

$$A(\mathbf{W}(i)) = \frac{1}{S} \sum_{\mathbf{x}_0(s) \in \mathbf{X}, s=1,2,\ldots,S} \sum_{t=0}^{H} U(i,t) \qquad (5)$$

The advantage of A over R is in faster computations of derivatives of A with respect to $\mathbf{W}(i)$ because the number of training trajectories per iteration is $S \ll N$, and the trajectory length is $H \ll T$. However, A must still be an adequate replacement of R and, possibly, I in order to improve the NN controller performance during its weight training. And of course A must also remain bounded over the iterations, otherwise the training process is not going to proceed successfully.

We assume that the NN weights are updated as follows

$$\mathbf{W}(i+1) = \mathbf{W}(i) + \mathbf{d}(i) \qquad (6)$$

where $\mathbf{d}(i)$ is an update vector. Employing the Taylor expansion of I around $\mathbf{W}(i)$ and neglecting terms higher than the first order yields

$$I(\mathbf{W}(i+1)) = I(\mathbf{W}(i)) + \frac{\partial I(i)}{\partial \mathbf{W}(i)}^T (\mathbf{W}(i+1) - \mathbf{W}(i)) \qquad (7)$$

Substituting for $(\mathbf{W}(i+1) - \mathbf{W}(i))$ from (6) yields

$$I(\mathbf{W}(i+1)) = I(\mathbf{W}(i)) + \frac{\partial I(i)}{\partial \mathbf{W}(i)}^T \mathbf{d}(i) \qquad (8)$$

The growth of I with iterations i is guaranteed if

$$\frac{\partial I(i)}{\partial \mathbf{W}(i)}^T \mathbf{d}(i) > 0 \qquad (9)$$

Alternatively, the decrease of I is assured if the inequality above is strictly negative; this is suitable for cost minimization problems, e.g., when $U(t) = (yr(t) - yp(t))^2$, which is popular in tracking problems.

It is popular to use gradients as the weight update:

$$\mathbf{d}(i) = \eta(i) \frac{\partial A(i)}{\partial \mathbf{W}(i)} \tag{10}$$

where $\eta(i) > 0$ is a learning rate. However, it is often much more effective to rely on updates computed with the help of second-order information; see Section 4 for details.

The condition (9) actually clarifies what it means for A to be an adequate substitute for R. The plant model is often required to train the NN controller. The model needs to provide accurate enough $\mathbf{d}$ such that (9) is satisfied. Interestingly, from the standpoint of NN controller training it is not critical to have a good match between plant outputs $\mathbf{yp}$ and their approximations by the model $\mathbf{ym}$. Coarse plant models which approximate well input-output sensitivities in the plant are sufficient. This has been noticed and successfully exploited by several researchers [58], [59], [60], [61].

In practice, of course it is not possible to guarantee that (9) always holds. This is especially questionable when even simpler approximations of R are employed, as is sometimes the case in practice, e.g., $S = 1$ and/or $H = 1$ in (5). However, if the behavior of $R(i)$ over the iterations i evolves towards its improvement, i.e., the trend is that R grows with i but not necessarily $R(i) < R(i+1), \forall i$, this would suggest that (9) does hold.

Our analysis above explains how the NN controller performance can be improved through training with imperfect models. It is in contrast with other studies, e.g., [62], [63], where the key emphasis is on proving the uniform ultimate boundedness (UUB) [64], which is not nearly as important in practice as the performance improvement because performance implies boundedness.

In terms of NN controller adaptation and in addition to the division of control to indirect and direct schemes, two adaptation extremes exist. The first is represented by the classic approach of fully adaptive NN controller which learns "on-the-fly", often without any prior knowledge; see, e.g., [65], [66]. This approach requires a detailed mathematical analysis of the plant and many assumptions, relegating NN to mere uncertainty compensators or look-up table replacement. Furthermore, the NN controller usually does not retain its long-term memory as reflected in the NN weights.

The second extreme is the approach employing NN controllers with weights fixed after training which relies on recurrent NN. It is known that RNN with fixed weights can imitate algorithms ([67], [68], [69], [70], [71], [72]) or adaptive systems [73] after proper training. Such RNN controllers are not supposed to require adaptation after deployment/in operation, thereby substantially reducing implementation cost especially in on-board applications. Figure 7 illustrates how a fixed-weight RNN can replace a set of controllers, each of which is designed for a specific operation mode of the time-varying plant. In this scheme the fixed-weight, trained RNN demonstrates its ability to generalize in the space of tasks, rather than just in the space of input-output vector pairs as non-recurrent networks do (see, e.g., [74]). As in the case of a properly trained non-recurrent NN which is very good at dealing with data similar to its training data, it is reasonable to expect that RNN can be trained to be good interpolators only in the space of tasks it has seen during training, meaning that significant extrapolation beyond training data is to be neither expected nor justified.

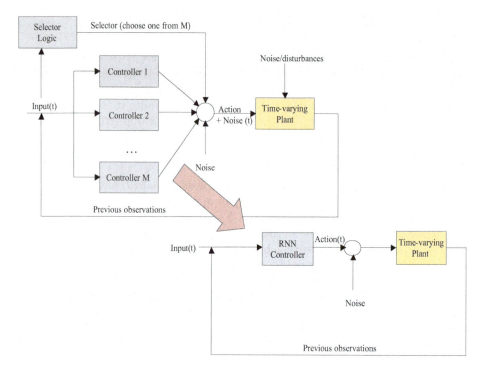

Fig. 7. A fixed-weight, trained RNN can replace a popular control scheme which includes a set of controllers specialized to handle different operating modes of the time-varying plant and a controller selector algorithm which chooses an appropriate controller based on the context of plant operation (input, feedback, etc.).

The fixed-weight approach is very suitable to such practically useful direction as training RNN off-line, i.e., on high-fidelity simulators of real systems, and preparing RNN through training to various sources of uncertainties and disturbances that can be encountered during system operation. And the performance of the trained RNN can also be verified on simulators to increase confidence in successful deployment of the RNN.

The fully adaptive approach is preferred if the plant may undergo very significant changes during its operation, e.g., when faults in the system force its performance to change permanently. Alternatively, the fixed-weight approach is more appropriate if the system may be repaired back to its normal state after the fault is corrected [32]. Various combinations of the two approaches above (hybrids of fully adaptive and fixed-weight approaches) are also possible [75].

Before concluding this section we would like to discuss on-line training implementation. On-line or continuous training occurs when the plant can not be returned to its initial state to begin another iteration of training, and it must be run continuously. This is in contrast with off-line training which assumes that the plant (its model in this case) can be reset to any specified state at any time.

On-line training can be done in a straightforward way by maintaining two distinct processes (see also [58]): foreground (network execution) and background (training). Figures 8 and 9 illustrate these processes.

The processes assume at least two groups of copies of the controller C labeled $C1$ and $C2$, respectively. The controller $C1$ is used in the foreground process which directly affects the plant P through the sequence of controller outputs **a1**.

The controller $C1$ weights are periodically replaced by those of the NN controller $C2$. The controller $C2$ is trained in the background process of Figure 9. The main difference from the previous figure is the replacement of the plant P with its model M. The model serves as a sensitivity pathway between utility U and controller $C2$ (cf. Figure 5), thereby enabling training $C2$ weights.

The model M could be trained as well, if necessary. For example, it can be done through adding another background process for training model of the plant. Of course, such process would have its own goal, e.g., minimization of the mean squared error between the model outputs $\mathbf{ym}(t+i)$ and the plant outputs $\mathbf{yp}(t+i)$. In general, simultaneous training of the model and the controller may result in training instability, and it is better to alternate cycles of model-controller training.

When referring to training NN in this and previous sections, we did not discuss possible training algorithms. This is done in the next section.

4 Training NN

Quite a variety of NN training methods exist (see, e.g., [13]). Here we provide an overview of selected methods illustrating diversity of NN training approaches, while referring the reader to detailed descriptions in appropriate references.

First, we discuss approaches that utilize derivatives. The two main methods for obtaining dynamic derivatives are real-time recurrent learning (RTRL) and backpropagation through time (BPTT) [76] or its truncated version BPTT(h) [77]. Often these are interpreted loosely as NN training methods, whereas they are merely the methods of obtaining derivatives to be combined subsequently with the NN weight update methods. (BPTT reduces to just BP when no dynamics needs to be accounted for in training.)

The RTRL algorithm was proposed in [78] for a fully connected recurrent layer of nodes. The name RTRL is derived from the fact that the weight updates of a recurrent network are performed concurrently with network execution. The term "forward method" is more appropriate to describe RTRL, since it better reflects the mechanics of the algorithm. Indeed, in RTRL, calculations of the derivatives

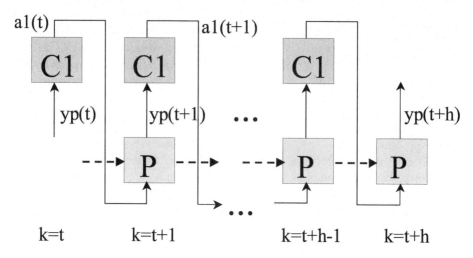

Fig. 8. The fixed-weight NN controller $C1$ influences the plant P through the controller outputs **a1** (actions) to optimize utility function U (not shown) in a temporal unfolding. The plant outputs **yp** are also shown. Note that this process in general continues for much longer than h time steps. The dashed lines symbolize temporal dependencies in the dynamic plant.

of node outputs with respect to weights of the network must be carried out during the forward propagation of signals in a network.

The computational complexity of the original RTRL scales as the fourth power of the number of nodes in a network (worst case of a fully connected RNN), with the space requirements (storage of all variables) scaling as the cube of the number of nodes [79]. Furthermore, RTRL for a RNN requires that the dynamic derivatives be computed at every time step for which that RNN is executed. Such coupling of forward propagation and derivative calculation is due to the fact that in RTRL both derivatives and RNN node outputs evolve recursively. This difficulty is independent of the weight update method employed, which might hinder practical implementation on a serial processor with limited speed and resources. Recently an effective RTRL method with quadratic scaling has been proposed [80] which approximates the full RTRL by ignoring derivatives not belonging to the same node.

Truncated backpropagation through time (BPTT(h), where h stands for the truncation depth) offers potential advantages relative to forward methods for obtaining sensitivity signals in NN training problems. The computational complexity scales as the product of h with the square of the number of nodes (for a fully connected NN). BPTT(h) often leads to a more stable computation of dynamic derivatives than do forward methods because its history is strictly finite. The use of BPTT(h) also permits training to be carried out asynchronously with the RNN

Preparation for controller training (background process)

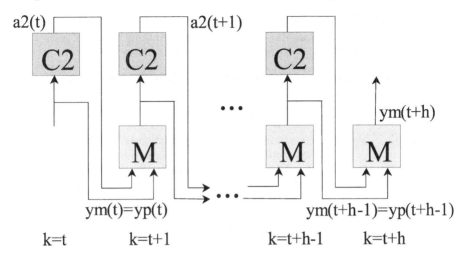

Fig. 9. Unlike the previous figure, another NN controller $C2$ and the plant model M are used here. It may be helpful to think of the current time step as step $t+h$, rather than step t. The controller $C2$ is a clone of $C1$ but their weights are different in general. The weights of $C2$ can be trained by an algorithm which requires that the temporal history of $h+1$ time steps be maintained. It is usually advantageous to align the model with the plant by forcing their outputs to match perfectly, especially if the model is sufficiently accurate for one-step-ahead predictions only. This is often called *teacher forcing* and shown here by setting $\mathbf{ym}(t+i) = \mathbf{yp}(t+i)$. Both $C2$ and M can be implemented as recurrent NN.

execution, as illustrated in Figures 8 and 9. This feature enabled testing a BPTT based approach on a real automotive hardware as described in [58].

As has been been observed some time ago [81], BPTT may suffer from the problem of vanishing gradients. This occurs because, in a typical RNN, the derivatives of sigmoidal nodes are less than the unity, while the RNN weights are often also less than the unity. Products of many of such quantities can become naturally very small, especially for large depths h. The RNN training would then become ineffective; the RNN would be "blind" and unable to associate target outputs with distant inputs.

Special RNN approaches such as those in [82] and [83] have been proposed to cope with the vanishing gradient problem. While we acknowledge that the problem may be indeed serious, it is not insurmountable. This is not just this author's opinion but also reflection on successful experience of Ford and Siemens NN Research (see, e.g., [84]).

In addition to calculation of derivatives of the performance measure with respect to the NN weights $\mathbf{W}$, we need to choose a weight update method. We can broadly classify weight update methods according to the amount of information used to perform an update. Still, the simple equation (6) holds, while the update $\mathbf{d}(i)$ may be determined in a much more complex process than the gradient method (10).

It is useful to summarize a typical BPTT(H) based training procedure for NN controllers because it highlights steps relevant to training NN with feedback in general.

1. Initiate states of each component of the system (e.g., RNN state): $\mathbf{x}(0) = \mathbf{x}_0(s)$, $s = 1, 2, ..., S$.
2. Run the system forward from time step $t = t_0$ to step $t = t_0 + H$, and compute U (see (5)) for all S trajectories.
3. For all S trajectories, compute dynamic derivatives of the relevant outputs with respect to NN controller weights, i.e., backpropagate to t_0. Usually backpropagating just $U(t_0 + H)$ is sufficient.
4. Adjust the NN controller weights according to the weight update $\mathbf{d}(i)$ using the derivatives obtained in step 3; increment i.
5. Move forward by one time step (run the closed-loop system forward from step $t = t_0 + H$ to step $t_0 + H + 1$ for all S trajectories), then increment t_0 and repeat the procedure beginning from step 3, etc., until the end of all trajectories ($t = T$) is reached.
6. Optionally, generate a new set of initial states and resume training from step 1.

The described procedure is similar to both model predictive control (MPC) with receding horizon (see, e.g., [85]) and optimal control based on the adjoint (Euler-Lagrange/Hamiltonian) formulation [86]. The most significant differences are that this scheme uses a parametric nonlinear representation for controller (NN) and that updates of NN weights are incremental, not "greedy" as in the receding-horizon MPC.

We henceforth assume that we deal with root-mean-squared (RMS) error minimization (corresponds to $-\frac{\partial A(i)}{\partial \mathbf{W}(i)}$ in (10)). Naturally, gradient descent is the simplest among all first-order methods of minimization for differentiable functions, and is the easiest to implement. However, it uses the smallest amount of information for performing weight updates. An imaginary plot of total error versus weight values, known as the error surface, is highly nonlinear in a typical neural network training problem, and the total error function may have many local minima. Relying only on the gradient in this case is clearly not the most effective way to update weights. Although various modifications and heuristics have been proposed to improve the effectiveness of the first-order methods, their convergence still remains quite slow due to the intrinsically ill-conditioned nature of training problems [13]. Thus, we need to utilize more information about the error surface to make the convergence of weights faster.

In differentiable minimization, the Hessian matrix, or the matrix of second-order partial derivatives of a function with respect to adjustable parameters, contains information that may be valuable for accelerated convergence. For instance, the minimum of a function quadratic in the parameters can be reached in one iteration, provided the inverse of the nonsingular positive definite Hessian matrix can be calculated. While such superfast convergence is only possible for quadratic functions, a great deal of experimental work has confirmed that much faster convergence is to be expected from weight update methods that use second-order information about error surfaces. Unfortunately, obtaining the inverse Hessian directly is practical only for small neural networks [15]. Furthermore, even if we can compute the inverse Hessian, it is frequently ill-conditioned and not positive definite, making it inappropriate for efficient minimization. For RNN, we have to rely on methods which build a positive definite estimate of the inverse Hessian without requiring its explicit knowledge. Such methods for weight updates belong to a family of second-order methods. For a detailed overview of the second-order methods, the reader is referred to [13]. If $\mathbf{d}(i)$ in (6) is a product of a specially created and maintained positive definite matrix, sometimes called the approximate inverse Hessian, and the vector $-\eta(i)\frac{\partial A(i)}{\partial \mathbf{W}(i)}$, we obtain the quasi-Newton method. Unlike first-order methods which can operate in either pattern-by-pattern or batch mode, most second-order methods employ batch mode updates (e.g., the popular Levenberg-Marquardt method [15]). In pattern-by-pattern mode, we update weights based on a gradient obtained for every instance in the training set, hence the term *instantaneous gradient*. In batch mode, the index i is no longer applicable to individual instances, and it becomes associated with a training iteration or epoch. Thus, the gradient is usually a sum of instantaneous gradients obtained for all training instances during the epoch i, hence the name *batch gradient*. The approximate inverse Hessian is recursively updated at the end of every epoch, and it is a function of the batch gradient and its history. Next, the best learning rate $\eta(i)$ is determined via a one-dimensional minimization procedure, called line search, which scales the vector $\mathbf{d}(i)$ depending on its influence on the total error. The overall scheme is then repeated until the convergence of weights is achieved.

Relative to first-order methods, effective second-order methods utilize more information about the error surface at the expense of many additional calculations for each training epoch. This often renders the overall training time to be comparable to that of a first-order method. Moreover, the batch mode of operation results in a strong tendency to move strictly downhill on the error surface. As a result, weight update methods that use batch mode have limited error surface exploration capabilities and frequently tend to become trapped in poor local minima. This problem may be particularly acute when training RNN on large and redundant training sets containing a variety of temporal patterns. In such a case, a weight update method that operates in pattern-by-pattern mode would be better, since it makes the search in the weight space *stochastic*. In other words, the training error can jump up and down, escaping from poor local minima. Of course, we are aware that no batch or sequential method, whether simple or sophisticated, provides a complete answer to the problem of multiple local minima. A reasonably small value of RMS error achieved on an independent testing set, not significantly larger than the RMS error obtained at the end of training, is a strong indication of success. Well known techniques, such as repeating a training exercise many times starting with different initial weights, are often useful to increase our confidence about solution quality and reproducibility.

Unlike weight update methods that originate from the field of differentiable function optimization, the extended Kalman filter (EKF) method treats supervised learning of a NN as a nonlinear sequential state estimation problem. The NN weights $\mathbf{W}$ are interpreted as states of the trivially evolving dynamic system, with the measurement equation described by the NN function $\mathbf{h}$:

$$\mathbf{W}(t+1) = \mathbf{W}(t) + \boldsymbol{\nu}(t) \tag{11}$$

$$\mathbf{y}^d(t) = \mathbf{h}(\mathbf{W}(t), \mathbf{i}(t), \mathbf{v}(t-1)) + \boldsymbol{\omega}(t) \tag{12}$$

where $\mathbf{y}^d(t)$ is the desired output vector, $\mathbf{i}(t)$ is the external input vector, $\mathbf{v}$ is the RNN state vector (internal feedback), $\boldsymbol{\nu}(t)$ is the process noise vector, and $\boldsymbol{\omega}(t)$ is the measurement noise vector. The weights $\mathbf{W}$ may be organized into g mutually exclusive weight groups. This trades off performance of the training method with its efficiency; a sufficiently effective and computationally efficient choice, termed node decoupling, has been to group together those weights that feed each node. Whatever the chosen grouping, the weights of group j are denoted by $\mathbf{W}_j$. The corresponding derivatives of network outputs with respect to weights $\mathbf{W}_j$ are placed in N_{out} columns of $\mathbf{H}_j$.

To minimize at time step t a cost function $cost = \sum_t \frac{1}{2}\boldsymbol{\xi}(t)^T \mathbf{S}(t)\boldsymbol{\xi}(t)$, where $\mathbf{S}(t) > 0$ is a weighting matrix and $\boldsymbol{\xi}(t)$ is the vector of errors, $\boldsymbol{\xi}(t) = \mathbf{y}^d(t) - \mathbf{y}(t)$, where $\mathbf{y}(t) = \mathbf{h}(\cdot)$ from (12), the decoupled EKF equations are as follows [58]:

$$\mathbf{A}^*(t) = \left[\frac{1}{\eta(t)}\mathbf{I} + \sum_{j=1}^{g} \mathbf{H}_j^*(t)^T \mathbf{P}_j(t)\mathbf{H}_j^*(t) \right]^{-1} \tag{13}$$

$$\mathbf{K}_j^*(t) = \mathbf{P}_j(t)\mathbf{H}_j^*(t)\mathbf{A}^*(t) \tag{14}$$

$$\mathbf{W}_j(t+1) = \mathbf{W}_j(t) + \mathbf{K}_j^*(t)\boldsymbol{\xi}^*(t) \tag{15}$$

$$\mathbf{P}_j(t+1) = \mathbf{P}_j(t) - \mathbf{K}_j^*(t)\mathbf{H}_j^*(t)^T \mathbf{P}_j(t) + \mathbf{Q}_j(t) \tag{16}$$

In these equations, the weighting matrix $\mathbf{S}(t)$ is distributed into both the derivative matrices and the error vector: $\mathbf{H}_j^*(t) = \mathbf{H}_j(t)\mathbf{S}(t)^{\frac{1}{2}}$ and $\boldsymbol{\xi}^*(t) = \mathbf{S}(t)^{\frac{1}{2}}\boldsymbol{\xi}(t)$. The matrices $\mathbf{H}_j^*(t)$ thus contain scaled derivatives of network (or the closed-loop system) outputs with respect to the jth group of weights; the concatenation of these matrices forms a global scaled derivative matrix $\mathbf{H}^*(t)$. A common global scaling matrix $\mathbf{A}^*(t)$ is computed with contributions from all g weight groups through the scaled derivative matrices $\mathbf{H}_j^*(t)$, and from all of the decoupled approximate error covariance matrices $\mathbf{P}_j(t)$. A user-specified learning rate $\eta(t)$ appears in this common matrix. (Components of the measurement noise matrix are inversely proportional to $\eta(t)$.) For each weight group j, a Kalman gain matrix $\mathbf{K}_j^*(t)$ is computed and used in updating the values of $\mathbf{W}_j(t)$ and in updating the group's approximate error covariance matrix $\mathbf{P}_j(t)$. Each approximate error covariance update is augmented by the addition of a scaled identity matrix $\mathbf{Q}_j(t)$ that represents additive data deweighting.

We often employ a multi-stream version of the algorithm above. A concept of multi-stream was proposed in [87] for improved training of RNN via EKF. It amounts to training N_s copies (N_s streams) of the same RNN with N_{out} outputs. Each copy has the same weights but different, separately maintained states. With each stream contributing its own set of outputs, every EKF weight update is based on information from all streams, with the total effective number of outputs increasing to $M = N_s N_{out}$. The multi-stream training may be especially effective for heterogeneous data sequences because it resists the tendency to improve local performance at the expense of performance in other regions.

The Stochastic Meta-Descent (SMD) is proposed in [88] for training nonlinear parameterizations including NN. The iterative SMD algorithm consists of two steps. First, we update the vector $\mathbf{p}$ of local learning rates

$$\mathbf{p}(t) = \mathtt{diag}(\mathbf{p}(t-1)) \\ \times \mathtt{max}(0.5, 1 + \mu \mathtt{diag}(\mathbf{v}(t))\boldsymbol{\nabla}(t)) \quad (17)$$

$$\mathbf{v}(t+1) = \gamma \mathbf{v}(t) + \mathtt{diag}(\mathbf{p}(t))(\boldsymbol{\nabla}(t) - \gamma \mathbf{C}\mathbf{v}(t)) \quad (18)$$

where γ is a forgetting factor, μ is a scalar meta-learning factor, $\mathbf{v}$ is an auxiliary vector, $\mathbf{C}\mathbf{v}(t)$ is the product of a curvature matrix $\mathbf{C}$ with $\mathbf{v}$, $\boldsymbol{\nabla}$ is a derivative of the instantaneous cost function with respect to $\mathbf{W}$ (e.g., the cost is $\frac{1}{2}\boldsymbol{\xi}(t)^T \mathbf{S}(t)\boldsymbol{\xi}(t)$; oftentimes $\boldsymbol{\nabla}$ is averaged over a short window of time steps).

The second step is the NN weight update:

$$\mathbf{W}(t+1) = \mathbf{W}(t) - \mathtt{diag}(\mathbf{p}(t))\boldsymbol{\nabla}(t) \quad (19)$$

In contrast to EKF which uses explicit approximation of the inverse curvature $\mathbf{C}^{-1}$ as the $\mathbf{P}$ matrix (16), the SMD calculates and stores the matrix-vector product $\mathbf{C}\mathbf{v}$, thereby achieving dramatic computational savings. Several efficient ways to obtain $\mathbf{C}\mathbf{v}$ are discussed in [88]. We utilize the product $\mathbf{C}\mathbf{v} = \boldsymbol{\nabla}\boldsymbol{\nabla}^T \mathbf{v}$ where we first compute the scalar product $\boldsymbol{\nabla}^T \mathbf{v}$, then scale the gradient $\boldsymbol{\nabla}$ by the result. The well adapted $\mathbf{p}$ allows the algorithm to behave as if it were a second-order method, with the dominant scaling *linear* in $\mathbf{W}$. This is clearly advantageous for problems requiring large NN.

Now we briefly discuss training methods which do not use derivatives.

ALOPEX, or ALgorithm Of Pattern EXtraction, is a correlation based algorithm proposed in [89]:

$$\Delta W_{ij}(n) = \eta \Delta W_{ij}(n-1) \Delta R(n) + r_i(n) \quad (20)$$

In terms of NN variables, $\Delta W_{ij}(n)$ is the difference between the current and previous value of weight W_{ij} at iteration n, $\Delta R(n)$ is the difference between the current and previous value of the NN performance function R (not necessarily in the form of (4)), η is the learning rate, and the stochastic term $r_i(n) \sim N(0, \sigma^2)$ (a non-Gaussian term is also possible) is added to help escaping poor local minima. Related correlation based algorithms are described in [90].

Another method of non-differential optimization is called particle swarm optimization (PSO) [91]. PSO is in principle a parallel search technique for finding solutions with the highest fitness. In terms of NN, it uses multiple weight vectors, or particles. Each particle has its own position $\mathbf{W}_i$ and velocity $\mathbf{V}_i$. The particle update equations are

$$V_{i,j}^{next} = \omega V_{i,j} + c_1 \phi_{i,j}^1 (W_{ibest,j} - W_{i,j}) + c_2 \phi_{i,j}^2 (W_{gbest,j} - W_{i,j}) \quad (21)$$
$$W_{i,j}^{next} = W_{i,j} + V_{i,j}^{next} \quad (22)$$

where the index i is the ith particle, j is its jth dimension (i.e., jth component of the weight vector), $\phi_{i,j}^1, \phi_{i,j}^2$ are uniform random numbers from zero to one, $\mathbf{W}_{ibest}$ is the best ith weight vector so far (in terms of evolution of the ith vector fitness), $\mathbf{W}_{gbest}$ is the overall best weight vector (in terms of fitness values of all weight vectors). The control parameters are termed the accelerations c_1, c_2 and the inertia ω. It is noteworthy that the first equation is to be done first for all pairs (i,j), followed by the second equation execution for all the pairs. It is also important to generate separate random numbers $\phi_{i,j}^1, \phi_{i,j}^2$ for each pair (i,j) (more common notation elsewhere omits the (i,j)-indexing, which may result in less effective PSO implementations if done literally).

The PSO algorithm is inherently a batch method. The fitness is to be evaluated over many data vectors to provide reliable estimates of NN performance.

Performance of the PSO algorithm above may be improved by combining it with particle ranking and selection according to their fitness [92], [93], [94], resulting in hybrids between PSO and evolutionary methods. In each generation, the PSO-EA hybrid ranks particles according to their fitness values and chooses the half of the particle population with the highest fitness for the PSO update, while discarding the second half of the population. The discarded half is replenished from the first half which is PSO-updated and then randomly mutated.

Simultaneous Perturbation Stochastic Approximation (SPSA) is also appealing due to its extreme simplicity and model-free nature. The SPSA algorithm has been tested on a variety of nonlinearly parameterized adaptive systems including neural networks [95].

A popular form of the gradient descent-like SPSA uses two cost evaluations *independent* of parameter vector dimensionality to carry out one update of each adaptive parameter. Each SPSA update can be described by two equations:

$$W_i^{next} = W_i - a G_i(\mathbf{W}) \quad (23)$$
$$G_i(\mathbf{W}) = \frac{cost(\mathbf{W} + c\boldsymbol{\Delta}) - cost(\mathbf{W} - c\boldsymbol{\Delta})}{2c\Delta_i} \quad (24)$$

where $\mathbf{W}^{next}$ is the updated value of the NN weight vector, $\boldsymbol{\Delta}$ is a vector of symmetrically distributed Bernoulli random variables generated anew for every update step (e.g., the ith component of $\boldsymbol{\Delta}$ denoted as Δ_i is either $+1$ or -1), c is size of a small perturbation step, and a is a learning rate.

Each SPSA update requires that two consecutive values of the cost function *cost* be computed, i.e., one value for the "positive" perturbation of weights $cost(\mathbf{W} + c\boldsymbol{\Delta})$ and another value for the "negative" perturbation $cost(\mathbf{W} - c\boldsymbol{\Delta})$ (in general, the cost function depends not only on $\mathbf{W}$ but also on other variables which are omitted for simplicity). This means that one SPSA update occurs no more than once every other time step. As in the case of the SMD algorithm (17)-(19), it may also be helpful to let the cost function represent changes of the cost over a short window of time steps, in which case each SPSA update would be even less frequent. Variations of the base SPSA algorithm are described in detail in [95].

Non-differential forms of KF have also been developed [96], [97], [98]. These replace backpropagation with many forward propagations of specially created test or *sigma* vectors. Such vectors are still only a small fraction of probing points required for high-accuracy approximations because it is easier to approximate a nonlinear transformation of a Gaussian density than an arbitrary nonlinearity itself. These truly nonlinear KF methods have been shown to result in more effective NN training than the EKF method ([99], [100], [101]), but at the price of significantly increased computational complexity.

Tremendous reductions in cost of the general-purpose computer memory and relentless increase in speed of processors have greatly relaxed implementation constraints for NN models. In addition, NN architectural innovations called liquid state machines (LSM) and echo state networks (ESN) have appeared recently (see, e.g., [102]), which reduce the recurrent NN training problem to that of training just the weights of the output nodes because other weights in the RNN are fixed. Recent advances in LSM/ESN are reported in [103].

5 RNN: A Motivating Example

Recurrent neural networks are capable to solve more complex problems than networks without feedback connections. We consider a simple example illustrating the need for RNN and propose an experimentally verifiable explanation for RNN behavior, referring the reader to other sources for additional examples and useful discussions [71], [104], [105], [106], [107], [108], [109], [110].

Figure 10 illustrates two different signals, all continued after 100 time steps at the *same* level of zero. An RNN is tasked with identifying two different signals by ascribing labels to them, e.g., +1 to one and −1 to another. It should be clear that only a recurrent NN is capable of solving this task. Only an RNN can retain potentially arbitrarily long memory of each input signal in the region where the two inputs are no longer distinguishable (the region beyond the first 100 time steps in Figure 10).

We chose an RNN with one input, one fully connected hidden layer of 10 recurrent nodes, and one bipolar sigmoid node as output. We employed the training based on BPTT(10) and EKF (see Section 4) with 150 time steps as the length of training trajectory, which turned out to be very quick due to simplicity of the task. Figure 11 illustrates results after training. The zero-signal segment is extended for additional 200 steps for testing, and the RNN still distinguishes the two signals clearly.

We examine the internal state (hidden layer) of the RNN. We can see clearly that all time series are different, depending on the RNN input; some node signals are very different, resembling the decision (output) node signal. For example,

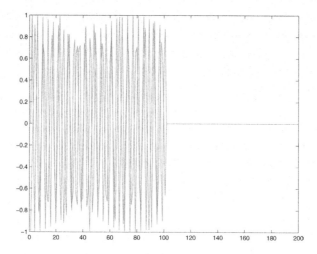

Fig. 10. Two inputs for the RNN motivating example. The blue curve is $sin(5t/\pi)$, where $t = [0:1:100]$, and the green curve is $sawtooth(t, 0.5)$ (Matlab notation).

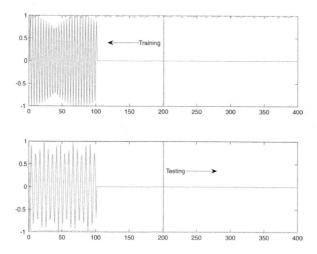

Fig. 11. The RNN results after training. The segment from 0 to 200 is for training, the rest is for testing.

Figure 12 shows the output of the hidden node 4 for both input signals. This hidden node could itself be used as the output node if the decision threshold is set at zero.

Our output node is non-recurrent. It is only capable of creating a separating hyperplane based on its inputs, or outputs of recurrent hidden nodes, and the bias node. The hidden layer behavior after training suggests that the RNN spreads the input signal into several dimensions such that in those dimensions the signal classification becomes easy.

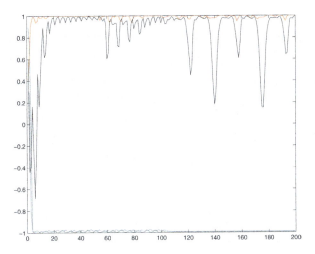

Fig. 12. The output of the hidden node 4 of the RNN responding to the first (black) and the second (green) input signals. The response of the output node is also shown in red and blue for the first and the second signal, respectively.

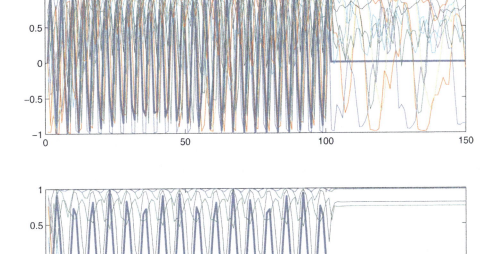

Fig. 13. The hidden node outputs of the RNN and the input signal (thick blue line) of the first (top) and the second (bottom) classes

The hidden node signals in the region where the input signal is zero do not have to converge to a fixed point. This is illustrated in Figure 13 for the segment where the input is zero (the top panel). It is sufficient that the hidden node behavior for each signal of a particular class belong to a distinct region of the hidden node state space, non-overlapping with regions for other classes. Thus, oscillatory or even chaotic behavior for hidden nodes is possible (and sometimes advantageous - see [110] and [109] for useful discussions), as long as a separating hyperplane exists for the output to make the classification decision. We illustrate in Figure 11 the long retention by testing the RNN on added 200-point segments of zero inputs to each of the training signals.

Though our example is for two classes of signals, it is straightforward to generalize it to multi-class problems. Clearly, not just classification problems but also regression problems can be solved, as demonstrated previously in [73], often with the addition of hidden (not necessarily recurrent) layers.

Though we employed the EKF algorithm for training of all RNN weights, other training methods can certainly be utilized. Furthermore, other researchers, e.g., [102], recently demonstrated that one might replace training RNN weights in the hidden layer with their random initializations, provided that the hidden layer nodes exhibit sufficiently diverse behavior. Only weights between the hidden nodes and the outputs would have to be trained, thereby greatly simplifying the training process. Indeed, it is plausible that even random weights in the RNN could sometimes result in sufficiently well separated responses to input signals of different classes, and this would also be consistent with our explanation for the trained RNN behavior observed in the example of this section.

6 Verification and Validation (V & V)

Verification and validation of performance of systems containing NN is a critical challenge of today and tomorrow [111], [112]. Proving mathematically that a NN will have the desired performance is possible, but such proofs are only as good as their assumptions. Sometimes too restrictive, hard to verify or not very useful assumptions are put forward just to create an appearance of mathematical rigor. For example, in many control papers a lot of efforts is spent on proving the uniform ultimate boundedness (UUB) property without due diligence demanded in practice by the need to control the value of that ultimate bound. Thus, stability becomes a proxy for performance, which is not often the case. In fact, physical systems in the automotive world (and in many other worlds too) are always bounded because of both physical limits and various safeguards.

As mentioned in Sections 3, it is reasonable to expect that a trained NN can do an adequate job *interpolating* to other sets of data it has not seen in training. *Extrapolation* significantly beyond the training set is not reasonable to expect. However, some automotive engineers and managers who are perhaps under pressure to deploy a NN system as quickly as possible may forget this and insist that the NN be tested on data which differs as much as possible from the training data, which clearly runs counter to the main principle of designing experiments with NN.

The inability to prove rigorously superior performance of systems with NN should not discourage automotive engineers from deploying such systems. Various high-fidelity simulators, HILS etc. are simplifying the work of performance verifiers. As such, these systems are already contributing to growing popularity of statistical methods for performance verification because other alternatives are simply not feasible [113], [114], [115], [116].

To illustrate statistical approach of performance verification of NN, we consider the following performance verification experiment. Assume that a NN is tested on N independent data sets. If the NN performance in terms of a performance measure m is better than m^d, then the experiment is considered successful, otherwise failed. The probability that a set of N experiments is successful is given by the classic formula of Bernoulli trials (see also [117]):

$$Prob = (1-p)^N \qquad (25)$$

where p is unknown true probability of failure. To keep the probability of observing no failures in N trials below κ even if $p \geq \epsilon$ requires

$$(1-\epsilon)^N \leq \kappa \qquad (26)$$

which means

$$N \geq \frac{\ln \kappa}{\ln(1-\epsilon)} = \frac{\ln \frac{1}{\kappa}}{\ln \frac{1}{1-\epsilon}} \approx \frac{1}{\epsilon} \ln \frac{1}{\kappa} \qquad (27)$$

If $\epsilon = \kappa = 10^{-6}$, then $N \geq 1.38 \cdot 10^7$. It would take less than four hours of testing (3.84 hours), assuming that a single verification experiment takes 1 msec.

Statistical performance verification illustrated above is applicable to other "black-box" approaches. It should be kept in mind that a NN is seldom the only component in the entire system. It may be useful and safer in practice to implement a hybrid system, i.e., a combination of a NN module ("black box") and a module whose functioning is more transparent than that of NN. The two modules together (and possibly the plant) form a system with desired properties. This approach is discussed in [8], which is the next chapter of the book.

References

[1] Jurgen, R.K.: Electronic Engine Control Technologies, 2nd edn., Society of Automotive Engineers (2004)
[2] Bryant, B.D., Marko, K.A.: Case example 2: Data analysis for diagnostics and process monitoring of automotive engines. In: Wang, B., Lee, J. (eds.) Computer-Aided Maintenance: Methodologies and Practices, pp. 281–301. Springer, Heidelberg (1999)

[3] Tascillo, A., Miller, R.: An in-vehicle virtual driving assistant using neural networks. In: Proceedings of the International Joint Conference on Neural Networks (IJCNN), vol. 3, pp. 2418–2423 (July 2003)
[4] Djurdjanovic, D., Liu, J., Marko, K.A., Ni, J.: Immune Systems Inspired Approach to Anomaly Detection and Fault Diagnosis for Engines. In: Proceedings of the International Joint Conference on Neural Networks (IJCNN), 2007, Orlando, FL, August 12-17, pp. 1375–1382 (2007)
[5] Chiu, S.: Developing commercial applications of intelligent control. IEEE Control Systems Magazine 17(2), 94–100 (1997)
[6] Kordon, A.K.: Application issues of industrial soft computing systems. In: Fuzzy Information Processing Society, 2005. NAFIPS 2005. Annual Meeting of the North American Publication Date: 26-28, pp. 110–115 (June 2005)
[7] Kordon, A.K.: Applied soft computing. Springer, Heidelberg (2008)
[8] Bloch, G., Lauer, F., Colin, G.: On Learning Machines for Engine Control. In: Prokhorov, D. (ed.) Comput. Intel. in Automotive Applications, ch.8, vol. 132. Springer, Heidelberg (2009)
[9] Marko, K.A., James, J., Dosdall, J., Murphy, J.: Automotive control system diagnostics using neural nets for rapid pattern classification of large data sets. In: Proceedings of the International Joint Conference on Neural Networks (IJCNN), Washington, DC, vol. 2, pp. 13–16 (July 1989)
[10] Puskorius, G.V., Feldkamp, L.A.: Neurocontrol of nonlinear dynamical systems with Kalman filter trained recurrent networks. IEEE Transactions on Neural Networks 5(2), 279–297 (1994)
[11] Okuda, N., Ishikawa, N., Kang, Z., Katayama, T., Nakai, T.: HILS Application for Hybrid System Development. SAE Technical Paper No. 2007-01-3469, Warrendale, PA (2007)
[12] Gusikhin, O.Y., Rychtyckyj, N., Filev, D.: Intelligent systems in the automotive industry: applications and trends. Knowl. Inf. Syst. 12(2), 147–168 (2007)
[13] Haykin, S.: Neural Networks: A Comprehensive Foundation, 2nd edn. Prentice Hall, Upper Saddle River (1999)
[14] Orr, G.B., Müller, K.-R. (eds.): NIPS-WS 1996, vol. 1524. Springer, Heidelberg (1998)
[15] Bishop, C.M.: Neural Networks for Pattern Recognition. Oxford University Press, Oxford (1995)
[16] Frigon, N.L., Matthews, D.: Practical Guide to Experimental Design. John Wiley and Sons, Chichester (1997)
[17] Meyer, S., Greff, A.: New Calibration Methods and Control Systems With Artificial Neural Networks. SAE Technical Paper No. 2002-01-1147, Warrendale, PA (2002)
[18] Schoggl, P., Koegeler, H.M., Gschweitl, K., Kokal, H., Williams, P., Hulak, K.: Automated EMS Calibration using Objective Driveability Assessment and Computer Aided Optimization Methods. SAE Technical Paper No. 2002-01-0849, Warrendale, PA (2002)
[19] Wu, B., Filipi, Z., Assanis, D., Kramer, D.M., Ohl, G.L., Prucka, M.J., DiValentin, E.: Using Artificial Neural Networks for Representing the Air Flow Rate through a 2.4 Liter VVT Engine. SAE Technical Paper No. 2004-01-3054, Warrendale, PA (2004)
[20] Schoop, U., Reeves, J., Watanabe, S., Butts, K.: Steady-State Engine Modeling for Calibration: A Productivity and Quality Study. In:' Proc. MathWorks Automotive Conference 2007, June 19-20, Dearborn, MI (2007)

[21] Wu, B., Filipi, Z.S., Prucka, R.G., Kramer, D.M., Ohl, G.L.: Cam-phasing Optimization Using Artificial Neural Networks as Surrogate Models - Fuel Consumption and NOx Emissions. SAE Technical Paper No. 2006-01-1512, Warrendale, PA (2006)

[22] Deignan, P.B., Meckl Jr., P.H., Franchek, M.A.: The MI - RBFN: Mapping for Generalization. In: Proceedings of the American Control Conference, Anchorage, AK, May 8-10, pp. 3840–3845 (2002)

[23] Papadimitriou, I., Warner, M.D., Silvestri, J.J., Lennblad, J., Tabar, S.: Neural Network Based Fast-Running Engine Models for Control-Oriented Applications. SAE Technical Paper No. 2005-01-0072, Warrendale, PA (2005)

[24] Specht, D.: Probabilistic neural networks. Neural Networks 3, 109–118 (1990)

[25] Wu, B., Filipi, Z.S., Prucka, R.G., Kramer, D.M., Ohl, G.L.: Cam-Phasing Optimization Using Artificial Neural Networks as Surrogate Models - Maximizing Torque Output. SAE Technical Paper No. 2005-01-3757, Warrendale, PA (2005)

[26] Ferrari, S., Stengel, R.F.: Smooth Function Approximation Using Neural Networks. IEEE Transactions on Neural Networks 16(1), 24–38 (2005)

[27] Gershenfeld, N.A.: Nature of Mathematical Modeling. MIT Press, Cambridge (1998)

[28] Gershenfeld, N., Schoner, B., Metois, E.: Cluster-weighted modelling for time-series analysis. Nature 397, 329–332 (1999)

[29] Prokhorov, D., Feldkamp, L., Feldkamp, T.: A New Approach to Cluster Weighted Modeling. In: Proc. of International Joint Conference on Neural Networks (IJCNN), Washington DC (July 2001)

[30] Hafner, M., Weber, M., Isermann, R.: Model-based control design for IC-engines on dynamometers: The toolbox "Optimot". In: Proc. 15th IFAC World Congress, July 21-26, Barcelona, Spain (2002)

[31] Torkzadeh, D., Baumann, J., Kiencke, U.: A Neuro Fuzzy Approach for Anti-Jerk Control. SAE Technical Paper 2003-01-0361, Warrendale, PA (2003)

[32] Prokhorov, D.: Toyota Prius HEV Neurocontrol and Diagnostics. Neural Networks 21, 458–465 (2008)

[33] Marko, K.A., James, J.V., Feldkamp, T.M., Puskorius, G.V., Feldkamp, L.A., Prokhorov, D.: Training recurrent networks for classification. In: Proceedings of the World Congress on Neural Networks, San Diego, pp. 845–850 (1996)

[34] Feldkamp, L.A., Prokhorov, D.V., Eagen, C.F., Yuan, F.: Enhanced multi-stream Kalman filter training for recurrent networks. In: Suykens, J., Vandewalle, J. (eds.) Nonlinear Modeling: Advanced Black-Box Techniques, pp. 29–53. Kluwer Academic Publishers, Dordrecht (1998)

[35] Feldkamp, L.A., Puskorius, G.V.: A Signal Processing Framework Based on Dynamic Neural Networks with Application to Problems in Adaptation, Filtering and Classification. In: Proceedings of the IEEE, vol. 86(11), pp. 2259–2277 (1998)

[36] Neural Network Competition at IJCNN, Washington D.C, GAC (2001), http://www.geocities.com/ijcnn/challenge.html

[37] Jesion, G., Gierczak, C.A., Puskorius, G.V., Feldkamp, L.A., Butler, J.W.: The application of dynamic neural networks to the estimation of feedgas vehicle emissions. In: Proc. World Congress on Computational Intelligence. International Joint Conference on Neural Networks, vol.1, pp. 69–73 (May 1998)

[38] Jarrett, R., Clark, N.N.: Weighting of Parameters in Artificial Neural Network Prediction of Heavy-Duty Diesel Engine Emissions. SAE Technical Paper No. 2002-01-2878, Warrendale, PA (2002)

[39] Brahma, I., Rutland, J.C.: Optimization of Diesel Engine Operating Parameters Using Neural Networks. SAE Technical Paper No. 2003-01-3228, Warrendale, PA (2003)
[40] Traver, M.L., Atkinson, R.J., Atkinson, C.M.: Neural Network-Based Diesel Engine Emissions Prediction Using In-Cylinder Combustion Pressure. SAE Technical Paper No. 1999-01-1532, Warrendale, PA (1999)
[41] del Re, L., Langthaler, P., Furtmüller, C., Winkler, S., Affenzeller, M.: NOx Virtual Sensor Based on Structure Identification and Global Optimization. SAE Technical Paper No. 2005-01-0050, Warrendale, PA (2005)
[42] Arsie, I., Pianese, C., Sorrentino, M.: Recurrent neural networks for AFR estimation and control in spark ignition automotive engines. In: Prokhorov, D. (ed.) Comput. Intel. in Automotive Applications, ch.9. Springer, Heidelberg (2009)
[43] Wickström, N., Larsson, M., Taveniku, M., Linde, A., Svensson, B.: Neural Virtual Sensors - Estimation of Combustion Quality in SI Engines using the Spark Plug. In: Proc. ICANN 1998 (1998)
[44] Howlett, R.J., Walters, S.D., Howson, P.A., Park, I.: Air-fuel ratio measurement in an internal combustion engine using a neural network. Advances in Vehicle Control and Safety (International Conference), AVCS 1998, July. Amiens, France (1998)
[45] Nareid, H., Grimes, M.R., Verdejo, J.R.: A Neural Network Based Methodology for Virtual Sensor Development. SAE Technical Paper No. 2005-01-0045
[46] Grimes, M.R., Verdejo, J.R., Bogden, D.M.: Development and Usage of a Virtual Mass Air Flow Sensor. SAE Technical Paper No. 2005-01-0074
[47] Miller III, W.T., Sutton, R.S., Werbos, P.J. (eds.): Neural Networks for Control. MIT Press, Cambridge (1990)
[48] Narendra, K.S.: Neural Networks for Control: Theory and Practice. Proceedings of the IEEE 84(10), 1385–1406 (1996)
[49] Suykens, J., Vandewalle, J., De Moor, B.: Artificial Neural Networks for Modeling and Control of Non-Linear Systems. Kluwer Academic, Dordrecht (1996)
[50] Hrycej, T.: Neurocontrol: Towards an Industrial Control Methodology. Wiley, Chichester (1997)
[51] Norgaard, M., Ravn, O., Poulsen, N.L., Hansen, L.K.: Neural Networks for Modelling and Control of Dynamic Systems. Springer, London (2000)
[52] Agarwal, M.: A systematic classification of neural-network-based control. IEEE Control Systems Magazine 17(2), 75–93 (1997)
[53] Sutton, R.S., Barto, A.G.: Reinforcement Learning: An Introduction. MIT Press, Cambridge (1998)
[54] Werbos, P.J.: Approximate Dynamic Programming for Real-Time Control and Neural Modeling. In: White, D.A., Sofge, D.A. (eds.) Handbook of Intelligent Control: Neural, Fuzzy, and Adaptive Approaches. Nostrand, New York (1992)
[55] Sutton, R.S., Barto, A.G., Williams, R.: Reinforcement learning is direct adaptive optimal control. IEEE Control Systems Magazine 12(2), 19–22
[56] Prokhorov, D.: Training Recurrent Neurocontrollers for Real-Time Applications. IEEE Transactions on Neural Networks 18(4), 1003–1015 (2007)
[57] Liu, D., Javaherian, H., Kovalenko, O., Huang, T.: Adaptive critic learning techniques for engine torque and air-fuel ratio control. IEEE Transactions on Systems, Man and Cybernetics-Part B: Cybernetics.
[58] Puskorius, G.V., Feldkamp, L.A., Davis Jr., L.I.: Dynamic neural network methods applied to on-vehicle idle speed control. Proceedings of the IEEE 84(10), 1407–1420 (1996)

[59] Prokhorov, D., Santiago, R.A., Wunsch, D.: Adaptive critic designs: A case study for neurocontrol. Neural Networks 8(9), 1367–1372
[60] Shannon, T.T.: Partial, Noisy and Qualitative Models for Adaptive Critic Based Neuro-control. In: Proc. of International Joint Conference on Neural Networks (IJCNN) 1999, Washington, DC (1999)
[61] Abbeel, P., Quigley, M., Ng, A.Y.: Using inaccurate models in reinforcement learning. In: Proceedings of the Twenty-third International Conference on Machine Learning, ICML (2006)
[62] He, P., Jagannathan, S.: Reinforcement Learning-Based Output Feedback Control of Nonlinear Systems With Input Constraints. IEEE Trans. SMC, Part B 35(1), 150–154 (2005)
[63] Sarangapani, J.: Neural Network Control of Nonlinear Discrete-Time Systems. CRC Press, Boca Raton (2006)
[64] Farrell, J.A., Polycarpou, M.M.: Adaptive Approximation Based Control. Wiley-Interscience, Hoboken (2006)
[65] Calise, A.J., Rysdyk, R.T.: Nonlinear adaptive flight control using neural networks. IEEE Control Systems Magazine 18(6), 14–25 (1998)
[66] Vance, J.B., Singh, A., Kaul, B.C., Jagannathan, S., Drallmeier, J.A.: Neural Network Controller Development and Implementation for Spark Ignition Engines With High EGR Levels. IEEE Trans. Neural Networks 18(4), 1083–1100 (2007)
[67] Schmidhuber, J.: A neural network that embeds its own meta-levels. In: Proc. of the IEEE International Conference on Neural Networks, San Francisco (1993)
[68] Siegelmann, H.T., Horne, B.G., Giles, C.L.: Computational capabilities of recurrent NARX neural networks. IEEE Trans. on Systems, Man and Cybernetics–Part B: Cybernetics 27(2), 208 (1997)
[69] Younger, S., Conwell, P., Cotter, N.: Fixed-weight on-line learning. IEEE Transaction on Neural Networks 10, 272–283
[70] Hochreiter, S., Younger, A.S., Conwell, P.R.: Learning to Learn Using Gradient Descent. In: Proceedings of the International Conference on Artificial Neural Networks (ICANN), August 21-25, pp.87–94 (2001)
[71] Feldkamp, L.A., Prokhorov, D.V., Feldkamp, T.M.: Simple and conditioned adaptive behavior from Kalman filter trained recurrent networks. Neural Networks 16(5-6), 683–689 (2003)
[72] Nishimoto, R., Tani, J.: Learning to generate combinatorial action sequences utilizing the initial sensitivity of deterministic dynamical systems. Neural Networks 17(7), 925–933 (2004)
[73] Feldkamp, L.A., Puskorius, G.V., Moore, P.C.: Adaptation from fixed weight dynamic networks. In: Proceedings of IEEE International Conference on Neural Networks, pp. 155–160 (1996)
[74] Feldkamp, L.A., Puskorius, G.V.: Fixed weight controller for multiple systems. In: Proceedings of the International Joint Conference on Neural Networks, vol. 2, pp.773–778 (1997)
[75] Prokhorov, D.: Toward effective combination of off-line and on-line training in ADP framework. In: Proceedings of the, IEEE Symposium on Approximate Dynamic Programming and Reinforcement Learning (ADPRL), Symposium Series on Computational Intelligence (SSCI), April 1-5, Honolulu, HI, pp. 268–271 (2007)
[76] Werbos, P.J.: Backpropagation through time: What it does and how to do it. Proceedings of the IEEE 78(10), 1550–1560 (1990)

[77] Williams, R.J., Peng, J.: An efficient gradient-based algorithm for on-line training of recurrent network trajectories. Neural Computation 2, 490–501 (1990)
[78] Williams, R.J., Zipser, D.: A learning algorithm for continually running fully recurrent neural networks. Neural Computation 1, 270–280 (1989)
[79] Williams, R.J., Zipser, D.: Gradient-based learning algorithms for recurrent networks and their computational complexity. In: Chauvin, Rumelhart (eds.) Backpropagation: Theory, Architectures and Applications, pp. 433–486. Erlbaum and Associates, New York (1995)
[80] Elhanany, I., Liu, Z.: A Fast and Scalable Recurrent Neural Network based on Stochastic Meta-Descent. IEEE Transactions on Neural Networks (2008)
[81] Bengio, Y., Simard, P., Frasconi, P.: Learning long-term dependencies with gradient descent is difficult. IEEE Transactions on Neural Networks 5(2), 157–166 (1994)
[82] Lin, T., Horne, B.G., Tino, P., Giles, C.L.: Learning long–term dependencies in NARX recurrent neural networks. IEEE Trans. on Neural Networks 7(6), 1329 (1996)
[83] Hochreiter, S., Schmidhuber, J.: Long Short-Term Memory. Neural Computation 9(8), 1735–1780 (1997)
[84] Zimmermann, H.G., Grothmann, R., Schäfer, A.M. tietz: Identification and Forecasting of Large Dynamical Systems by Dynamical Consistent Neural Networks. In: Haykin, S., Principe, J., Sejnowski, T., Mc Whirter, J. (eds.) New Directions in Statistical Signal Processing: From Systems to Brain. MIT Press, Cambridge (2006)
[85] Allgöwer, F., Zheng, A.: Progress in systems and Control Theory Series, vol. 26. Birkhauser Verlag, Basel (2000)
[86] Stengel, R.: Optimal Control and Estimation. Dover (1994)
[87] Feldkamp, L.A., Puskorius, G.V.: Training controllers for robustness: Multistream DEKF. In: Proceedings of the IEEE International Conference on Neural Networks, Orlando, pp. 2377–2382 (1994)
[88] Schraudolph, N.N.: Fast Curvature Matrix-Vector Products for Second-Order Gradient Descent. Neural Computation 14, 1723–1738 (2002)
[89] Harth, E., Tzanakou, E.: Alopex: A stochastic method for determining visual receptive fields. Vision Research 14, 1475–1482 (1974)
[90] Haykin, S., Chen, Z., Becker, S.: Stochastic correlative learning algorithms. IEEE Trans. Signal Processing 52(8), 2200–2209 (2004)
[91] Kennedy, J., Shi, Y.: Swarm Intelligence. Morgan Kaufmann/Academic Press (April 2001)
[92] Juang, C.-F.: A hybrid of genetic algorithm and particle swarm optimization for recurrent network design. IEEE Transactions on Systems, Man, and Cybernetics, Part B 34(2), 997–1006 (2004)
[93] Das, S., Abraham, A.: Synergy of Particle Swarm Optimization with Differential Evolution Algorithms for Intelligent Search and Optimization. In: Bajo, J., et al. (eds.) Proceedings of the Hybrid Artificial Intelligence Systems Workshop (HAIS 2006), Salamanca, Spain, pp. 89–99 (2006)
[94] Cai, X., Zhang, N., Venayagamoorthy, G.K., Wunsch, D.C.: Time series prediction with recurrent neural networks trained by a hybrid PSO-EA algorithm. Neurocomputing 70(13-15), 2342–2353 (2007)

[95] Spall, J.C., Cristion, J.A.: A Neural Network Controller for Systems with Unmodeled Dynamics with Applications to Wastewater Treatment. IEEE Trans. Systems, Man and Cybernetics, Part B 27(3), 369–375 (1997)
[96] Julier, S.J., Uhlmann, J.K., Durrant-Whyte, H.F.: A New Approach for Filtering Nonlinear Systems. In: Proceedings of the American Control Conference, Seattle WA, USA, pp. 1628–1632 (1995)
[97] Norgaard, M., Poulsen, N.K., Ravn, O.: New Developments in State Estimation for Nonlinear Systems. Automatica 36, 1627–1638 (2000)
[98] Arasaratnam, I., Haykin, S., Elliott, R.J.: Discrete-Time NonLinear Filtering Algorithms Using Gauss-Hermite Quadrature. Proceedings of the IEEE 95, 953–977 (2007)
[99] Wan, E.A., van der Merwe, R.: The Unscented Kalman Filter for Nonlinear Estimation. In: Proceedings of the IEEE Symposium 2000 on Adaptive Systems for Signal Processing, Communication and Control (AS-SPCC), Lake Louise, Alberta, Canada (October 2000)
[100] Feldkamp, L.A., Feldkamp, T.M., Prokhorov, D.V.: Neural Network Training with the nprKF. In: Proceedings of International Joint Conference on Neural Networks 2001, Washington, D.C., pp. 109–114 (2001)
[101] Prokhorov, D.: Training Recurrent Neurocontrollers for Robustness with Derivative-Free Kalman Filter. IEEE Trans. Neural Networks, 1606–1616 (November 2006)
[102] Jaeger, H., Haas, H.: Harnessing nonlinearity: predicting chaotic systems and saving energy in wireless telecommunications. Science, pp. 78–80, April 2 (2004)
[103] Jaeger, H., Maass, W., Principe, J. (eds.) : Neural Networks, vol. 20(3) (April 2007) Special issue on echo state networks and liquid state machines
[104] Mandic, D., Chambers, J.: Recurrent Neural Networks for Prediction. Wiley, Chichester (2001)
[105] Kolen, J., Kremer, S.: A Field Guide to Dynamical Recurrent Networks. IEEE Press, Los Alamitos (2001)
[106] Schmidhuber, J., Wierstra, D., Gagliolo, M., Gomez, F.: Training Recurrent Networks by Evolino. Neural Computation 19(3), 757–779 (2007)
[107] Back, A.D., Chen, T.: Universal Approximation of Multiple Nonlinear Operators by Neural Networks. Neural Computation 14(11), 2561–2566 (2002)
[108] Santiago, R.A., Lendaris, G.G.: Context Discerning Multifunction Networks: Reformulating fixed weight neural networks. In: Proceedings of the International Joint Conference on Neural Networks (IJCNN), Budapest, Hungary (2004)
[109] Molter, C., Salihoglu, U., Bersini, H.: The road to chaos by hebbian learning in recurrent neural networks. Neural Computation 19(1) (January 2007)
[110] Tyukin, I., Prokhorov, D., van Leeuwen, C.: Adaptive classification of temporal signals in fixed-weights recurrent neural networks: an existence proof. Neural Computation (2008)
[111] Taylor, B.J. (ed.): Methods and Procedures for the Verification and Validation of Artificial Neural Networks. Springer, Heidelberg (2005)
[112] Pullum, L.L., Taylor, B.J., Darrah., M.A.: Guidance for the Verification and Validation of Neural Networks. Wiley-IEEE Computer Society Press (2007)
[113] Vidyasagar, M.: Statistical learning theory and randomized algorithms for control. IEEE Control Systems Magazine 18(6), 69–85 (1998)
[114] Zakrzewski, R.R.: Verification of a trained neural network accuracy. In: Proceedings of International Joint Conference on Neural Networks (IJCNN), vol. 3, pp. 1657–1662 (2001)

[115] Samad, T., Cofer, D.D., Ha, V., Binns, P.: High-confidence control: Ensuring reliability in high-performance real-time systems. Int. J. Intell. Syst. 19(4), 315–326 (2004)

[116] Schumann, J., Gupta, P.: Monitoring the Performance of a Neuro-Adaptive Controller. In: Proc. MAXENT, American Institute of Physics Conference Proceedings 735, pp. 289–296 (2004)

[117] Zakrzewski, R.R.: Randomized approach to verification of neural networks. In: Proceedings of International Joint Conference on Neural Networks (IJCNN), vol. 4, pp. 2819–2824 (2004)

On Learning Machines for Engine Control

Gérard Bloch[1], Fabien Lauer[1], and Guillaume Colin[2]

[1] Centre de Recherche en Automatique de Nancy (CRAN), Nancy-University, CNRS,
2 rue Jean Lamour, 54519 Vandoeuvre lès Nancy, France
gerard.bloch@esstin.uhp-nancy.fr, fabien.lauer@esstin.uhp-nancy.fr
[2] Laboratoire de Mécanique et d'Energétique (LME), University of Orléans,
8 rue Léonard de Vinci, 45072 Orléans Cedex 2, France
guillaume.colin@univ-orleans.fr

Summary. The chapter deals with neural networks and learning machines for engine control applications, particularly in modeling for control. In the first section, basic features of engine control in a layered engine management architecture are reviewed. The use of neural networks for engine modeling, control and diagnosis is then briefly described. The need for descriptive models for model-based control and the link between physical models and black box models are emphasized by the grey box approach discussed in this chapter. The second section introduces the neural models frequently used in engine control, namely, MultiLayer Perceptrons (MLP) and Radial Basis Function (RBF) networks. A more recent approach, known as Support Vector Regression (SVR), to build models in kernel expansion form is also presented. The third section is devoted to examples of application of these models in the context of turbocharged Spark Ignition (SI) engines with Variable Camshaft Timing (VCT). This specific context is representative of modern engine control problems. In the first example, the airpath control is studied, where open loop neural estimators are combined with a dynamical polytopic observer. The second example considers modeling the in-cylinder residual gas fraction by Linear Programming SVR (LP-SVR) based on a limited amount of experimental data and a simulator built from prior knowledge. Each example demonstrates that models based on first principles and neural models must be joined together in a grey box approach to obtain effective and acceptable results.

1 Introduction

The following gives a short introduction on learning machines in engine control. For a more detailed introduction on engine control in general, the reader is referred to (20). After a description of the common features in engine control (Sect. 1.1), including the different levels of a general control strategy, an overview of the use of neural networks in this context is given in Sect. 1.2. Section 1 ends with the presentation of the grey box approach considered in this chapter. Then, in Section 2, the neural models that will be used in the illustrative applications of Section 3, namely, the MultiLayer Perceptron (MLP), the Radial Basis Function Network (RBFN) and a kernel model trained by Support Vector Regression (SVR) are exposed. The examples of Section 3 are taken from a context representative of modern engine control problems, such as airpath control of a turbocharged Spark Ignition (SI) engine with Variable Camshaft Timing

(VCT) (Sect. 3.2) and modeling of the in-cylinder residual gas fraction based on very few samples in order to limit the experimental costs (Sect. 3.3).

1.1 Common Features in Engine Control

The main function of the engine is to ensure the vehicle mobility by providing the power to the vehicle transmission. Nevertheless, the engine torque is also used for peripheral devices such as the air conditioning or the power steering. In order to provide the required torque, the engine control manages the engine actuators, such as ignition coils, injectors and air path actuators for a gasoline engine, pump and valve for diesel engine. Meanwhile, over a wide range of operating conditions, the engine control must satisfy some constraints: driver pleasure, fuel consumption and environmental standards.

In (13), a hierarchical (or stratified) structure, shown in figure 1, is proposed for engine control. In this framework, the engine is considered as a torque source (18) with constraints on fuel consumption and pollutant emission. From the global characteristics of the vehicle, the *Vehicle layer* controls driver strategies and manages the links with other devices (gear box, ...). The *Engine layer* receives from the *Vehicle layer* the effective torque set point (with friction) and translates it into an indicated torque set point (without friction) for the combustion by using an internal model (often a map). The *Combustion layer* fixes the set points for the in-cylinder masses while taking into account the constraints on pollutant emissions. The *Energy layer* ensures the engine load with e.g. the Air to Fuel Ratio (AFR) control and the turbo control. The lower level, specific for a given engine, is the *Actuator layer*, which controls, for instance, the throttle position, the injection and the ignition.

With the multiplication of complex actuators, advanced engine control is necessary to obtain an efficient torque control. This notably includes the control of the ignition coils, fuel injectors and air actuators (throttle, Exhaust Gas Recirculation (EGR), Variable Valve Timing (VVT), turbocharger...). The air actuator

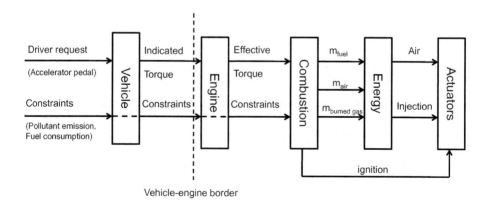

Fig. 1. Hierarchical torque control adapted from (13)

controllers generally used are PID controllers which are difficult to tune. Moreover, they often produce overshooting and bad set point tracking because of the system nonlinearities. Only model-based control can enhance engine torque control.

Several common characteristics can be found in engine control problems. First of all, the descriptive models are dynamic and nonlinear. They require a lot of work to be determined, particularly to fix the parameters specific to each engine type ("mapping"). For control, a sampling period depending on the engine speed (very short in the worst case) must be considered. The actuators present strong saturations. Moreover, many internal state variables are not measured, partly because of the physical impossibility of measuring and the difficulties in justifying the cost of setting up additional sensors. At a higher level, the control must be multi-objective in order to satisfy contradictory constraints (performance, comfort, consumption, pollution). Lastly, the control must be implemented in on-board computers (Electronic Control Units, ECU), whose computing power is increasing, but remains limited.

1.2 Neural Networks in Engine Control

Artificial neural networks have been the focus of a great deal of attention during the last two decades, due to their capabilities to solve nonlinear problems by learning from data. Although a broad range of neural network architectures can be found, MultiLayer Perceptrons (MLP) and Radial Basis Function Networks (RBFN) are the most popular neural models, particularly for system modeling and identification (47). The universal approximation and flexibility properties of such models enable the development of modeling approaches, and then control and diagnosis schemes, which are independent of the specifics of the considered systems. As an example, the linearized neural model predictive control of a turbocharger is described in (12). They allow the construction of nonlinear global models, static or dynamic. Moreover, neural models can be easily and generically differentiated so that a linearized model can be extracted at each sample time and used for the control design. Neural systems can then replace a combination of control algorithms and look-up tables used in traditional control systems and reduce the development effort and expertise required for the control system calibration of new engines. Neural networks can be used as observers or software sensors, in the context of a low number of measured variables. They enable the diagnosis of complex malfunctions by classifiers determined from a base of signatures.

First use of neural networks for automotive application can be traced back to early 90's. In 1991, Marko tested various neural classifiers for online diagnosis of engine control defects (misfires) and proposed a direct control by inverse neural model of an active suspension system (32). In (40), Puskorius and Feldkamp, summarizing one decade of research, proposed neural nets for various subfunctions in engine control: AFR and idle speed control, misfire detection, catalyst monitoring, prediction of pollutant emissions. Indeed, since the beginning of the 90's, neural approaches have been proposed by numerous authors, for example, for

- *vehicle control.* Anti-lock braking system (ABS), active suspension, steering, speed control;

- *engine modeling*. Manifold pressure, air mass flow, volumetric efficiency, indicated pressure into cylinders, AFR, start-of-combustion for Homogeneous Charge Compression Ignition (HCCI), torque or power;
- *engine control*. Idle speed control, AFR control, transient fuel compensation (TFC), cylinder air charge control with VVT, ignition timing control, throttle, turbocharger, EGR control, pollutants reduction;
- *engine diagnosis*. Misfire and knock detection, spark voltage vector recognition systems.

The works are too numerous to be referenced here. Nevertheless, the reader can consult the publications (45; 1; 4; 5; 39) and the references therein, for an overview.

More recently, Support Vector Machines (SVMs) have been proposed as another approach for nonlinear black box modeling (24; 53; 41) or monitoring (43) of automotive engines.

1.3 Grey Box Approach

Let us now focus on the development cycle of engine control, presented in Figure 2, and the different models that are used in this framework. The design process is the following:

1. Building of an engine simulator mostly based on prior knowledge,
2. First identification of control models from data provided by the simulator,
3. Control scheme design,
4. Simulation and pre-calibration of the control scheme with the simulator,
5. Control validation with the simulator,
6. Second identification of control models from data gathered on the engine,
7. Calibration and final test of the control with the engine.

This shows that, in current practice, more or less complex simulation environments based on physical relations are built for internal combustion engines. The great amount of knowledge that is included is consequently available. These simulators are built to be accurate, but this accuracy depends on many physical parameters which must be fixed. In any case, these simulation models cannot be used online, contrary to real time control models. Such control models, e.g. neural models, must be identified first from the simulator and then re-identified or adapted from experimental data. If the modeling process is improved, much gain can be expected for the overall control design process.

Relying in the control design on meaningful physical equations has a clear justification. This partially explains that the fully black box modeling approach has a difficult penetration in the engine control engineering community. Moreover the fully black box (e.g. neural) model based control solutions have still to practically prove their efficiency in terms of robustness, stability and real time applicability. This issue motivates the material presented in this chapter, which concentrates on developing modeling and control solutions, through several examples, mixing physical models and nonlinear black box models in a grey

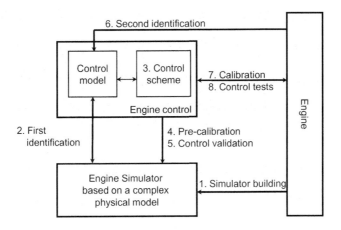

Fig. 2. Engine control development cycle

box approach. In short, *use neural models whenever needed, i. e. whenever first-principles models are not sufficient.* In practice, this can be expressed in two forms.

- Neural models should be used to enhance – not replace – physical models, particularly by extending two dimensional static maps or by correcting physical models when applied to real engines. This is developed in section 3.2.
- Physical insights should be incorporated as prior knowledge into the learning of the neural models. This is developed in section 3.3.

2 Neural Models

This section provides the necessary background on standard MultiLayer Perceptron (MLP) and Radial Basis Function (RBF) neural models, before presenting kernel models and support vector regression.

2.1 Two Neural Networks

As depicted in (47), a general neural model with a single output may be written as a function expansion of the form

$$f(\boldsymbol{\varphi}, \boldsymbol{\theta}) = \sum_{k=1}^{n} \alpha_k g_k(\boldsymbol{\varphi}) + \alpha_0 , \qquad (1)$$

where $\boldsymbol{\varphi} = [\varphi_1 \ \ldots \ \varphi_i \ \ldots \ \varphi_p]^T$ is the regression vector and $\boldsymbol{\theta}$ is the parameter vector.

The restriction of the multilayer perceptron to only one hidden layer and to a linear activation function at the output corresponds to a particular choice, the sigmoid function, for the basis function g_k, and to a "ridge" construction for the inputs in model (1). Although particular, this model will be called MLP in this chapter. Its form is given, for a single output f_{nn}, by

$$f_{nn}(\boldsymbol{\varphi}, \boldsymbol{\theta}) = \sum_{k=1}^{n} w_k^2 \, g\left(\sum_{j=1}^{p} w_{kj}^1 \varphi_j + b_k^1\right) + b^2, \qquad (2)$$

where $\boldsymbol{\theta}$ contains all the weights w_{kj}^1 and biases b_k^1 of the n hidden neurons together with the weights and bias w_k^2, b^2 of the output neuron, and where the activation function g is a sigmoid function (often the hyperbolic tangent $g(x) = 2/(1 + e^{-2x}) - 1$).

On the other hand, choosing a Gaussian function $g(x) = \exp\left(-x^2/\sigma^2\right)$ as basis function and a radial construction for the inputs leads to the radial basis function network (RBFN) (38), of which the output is given by

$$f(\boldsymbol{\varphi}, \boldsymbol{\theta}) = \sum_{k=1}^{n} \alpha_k g\left(\|\boldsymbol{\varphi} - \boldsymbol{\gamma}_k\|_{\boldsymbol{\sigma}_k}\right) + \alpha_0 \qquad (3)$$

$$= \sum_{k=1}^{n} \alpha_k \exp\left(-\frac{1}{2} \sum_{j=1}^{p} \frac{(\varphi_j - \gamma_{kj})^2}{\sigma_{kj}^2}\right) + \alpha_0 \,,$$

where $\boldsymbol{\gamma}_k = [\gamma_{k1} \, \ldots \, \gamma_{kp}]^T$ is the "center" or "position" of the kth Gaussian and $\boldsymbol{\sigma}_k = [\sigma_{k1} \, \ldots \, \sigma_{kp}]^T$ its "scale" or "width", most of the time with $\sigma_{kj} = \sigma_k$, $\forall j$, or even $\sigma_{kj} = \sigma$, $\forall j, k$.

The process of approximating nonlinear relationships from data with these models can be decomposed in several steps:

- determining the structure of the regression vector $\boldsymbol{\varphi}$ or selecting the inputs of the network, see e.g. (46) for dynamic system identification;
- choosing the nonlinear mapping f or, in the neural network terminology, selecting an internal network architecture, see e.g. (42) for MLP's pruning or (37) for RBFN's center selection;
- estimating the parameter vector $\boldsymbol{\theta}$, i.e., (weight) "learning" or "training";
- validating the model.

This approach is similar to the classical one for linear system identification (29), the selection of the model structure being, nevertheless, more involved. For a more detailed description of the training and validation procedures, see (7) or (36).

Among the numerous nonlinear models, neural or not, which can be used to estimate a nonlinear relationship, the advantages of the one hidden layer perceptron, as well as those of the radial basis function network, can be summarized as follows: they are *flexible and parsimonious nonlinear black box models, with universal approximation capabilities* (6).

2.2 Kernel Expansion Models and Support Vector Regression

In the past decade, kernel methods (44) have attracted much attention in a large variety of fields and applications: classification and pattern recognition, regression, density estimation etc. Indeed, using kernel functions, many linear methods can be extended to the nonlinear case in an almost straightforward manner, while avoiding the curse of dimensionality by transposing the focus from the data dimension to the number of data. In particular, Support Vector Regression (SVR), stemming from statistical learning theory (52) and based on the same concepts as the Support Vector Machine (SVM) for classification, offers an interesting alternative both for nonlinear modeling and system identification (16; 33; 54).

SVR originally consists in finding the kernel model that has at most a deviation ε from the training samples with the smallest complexity (48). Thus, SVR amounts to solving a constrained optimization problem known as a quadratic program (QP), where both the ℓ_1-norm of the errors larger than ε and the ℓ_2-norm of the parameters are minimized. Other formulations of the SVR problem minimizing the ℓ_1-norm of the parameters can be derived to yield linear programs (LP) (49; 31). Some advantages of this latter approach can be noticed compared to the QP formulation such as an increased sparsity of support vectors or the ability to use more general kernels (30). The remaining of this chapter will thus focus on the LP formulation of SVR (LP-SVR).

Nonlinear Mapping and Kernel Functions

A kernel model is an expansion of the inner products by the N training samples $\mathbf{x}_i \in \mathbb{R}^p$ mapped in a higher dimensional feature space. Defining the *kernel function* $k(\mathbf{x}, \mathbf{x}_i) = \mathbf{\Phi}(\mathbf{x})^T \mathbf{\Phi}(\mathbf{x}_i)$, where $\mathbf{\Phi}(\mathbf{x})$ is the image of the point $\mathbf{x}$ in that feature space, allows to write the model as a kernel expansion

$$f(\mathbf{x}) = \sum_{i=1}^{N} \alpha_i k(\mathbf{x}, \mathbf{x}_i) + b = \mathbf{K}(\mathbf{x}, \mathbf{X}^T)\boldsymbol{\alpha} + b \, , \tag{4}$$

where $\boldsymbol{\alpha} = [\alpha_1 \ \ldots \ \alpha_i \ \ldots \ \alpha_N]^T$ and b are the parameters of the model, the data $(\mathbf{x}_i, y_i)$, $i = 1, \ldots, N$, are stacked as rows in the matrix $\mathbf{X} \in \mathbb{R}^{N \times p}$ and the vector $\mathbf{y}$, and $\mathbf{K}(\mathbf{x}, \mathbf{X}^T)$ is a vector defined as follows. For $\mathbf{A} \in \mathbb{R}^{p \times m}$ and $\mathbf{B} \in \mathbb{R}^{p \times n}$ containing p-dimensional sample vectors, the "kernel" $\mathbf{K}(\mathbf{A}, \mathbf{B})$ maps $\mathbb{R}^{p \times m} \times \mathbb{R}^{p \times n}$ in $\mathbb{R}^{m \times n}$ with $\mathbf{K}(\mathbf{A}, \mathbf{B})_{i,j} = k(\mathbf{A}_i, \mathbf{B}_j)$, where $\mathbf{A}_i$ and $\mathbf{B}_j$ are the ith and jth columns of $\mathbf{A}$ and $\mathbf{B}$. Typical kernel functions are the linear ($k(\mathbf{x}, \mathbf{x}_i) = \mathbf{x}^T \mathbf{x}_i$), Gaussian RBF ($k(\mathbf{x}, \mathbf{x}_i) = \exp(-\|\mathbf{x} - \mathbf{x}_i\|_2^2 / 2\sigma^2)$) and polynomial ($k(\mathbf{x}, \mathbf{x}_i) = (\mathbf{x}^T \mathbf{x}_i + 1)^d$) kernels. The kernel function defines the feature space $\mathcal{F}$ in which the data are implicitly mapped. The higher the dimension of $\mathcal{F}$, the higher the approximation capacity of the function f, up to the universal approximation capacity obtained for an infinite feature space, as with Gaussian RBF kernels.

Support Vector Regression by Linear Programming

In Linear Programming Support Vector Regression (LP-SVR), the model complexity, measured by the ℓ_1-norm of the parameters $\boldsymbol{\alpha}$, is minimized together with the error on the data, measured by the ε-insensitive loss function l, defined by (52) as

$$l(y_i - f(\mathbf{x}_i)) = \begin{cases} 0 & \text{if } |y_i - f(\mathbf{x}_i)| \leq \varepsilon \, , \\ |y_i - f(\mathbf{x}_i)| - \varepsilon & \text{otherwise} \, . \end{cases} \tag{5}$$

Minimizing the complexity of the model allows to control its generalization capacity. In practice, this amounts to penalizing non-smooth functions and implements the general smoothness assumption that two samples close in input space tend to give the same output.

Following the approach of (31), two sets of optimization variables, in two positive slack vectors **a** and **ξ**, are introduced to yield a linear program solvable by standard optimization routines such as the MATLAB *linprog* function. In this scheme, the LP-SVR problem may be written as

$$\min_{(\alpha,b,\boldsymbol{\xi}\geq 0,\mathbf{a}\geq 0)} \mathbf{1}^T\mathbf{a} + C\mathbf{1}^T\boldsymbol{\xi}$$

$$\text{s.t.} \quad -\boldsymbol{\xi} \leq \mathbf{K}(\mathbf{X},\mathbf{X}^T)\boldsymbol{\alpha} + b\mathbf{1} - \mathbf{y} \leq \boldsymbol{\xi} \tag{6}$$
$$0 \leq \mathbf{1}\varepsilon \leq \boldsymbol{\xi}$$
$$-\mathbf{a} \leq \boldsymbol{\alpha} \leq \mathbf{a},$$

where a hyperparameter C is introduced to tune the trade-off between the minimization of the model complexity and the minimization of the error. The last set of constraints ensures that $\mathbf{1}^T\mathbf{a}$, which is minimized, bounds $\|\boldsymbol{\alpha}\|_1$. In practice, sparsity is obtained as a certain number of parameters α_i will tend to zero. The input vectors $\mathbf{x}_i$ for which the corresponding α_i are non-zero are called *support vectors* (SVs).

2.3 Link between Support Vector Regression and RBFNs

For a Gaussian kernel, the kernel expansion (4) can be interpreted as a RBFN with N neurons in the hidden layer centered at the training samples $\mathbf{x}_i$ and with a unique width $\boldsymbol{\sigma}_k = [\sigma \ \ldots \ \sigma]^T$, $k = 1, \ldots, N$. Compared to neural networks, SVR has the following advantages: automatic selection and sparsity of the model, intrinsic regularization, no local minima (convex problem with a unique solution), and good generalization ability from a limited amount of samples.

It seems though that least squares estimates of the parameters or standard RBFN training algorithms are most of the time satisfactory, particularly when a sufficiently large number of samples corrupted by Gaussian noise is available. Moreover, in this case, standard center selection algorithms may be faster and yield a sparser model than SVR. However, in difficult cases, the good generalization capacity and the better behavior with respect to outliers of SVR may help. Even if non-quadratic criteria have been proposed to train (9) or prune neural networks (51; 25), the SVR loss function is intrinsically robust and thus allows accommodation to non-Gaussian noise probability density functions. In practice, it is advised to employ SVR in the following cases.

- Few data points are available.
- The noise is non-Gaussian.
- The training set is corrupted by outliers.

Finally, the computational framework of SVR allows for easier extensions such as the one described in this chapter, namely, the inclusion of prior knowledge.

3 Engine Control Applications

3.1 Introduction

The application treated here, the control of the turbocharged Spark Ignition engine with Variable Camshaft Timing, is representative of modern engine control problems. Indeed, such an engine presents for control the common characteristics mentioned in the introduction 1.1 and comprises several air actuators and therefore several degrees of freedom for airpath control.

More stringent standards are being imposed to reduce fuel consumption and pollutant emissions for Spark Ignited (SI) engines. In this context, downsizing appears as a major way for reducing fuel consumption while maintaining the advantage of low emission capability of three-way catalytic systems and combining several well known technologies (28). (Engine) downsizing is the use of a smaller capacity engine operating at higher specific engine loads, i.e. at better efficiency points. In order to feed the engine, a well-adapted turbocharger seems to be the best solution. Unfortunately, the turbocharger inertia involves long torque transient responses (28). This problem can be partially solved by combining turbocharging and Variable Camshaft Timing (VCT) which allows air scavenging from the intake to the exhaust.

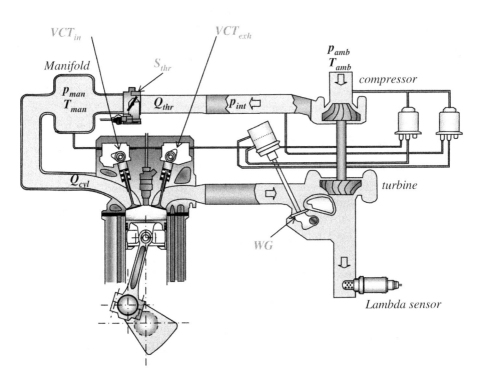

Fig. 3. Airpath of a Turbocharged SI Engine with VCT

The air intake of a turbocharged SI Engine with VCT, represented in Figure 3, can be described as follows. The compressor (pressure p_{int}) produces a flow from the ambient air (pressure p_{amb} and temperature T_{amb}). This air flow Q_{th} is adjusted by the intake throttle (section S_{th}) and enters the intake manifold (pressure p_{man} and temperature T_{man}). The flow that goes into the cylinders Q_{cyl} passes through the intake valves, whose timing is controlled by the intake Variable Camshaft Timing VCT_{in} actuator. After the combustion, the gases are expelled into the exhaust manifold through the exhaust valve, controlled by the exhaust Variable Camshaft Timing VCT_{exh} actuator. The exhaust flow is split into turbine flow and wastegate flow. The turbine flow powers up the turbine and drives the compressor through a shaft. Thus, the supercharged pressure p_{int} is adjusted by the turbine flow which is controlled by the wastegate WG.

The effects of Variable Camshaft Timing (VCT) can be summarized as follows. On the one hand, cam timing can inhibit the production of nitrogen oxides (NO_x). Indeed, by acting on the cam timing, combustion products which would otherwise be expelled during the exhaust stroke are retained in the cylinder during the subsequent intake stroke. This dilution of the mixture in the cylinder reduces the combustion temperature and limits the NO_x formation. Therefore, it is important to estimate and control the back-flow of burned gases in the cylinder. On the other hand, with camshaft timing, air scavenging can appear, that is air passing directly from the intake to the exhaust through the cylinder. For that, the intake manifold pressure must be greater than the exhaust pressure when the exhaust and intake valves are opened together. In that case, the engine torque dynamic behavior is improved, i.e. the settling times decreased. Indeed, the flow which passes through the turbine is increased and the corresponding energy is transmitted to the compressor. In transient, it is also very important to estimate and control this scavenging for torque control.

For such an engine, the following presents the inclusion of neural models in various modeling and control schemes in two parts: an air path control based on an in-cylinder air mass observer, and an in-cylinder residual gas estimation. In the first example, the air mass observer will be necessary to correct the manifold pressure set point. The second example deals with the estimation of residual gases for a single cylinder naturally-aspirated engine. In this type of engine, no scavenging appears, so that the estimation of burned gases and air scavenging of the first example are simplified into a residual gas estimation.

3.2 Airpath Observer Based Control

Control Scheme

The objective of engine control is to supply the torque requested by the driver while minimizing the pollutant emissions. For a SI engine, the torque is directly linked to the air mass trapped in the cylinder for a given engine speed N_e and an efficient control of this air mass is then required. The air path control, i.e. throttle, turbocharger and variable camshaft timing (VCT) control, can be divided in two main parts: the air mass control by the throttle and the turbocharger and the control of the gas mix by the variable camshaft timing (see (12) for further details on VCT control). The structure of the air mass control scheme, described in figure 4, is now detailed block by block. The supervisor, that corresponds to a part of the *Combustion layer* of figure 1, builds the in-cylinder air mass set point from the indicated torque set point, computed by the *Engine layer*. The determination of manifold pressure set points is presented at the end of the section. The general control structure uses an in-cylinder air mass observer discussed below that corrects the errors of the manifold pressure model. The remaining blocks are not described in this chapter but an Internal Model Control (IMC) of the throttle is proposed in (12) and a linearized neural Model Predictive Control (MPC) of the turbocharger can be found in (11; 12). The IMC scheme relies on a grey box model, which includes a neural static estimator. The MPC scheme is based on a dynamical neural model of the turbocharger.

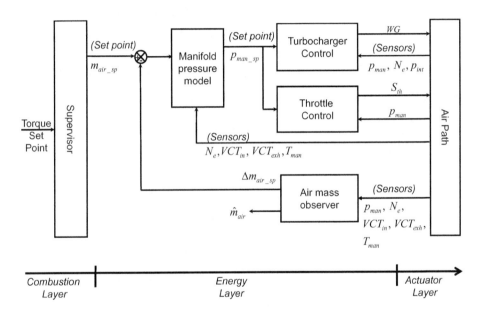

Fig. 4. General control scheme

Observation Scheme

Here two nonlinear estimators of the air variables, the recirculated gas mass RGM and the in-cylinder air mass m_{air}, are presented. Because these variables are not measured, data provided by a complex but accurate high frequency engine simulator (27) are used to build the corresponding models.

Because scavenging and burned gas back-flow correspond to associated flow phenomena, only one variable, the Recirculated Gas Mass (RGM), is defined:

$$RGM = \begin{cases} m_{bg} & \text{, if } m_{bg} > m_{sc} \\ -m_{sc} & \text{, otherwise,} \end{cases} \quad (7)$$

where m_{bg} is the in-cylinder burned gas mass and m_{sc} is the scavenged air mass. Note that, when scavenging from the intake to the exhaust occurs, the burned gases are insignificant. The recirculated gas mass RGM estimator is a neural model entirely obtained from the simulated data.

Considering in-cylinder air mass observation, a lot of references are available especially for air-fuel ratio (AFR) control in a classical engine (21). More recently, (50) uses an "input observer" to determine the engine cylinder flow and (3) uses a Kalman filter to reconstruct the air mass for a turbocharged SI engine.

A novel observer for the in-cylinder air mass m_{air} is presented below. Contrary to the references above, it takes into account a non measured phenomenon (scavenging), and can thus be applied with advanced engine technology (turbocharged VCT engine). Moreover, its on-line computational load is low. As

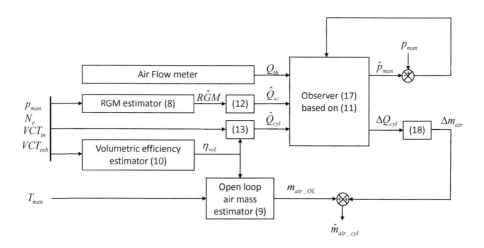

Fig. 5. Air mass observer scheme

presented in Figure 5, this observer combines open loop nonlinear neural based statical estimators of RGM and m_{air}, and a "closed loop" polytopic observer. The observer is built from the Linear Parameter Varying model of the intake manifold and dynamically compensates for the residual error ΔQ_{cyl} committed by one of the estimators, based on a principle similar to the one presented in (2).

Open Loop Estimators

Recirculated gas mass model

Studying the RGM variable (7) is complex because it cannot be measured on-line. Consequently, a static model is built from data provided by the engine simulator. The perceptron with one hidden layer and a linear output unit (2) is chosen with a hyperbolic tangent activation function g.

The choice of the regressors φ_j is based on physical considerations and the estimated Recirculated Gas Mass $\widehat{RGM}$ is given by

$$\widehat{RGM} = f_{nn}(p_{man}, N_e, VCT_{in}, VCT_{exh}), \qquad (8)$$

where p_{man} is the intake manifold pressure, N_e the engine speed, VCT_{in} the intake camshaft timing, and VCT_{exh} the exhaust camshaft timing.

Open loop air mass estimator

The open loop model m_{air_OL} of the in-cylinder air mass is based on the volumetric efficiency equation:

$$m_{air\_OL} = \eta_{vol} \frac{p_{amb} V_{cyl}}{r T_{man}}, \qquad (9)$$

where T_{man} is the manifold temperature, p_{amb} the ambient pressure, V_{cyl} the displacement volume, r the perfect gas constant, and where the volumetric efficiency η_{vol} is described by the static nonlinear function f of four variables: p_{man}, N_e, VCT_{in} and VCT_{exh}.

In (15), various black box models, such as polynomial, spline, MLP and RBFN models, are compared for the static prediction of the volumetric efficiency. In (10), three models of the function f, obtained from engine simulator data, are compared: a polynomial model linear in manifold pressure proposed by Jankovic (23) $f_1(N_e, VCT_{in}, VCT_{exh})p_{man} + f_2(N_e, VCT_{in}, VCT_{exh})$, where f_1 et f_2 are 4th order polynomials, complete with 69 parameters, then reduced by stepwise regression to 43 parameters; a standard 4th order polynomial model $f_3(p_{man}, N_e, VCT_{in}, VCT_{exh})$, complete with 70 parameters then reduced to 58 parameters; and a MLP model with 6 hidden neurons (37 parameters)

$$\eta_{vol} = f_{nn}(p_{man}, N_e, VCT_{in}, VCT_{exh}) . \qquad (10)$$

Training of the neural model has been performed by minimizing the mean squared error, using the Levenberg-Marquardt algorithm. The behavior of these models is similar, and the most important errors are committed at the same operating points. Nevertheless, the neural model, that involves the smallest number of parameters and yields slightly better approximation results, is chosen as the static model of the volumetric efficiency. These results illustrate the parsimony property of the neural models.

Air Mass Observer

Principle

The air mass observer is based on the flow balance in the intake manifold. As shown in figure 6, a flow Q_{th} enters the manifold and two flows leave it: the flow that is captured in the cylinder Q_{cyl} and the flow scavenged from the intake to the exhaust Q_{sc}. The flow balance in the manifold can thus be written as

$$\dot{p}_{man}(t) = \frac{rT_{man}(t)}{V_{man}} \left(Q_{th}(t) - Q_{cyl}(t) - \Delta Q_{cyl}(t) - Q_{sc}(t) \right), \qquad (11)$$

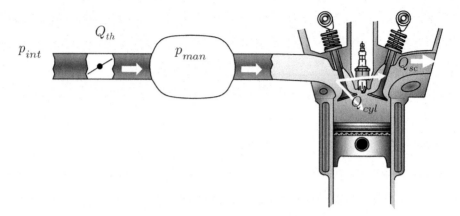

Fig. 6. Intake manifold and cylinder. From the intake manifold, the throttle air flow Q_{th} is divided into in-cylinder air flow Q_{cyl} and air scavenged flow Q_{sc}.

where, for the intake manifold, p_{man} is the pressure to be estimated (in Pa), T_{man} is the temperature (K), V_{man} is the volume (m^3), supposed to be constant and r is the ideal gas constant. In (11), Q_{th} can be measured by an air flow meter (kg/s). On the other hand, Q_{sc} (kg/s) and Q_{cyl} (kg/s) are respectively estimated by differentiating the Recirculated Gas Mass $\widehat{RGM}$ (8):

$$\hat{Q}_{sc} = min(-\widehat{RGM}, 0)/t_{tdc}, \tag{12}$$

where $t_{tdc} = \frac{2 \times 60}{N_e \, n_{cyl}}$ is the variable sampling period between two intake top dead center (TDC), and by

$$\hat{Q}_{cyl}(t) = \eta_{vol}(t) \frac{p_{amb}(t) \, V_{cyl} \, N_e(t) \, n_{cyl}}{r T_{man}(t) 2 \times 60}, \tag{13}$$

where η_{vol} is given by the neural model (10), p_{amb} (Pa) is the ambient pressure, V_{cyl} (m^3) is the displacement volume, N_e (rpm) is the engine speed and n_{cyl} is the number of cylinders. The remaining term in (11), ΔQ_{cyl}, is the error made by the model (13).

Considering slow variations of ΔQ_{cyl}, i.e. $\dot{\Delta Q}_{cyl}(t) = 0$, and after discretization at each top dead center (TDC), thus with a variable sampling period $t_{tdc}(k) = \frac{2 \times 60}{N_e(k) \, n_{cyl}}$, the corresponding state space representation can be written as

$$\begin{cases} \mathbf{x}_{k+1} = \mathbf{A} \, \mathbf{x}_k + \mathbf{B} \, \mathbf{u}_k \\ y_k = \mathbf{C} \, \mathbf{x}_k \, , \end{cases} \tag{14}$$

where

$$\mathbf{x}_k = \begin{bmatrix} p_{man}(k) \\ \Delta Q_{cyl}(k) \end{bmatrix}, \quad \mathbf{u}_k = \begin{bmatrix} Q_{th}(k) \\ Q_{cyl}(k) \\ Q_{sc}(k) \end{bmatrix}, \quad y_k = p_{man}(k), \tag{15}$$

and, defining $\rho(k) = -\frac{r \, T_{man}(k)}{V_{man}} t_{tdc}(k)$, where

$$\mathbf{A} = \begin{bmatrix} 1 & \rho(k) \\ 0 & 1 \end{bmatrix}, \quad \mathbf{B} = \begin{bmatrix} -\rho(k) & \rho(k) & \rho(k) \\ 0 & 0 & 0 \end{bmatrix}. \tag{16}$$

Note that this system is Linear Parameter Varying (LPV), because the matrices $\mathbf{A}$ and $\mathbf{B}$ depend linearly on the (measured) parameter $\rho(k)$, which depends on the manifold temperature $T_{man}(k)$ and the engine speed $N_e(k)$.

The state reconstruction for system (14) can be achieved by resorting to the so-called polytopic observer of the form

$$\begin{cases} \hat{\mathbf{x}}_{k+1} = \mathbf{A}(\rho_k)\hat{\mathbf{x}}_k + \mathbf{B}(\rho_k)\mathbf{u}_k + \mathbf{K}(y_k - \hat{y}_k) \\ \hat{y}_k = \mathbf{C}\hat{\mathbf{x}}_k, \end{cases} \tag{17}$$

with a constant gain $\mathbf{K}$.

This gain is obtained by solving a Linear Matrix Inequality (LMI). This LMI ensures the convergence towards zero of the reconstruction error for the whole operating domain of the system based on its polytopic decomposition. This ensures the global convergence of the observer. See (35), (34) and (14) for further details.

Then, the state ΔQ_{cyl} is integrated (i.e. multiplied by t_{tdc}) to give the air mass bias

$$\Delta m_{air} = \Delta Q_{cyl} \times t_{tdc}. \tag{18}$$

Finally, the in-cylinder air mass can be estimated by correcting the open loop estimator (9) with this bias as

$$\hat{m}_{air\_cyl} = m_{air\_OL} + \Delta m_{air}. \tag{19}$$

Results

Some experimental results, normalized between 0 and 1, obtained on a 1.8 Liter turbocharged 4 cylinder engine with Variable Camshaft Timing are given in figure 7.

A measurement of the in-cylinder air mass, only valid in steady state, can be obtained from the measurement of Q_{th} by an air flow meter. Indeed, in steady state with no scavenging, the air flow that gets into the cylinder Q_{cyl} is equal to the flow that passes through the throttle Q_{th} (see Figure 6). In consequence, this air mass measurement is obtained by integrating Q_{th} (i.e. multiplying by t_{tdc}). Figure 7 compares this measurement, the open loop neural estimator ((9) with a neural model (10)), an estimation not based on this neural model (observer (17) based on model (11) but with $Q_{cyl} = Q_{sc} = 0$), the proposed estimation ((19) combining the open loop neural estimator (9) and the polytopic observer (17) based on model (11) with Q_{cyl} given by (13) using the neural model (10) and Q_{sc} given by (12) using (8)).

For steps of air flow, the open loop neural estimator tracks very quickly the measurement changes, but a small steady state error can be observed (see for example between 32 s and 34 s). Conversely, the closed loop observer which does not take into account this feedforward estimator involves a long transient error while guaranteeing the convergence in steady state. Finally, the proposed

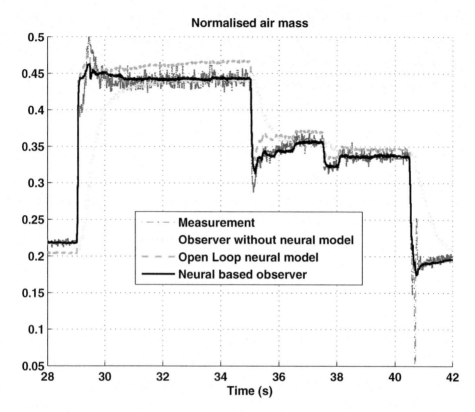

Fig. 7. Air mass observer results (mg) vs. time (s) on an engine test bench

estimator, including feedforward statical estimators and a polytopic observer, combines both the advantages: very fast tracking and no steady state error. This observer can be used to design and improve the engine supervisor of figure 5 by determining the air mass set points.

Computing the Manifold Pressure Set Points

To obtain the desired torque of a SI engine, the air mass trapped in the cylinder must be precisely controlled. The corresponding measurable variable is the manifold pressure. Without Variable Camshaft Timing (VCT), this variable is linearly related to the trapped air mass, whereas with VCT, there is no more one-to-one correspondence. Figure 8 shows the relationship between the trapped air mass and the intake manifold pressure at three particular VCT positions for a fixed engine speed.

Thus, it is necessary to model the intake manifold pressure p_{man}. The chosen static model is a perceptron with one hidden layer (2). The regressors have been chosen from physical considerations: air mass m_{air} (corrected by the intake manifold temperature T_{man}), engine speed N_e, intake VCT_{in} and exhaust VCT_{exh}

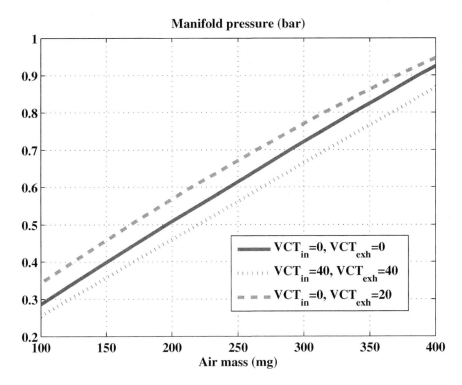

Fig. 8. Relationship between the manifold pressure (in bar) and the air mass trapped (in mg) for a SI engine with VCT at 2000 rpm

camshaft timing. The intake manifold pressure model is thus given by

$$p_{man} = f_{nn}(m_{air}, N_e, VCT_{in}, VCT_{exh}) . \qquad (20)$$

Training of the neural model from engine simulator data has been performed by minimizing the mean squared error, using the Levenberg-Marquardt algorithm.

The supervisor gives an air mass set point m_{air_sp} from the torque set point (figure 4). The intake manifold pressure set point, computed by model (20), is corrected by the error Δm_{air} (18) to yield the final set point p_{man_sp} as

$$p_{man\_sp} = f_{nn}(m_{air\_sp} - \Delta m_{air\_sp}, N_e, VCT_{in}, VCT_{exh}) . \qquad (21)$$

Engine Test Bench Results

The right part of figure 9 shows an example of results for air mass control, in which the VCT variations are not taken into account. Considerable air mass variations (nearly ±25% of the set point) can be observed. On the contrary, the left part shows the corresponding results for the proposed air mass control. The air mass is almost constant (nearly ±2% of variation), illustrating that the manifold pressure set point is well computed with (21). This allows to reduce the pollutant emissions without degrading the torque set point tracking.

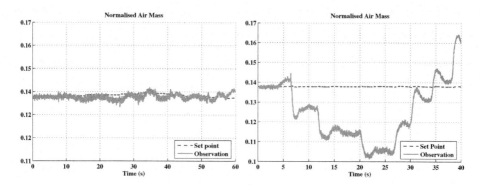

Fig. 9. Effect of the variation of VCT's on air mass with the proposed control scheme (left) and without taking into account the variation of VCT's in the control scheme (right)

3.3 Estimation of In-Cylinder Residual Gas Fraction

The application deals with the estimation of residual gases in the cylinders of Spark Ignition (SI) engines with Variable Camshaft Timing (VCT) by Support Vector Regression (SVR) (8). More precisely, we are interested in estimating the residual gas mass fraction by incorporating prior knowledge in the SVR learning with the general method proposed in (26). Knowing this fraction allows to control torque as well as pollutant emissions. The residual gas mass fraction χ_{res} can be expressed as a function of the engine speed N_e, the ratio p_{man}/p_{exh}, where p_{man} and p_{exh} are respectively the (intake) manifold pressure and the exhaust pressure, and an overlapping factor OF (in $°/m$) (17), related to the time during which the valves are open together.

The available data are provided, on one hand, from the modeling and simulation environment Amesim (22), which uses a high frequency zero-dimensional thermodynamic model and, on the other hand, from off line measurements, which are accurate, but complex and costly to obtain, by direct in-cylinder sampling (19). The problem is this. How to obtain a simple, embeddable, black box model with a good accuracy and a large validity range for the real engine, from precise real measurements as less numerous as possible and a representative, but possibly biased, prior simulation model? The problem thus posed, although particular, is very representative of numerous situations met in engine control, and more generally in engineering, where complex models, more or less accurate, exist and where the experimental data which can be used for calibration are difficult or expensive to obtain.

The simulator being biased but approximating rather well the overall shape of the function, the prior knowledge will be incorporated in the derivatives. Prior knowledge of the derivatives of a SVR model can be enforced in the training by noticing that the kernel expansion (4) is linear in the parameters $\boldsymbol{\alpha}$, which allows to write the derivative of the model output with respect to the scalar input x^j as

$$\frac{\partial f(\mathbf{x})}{\partial x^j} = \sum_{i=1}^{N} \alpha_i \frac{\partial k(\mathbf{x}, \mathbf{x}_i)}{\partial x^j} = \mathbf{r}_j(\mathbf{x})^T \boldsymbol{\alpha} , \qquad (22)$$

where $\mathbf{r}_j(\mathbf{x}) = [\partial k(\mathbf{x}, \mathbf{x}_1)/\partial x^j \ldots \partial k(\mathbf{x}, \mathbf{x}_i)/\partial x^j \ldots \partial k(\mathbf{x}, \mathbf{x}_N)/\partial x^j]^T$ is of dimension N. The derivative (22) is linear in $\boldsymbol{\alpha}$. In fact, the form of the kernel expansion implies that the derivatives of any order with respect to any component are linear in $\boldsymbol{\alpha}$. Prior knowledge of the derivatives can thus be formulated as linear constraints.

The proposed model is trained by a variant of algorithm (6) with additional constraints on the derivatives at the points $\tilde{\mathbf{x}}_p$ of the simulation set. In the case where the training data do not cover the whole input space, extrapolation occurs, which can become a problem when using local kernels such as the RBF kernel. To avoid this problem, the simulation data $\tilde{\mathbf{x}}_p$, $p = 1, \ldots, N^{pr}$, are introduced as potential support vectors (SVs). The resulting global model is now

$$f(\mathbf{x}) = \mathbf{K}(\mathbf{x}, [\mathbf{X}^T \ \tilde{\mathbf{X}}^T])\boldsymbol{\alpha} + b , \qquad (23)$$

where $\boldsymbol{\alpha} \in \mathbb{R}^{N+N^{pr}}$ and $[\mathbf{X}^T \ \tilde{\mathbf{X}}^T]$ is the concatenation of the matrices $\mathbf{X}^T = [\mathbf{x}_1 \ldots \mathbf{x}_i \ldots \mathbf{x}_N]$, containing the real data, and $\tilde{\mathbf{X}}^T = [\tilde{\mathbf{x}}_1 \ldots \tilde{\mathbf{x}}_p \ldots \tilde{\mathbf{x}}_{N^{pr}}]$, containing the simulation data. Defining the $N^{pr} \times (N + N^{pr})$-dimensional matrix $\mathbf{R}(\tilde{\mathbf{X}}^T, [\mathbf{X}^T \ \tilde{\mathbf{X}}^T]) = [\mathbf{r}_1(\tilde{\mathbf{x}}_1) \ldots \mathbf{r}_1(\tilde{\mathbf{x}}_p) \ldots \mathbf{r}_1(\tilde{\mathbf{x}}_{N^{pr}})]^T$, where $\mathbf{r}_1(\mathbf{x})$ corresponds to the derivative of (23) with respect to the input $x^1 = p_{man}/p_{exh}$, the proposed model is obtained by solving

$$\min_{(\boldsymbol{\alpha}, b, \boldsymbol{\xi}, \mathbf{a}, \mathbf{z})} \frac{1}{N+N^{pr}} \mathbf{1}^T \mathbf{a} + \frac{C}{N} \mathbf{1}^T \boldsymbol{\xi} + \frac{\lambda}{N^{pr}} \mathbf{1}^T \mathbf{z}$$

$$\text{s.t.} \quad \begin{array}{rcl} -\boldsymbol{\xi} \leq & \mathbf{K}(\mathbf{X}^T, [\mathbf{X}^T \ \tilde{\mathbf{X}}^T])\boldsymbol{\alpha} + b\mathbf{1} - \mathbf{y} & \leq \boldsymbol{\xi} \\ 0 \leq & 1\varepsilon & \leq \boldsymbol{\xi} \\ -\mathbf{a} \leq & \boldsymbol{\alpha} & \leq \mathbf{a} \\ -\mathbf{z} \leq & \mathbf{R}(\tilde{\mathbf{X}}^T, [\mathbf{X}^T \ \tilde{\mathbf{X}}^T])\boldsymbol{\alpha} - \mathbf{y}' & \leq \mathbf{z} , \end{array} \qquad (24)$$

where $\mathbf{y}'$ contains the N^{pr} known values of the derivative with respect to the input p_{man}/p_{exh}, at the points $\tilde{\mathbf{x}}_p$ of the simulation set. In order to be able to evaluate these values, a *prior model*, $\tilde{f}(\mathbf{x}) = \sum_{p=1}^{N^{pr}} \tilde{\alpha}_p k(\mathbf{x}, \tilde{\mathbf{x}}_p) + \tilde{b}$, is first trained on the N^{pr} simulation data $(\tilde{\mathbf{x}}_p, \tilde{y}_p)$, $p = 1, \ldots, N^{pr}$, only. This prior model is then used to provide the prior derivatives $\mathbf{y}' = [\partial \tilde{f}(\tilde{\mathbf{x}}_1)/\partial \tilde{x}^1 \ldots \partial \tilde{f}(\tilde{\mathbf{x}}_p)/\partial \tilde{x}^1 \ldots \partial \tilde{f}(\tilde{\mathbf{x}}_{N^{pr}})/\partial \tilde{x}^1]^T$.

Note that the knowledge of the derivatives is included by soft constraints, thus allowing to tune the trade-off between the data and the prior knowledge. The weighting hyperparameters are set to C/N and λ/N^{pr} in order to maintain the same order of magnitude between the regularization, error and prior knowledge terms in the objective function. This allows to ease the choice of C and λ based on the application goals and confidence in the prior knowledge. Hence, the hyperparameters become problem independent.

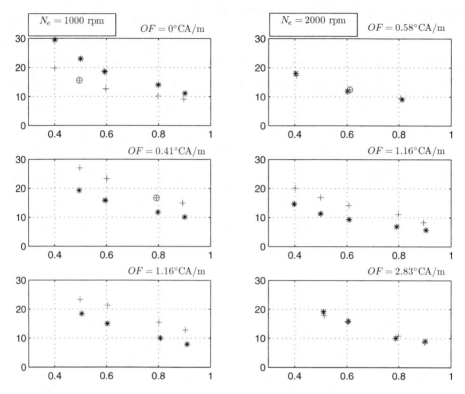

Fig. 10. Residual gas mass fraction χ_{res} in percentages as a function of the ratio p_{man}/p_{exh} for two engine speeds N_e and different overlapping factors OF. The 26 experimental data are represented by plus signs (+) with a superposed circle (⊕) for the 3 points retained as training samples. The 26 simulation data appear as asterisks (∗).

The method is now evaluated on the in-cylinder residual gas fraction application. In this experiment, three data sets are built from the available data composed of 26 experimental samples plus 26 simulation samples:

- the training set $(\mathbf{X}, \mathbf{y})$ composed of a limited amount of real data (N samples),
- the test set composed of independent real data ($26 - N$ samples),
- the simulation set $(\tilde{\mathbf{X}}, \tilde{\mathbf{y}})$ composed of data provided by the simulator ($N^{pr} = 26$ samples).

The test samples are assumed to be unknown during the training and are retained for testing only. It must be noted that the inputs of the simulation data do not exactly coincide with the inputs of the experimental data as shown in Figure 10 for $N = 3$.

The comparison is performed between four models.

- The *experimental model* trained by (6) on the real data set $(\mathbf{X}, \mathbf{y})$ only,
- the *prior model* trained by (6) on the simulation data $(\tilde{\mathbf{X}}, \tilde{\mathbf{y}})$ only,

Table 1. Errors on the residual gas mass fraction with the number of real and simulation data used for training. '–' appears when the result is irrelevant (model mostly constant).

Model	# of real data N	# of simulation data N^{pr}	RMSE test	RMSE	MAE
experimental model	6		6.84	6.00	15.83
prior model		26	4.86	4.93	9.74
mixed model	6	26	4.85	4.88	9.75
proposed model	6	26	**2.44**	**2.15**	**5.94**
experimental model	3		–	–	–
prior model		26	4.93	4.93	9.74
mixed model	3	26	4.89	4.86	9.75
proposed model	3	26	**2.97**	**2.79**	**5.78**

- the *mixed model* trained by (6) on the real data set simply extended with the simulation data $([\mathbf{X}^T \ \tilde{\mathbf{X}}^T]^T, [\mathbf{y}^T \ \tilde{\mathbf{y}}^T]^T)$ (the training of this model is close in spirit to the virtual sample approach, where extra data are added to the training set),
- the *proposed model* trained by (24) and using both the real data $(\mathbf{X}, \mathbf{y})$ and the simulation data $(\tilde{\mathbf{X}}, \tilde{\mathbf{y}})$.

These models are evaluated on the basis of three indicators: the root mean square error (RMSE) on the test set (*RMSE test*), the RMSE on all experimental data (*RMSE*) and the maximum absolute error on all experimental data (*MAE*).

Before training, the variables are normalized with respect to their mean and standard deviation. The different hyperparameters are set according to the following heuristics. One goal of the application is to obtain a model that is accurate on both the training and test samples (the training points are part of the performance index *RMSE*). Thus C is set to a large value ($C = 100$) in order to ensure a good approximation of the training points. Accordingly, ε is set to 0.001 in order to approximate the real data well. The trade-off parameter λ of the proposed method is set to 100, which gives as much weight to both the training data and the prior knowledge. Since all standard deviations of the inputs are equal to 1 after normalization, the RBF kernel width σ is set to 1.

Two sets of experiments are performed for very low numbers of training samples $N = 6$ and $N = 3$. The results in Table 1 show that both the *experimental* and the *mixed* models cannot yield a better approximation than the *prior model* with so few training data. Moreover, for $N = 3$, the *experimental model* yields a quasi-constant function due to the fact that the model has not enough free parameters (only 3 plus a bias term) and thus cannot model the data. In this case, the RMSE is irrelevant. On the contrary, the *proposed model* does not suffer from a lack of basis functions, thanks to the inclusion of the simulation data as potential support vectors. This model yields good results from very few training samples. Moreover, the performance decreases only slightly when reducing the training set size from 6 to 3. Thus, the proposed method seems to be a promising alternative to obtain a simple black box model with a good accuracy from a limited number of experimental data and a prior simulation model.

4 Conclusion

The chapter exposed learning machines for engine control applications. The two neural models most used in modeling for control, the MultiLayer Perceptron and the Radial Basis Function network, have been described, along with a more recent approach, known as Support Vector Regression. The use of such black box models has been placed in the design cycle of engine control, where the modeling steps constitute the bottleneck of the whole process. Application examples have been presented for a modern engine, a turbocharged Spark Ignition engine with Variable Camshaft Timing. In the first example, the airpath control was studied, where open loop neural estimators are combined with a dynamical polytopic observer. The second example considered modeling a variable which is not measurable on-line, from a limited amount of experimental data and a simulator built from prior knowledge.

The neural black box approach for modeling and control allows to develop generic, application independent, solutions. The price to pay is the loss of the physical interpretability of the resulting models. Moreover, the fully black box (e.g. neural) model based control solutions have still to practically prove their efficiency in terms of robustness or stability. On the other hand, models based on first principles (white box models) are completely meaningful and many control approaches with good properties have been proposed, which are well understood and accepted in the engine control community. However, these models are often inadequate, too complex or too difficult to parametrize, as real time control models. Therefore, intermediate solutions, involving grey box models, seem to be preferable for engine modeling, control and, at a higher level, optimization. In this framework, two approaches can be considered. First, besides first principles models, black box neural sub-models are chosen for variables difficult to model (e.g. volumetric efficiency, pollutant emissions). Secondly, black box models can be enhanced thanks to physical knowledge. The examples presented in this chapter showed how to implement these grey box approaches in order to obtain efficient and acceptable results.

References

[1] Alippi, C., De Russis, C., Piuri, V.: A neural-network based control solution to air fuel ratio for automotive fuel injection system. IEEE Trans. on System Man and Cybernetics - Part C 33(2), 259–268 (2003)

[2] Andersson, P.: Intake Air Dynamics on a Turbocharged SI Engine with Wastegate. PhD thesis, Linköping University, Sweden (2002)

[3] Andersson, P., Eriksson, L.: Mean-value observer for a turbocharged SI engine. In: Proc. of the IFAC Symp. on Advances in Automotive Control, Salerno, Italy, pp. 146–151 (April 2004)

[4] Arsie, I., Pianese, C., Sorrentino, M.: A procedure to enhance identification of recurrent neural networks for simulating air-fuel ratio dynamics in SI engines. Engineering Applications of Artificial Intelligence 19(1), 65–77 (2006)

[5] Arsie, I., Pianese, C., Sorrentino, M.: Recurrent neural networks for AFR estimation and control in spark ignition automotive engines. In: Prokhorov, D. (ed.) Comput. Intel. in Automotive Applications, ch. 9. Springer, Heidelberg (2009)

[6] Barron, A.R.: Universal approximation bounds for superpositions of a sigmoidal function. IEEE Trans. on Information Theory 39(3), 930–945 (1993)

[7] Bloch, G., Denoeux, T.: Neural networks for process control and optimization: two industrial applications. ISA Transactions 42(1), 39–51 (2003)

[8] Bloch, G., Lauer, F., Colin, G., Chamaillard, Y.: Combining experimental data and physical simulation models in support vector learning. In: Iliadis, L., Margaritis, K. (eds.) Proc. of the 10th Int. Conf. on Engineering Applications of Neural Networks (EANN) CEURWorkshop Proceedings, Thessaloniki, Greece, vol. 284, pp. 284–295 (2007)

[9] Bloch, G., Sirou, F., Eustache, V., Fatrez, P.: Neural intelligent control of a steel plant. IEEE Trans. on Neural Networks 8(4), 910–918 (1997)

[10] Colin, G.: Contrôle des systèmes rapides non linéaires – Application au moteur à allumage commandé turbocompressé à distribution variable. PhD thesis, University of Orléans, France (2006)

[11] Colin, G., Chamaillard, Y., Bloch, G., Charlet, A.: Exact and linearised neural predictive control - a turbocharged SI engine example. Journal of Dynamic Systems, Measurement and Control - Trans. of the ASME 129(4), 527–533 (2007)

[12] Colin, G., Chamaillard, Y., Bloch, G., Corde, G.: Neural control of fast nonlinear systems - Application to a turbocharged SI engine with VCT. IEEE Trans. on Neural Networks 18(4), 1101–1114 (2007)

[13] Corde, G.: Le contrôle moteur. In: Gissinger, G., Le Fort Piat, N. (eds.) Contrôle commande de la voiture, Hermès (2002)

[14] Daafouz, J., Bernussou, J.: Parameter dependent Lyapunov functions for discrete time systems with time varying parametric uncertainties. Systems & Control Letters 43(5), 355–359 (2001)

[15] De Nicolao, G., Scattolini, R., Siviero, C.: Modelling the volumetric efficiency of IC engines: parametric, non-parametric and neural techniques. Control Engineering Practice 4(10), 1405–1415 (1996)

[16] Drezet, P.M.L., Harrison, R.F.: Support vector machines for system identification. In: Proc. of the UKACC Int. Conf. on Control, Swansea, UK, vol. 1, pp. 688–692 (1998)

[17] Fox, J.W., Cheng, W.K., Heywood, J.B.: A model for predicting residual gas fraction in sparkignition engines. SAE Technical Papers (931025) (1993)

[18] Gerhardt, J., Hönniger, H., Bischof, H.: A new approach to functionnal and software structure for engine management systems - BOSCH ME7. SAE Technical Papers (980801) (1998)

[19] Giansetti, P., Colin, G., Higelin, P., Chamaillard, Y.: Residual gas fraction measurement and computation. International Journal of Engine Research 8(4), 347–364 (2007)

[20] Guzzella, L., Onder, C.H.: Introduction to Modeling and control of Internal Combustion Engine Systems. Springer, Heidelberg (2004)

[21] Hendricks, E., Luther, J.: Model and observer based control of internal combustion engines. In: Proc. of the 1st Int. Workshop on Modeling Emissions and Control in Automotive Engines (MECA), Salerno, Italy, pp. 9–20, Salerno, Italy (2001)

[22] Imagine. Amesim web site (2006), www.amesim.com
[23] Jankovic, M., Magner, S.W.: Variable Cam Timing: Consequences to Automotive Engine Control Design. In: Proc. of the 15th Triennial IFAC World Congress, Barcelona, Spain, pp. 271–276 (2002)
[24] Kolmanovsky, I.: Support vector machine-based determination of gasoline direct-injected engine admissible operating envelope. SAE Technical Papers (2002-01-1301) (2002)
[25] Lairi, M., Bloch, G.: A neural network with minimal structure for maglev system modeling and control. In: Proc. of the IEEE Int. Symp. on Intelligent Control / Intelligent Systems & Semiotics, Cambridge, MA, USA, pp. 40–45 (1999)
[26] Lauer, F., Bloch, G.: Incorporating prior knowledge in support vector regression. Machine Learning 70(1), 89–118 (2008)
[27] Le Berr, F., Miche, M., Colin, G., Le Solliec, G., Lafossas, F.: Modelling of a Turbocharged SI Engine with Variable Camshaft Timing for Engine Control Purposes. SAE Technical Paper (2006-01-3264) (2006)
[28] Lecointe, B., Monnier, G.: Downsizing a Gasoline Engine Using Turbocharging with Direct Injection. SAE Technical Paper (2003-01-0542) (2003)
[29] Ljung, L.: System identification: Theory for the user, 2nd edn. Prentice-Hall, Englewood Cliffs (1999)
[30] Mangasarian, O.: Generalized support vector machines. In: Smola, A., Bartlett, P., Schölkopf, B., Schuurmans, D. (eds.) Advances in Large Margin Classifiers, pp. 135–146. MIT Press, Cambridge (2000)
[31] Mangasarian, O.L., Musicant, D.R.: Large scale kernel regression via linear programming. Machine Learning 46(1-3), 255–269 (2002)
[32] Marko, K.A.: Neural network application to diagnostics and control of vehicle control systems. In: Lippmann, R., Moody, J.E., Touretzky, D.S. (eds.) Advances in Neural Information Processing Systems, vol. 3, pp. 537–543. Morgan Kaufmann, San Francisco (1991)
[33] Mattera, D., Haykin, S.: Support vector machines for dynamic reconstruction of a chaotic system. In: Schölkopf, B., Burges, C.J.C., Smola, A.J. (eds.) Advances in kernel methods: support vector learning, pp. 211–241. MIT Press, Cambridge (1999)
[34] Millérioux, G., Anstett, F., Bloch, G.: Considering the attractor structure of chaotic maps for observer-based synchronization problems. Mathematics and Computers in Simulation 68(1), 67–85 (2005)
[35] Millérioux, G., Rosier, L., Bloch, G., Daafouz, J.: Bounded state reconstruction error for LPV systems with estimated parameters. IEEE Trans. on Automatic Control 49(8), 1385–1389 (2004)
[36] Nelles, O.: Nonlinear System Identification: From Classical Approaches to Neural Networks and Fuzzy Models. Springer, Berlin (2001)
[37] Orr, M.J.L.: Recent advances in radial basis function networks. Technical report, Edinburgh University, UK (1999)
[38] Poggio, T., Girosi, F.: Networks for approximation and learning. Proc. IEEE 78(10), 1481–1497 (1990)
[39] Prokhorov, D.V.: Neural networks in automotive applications. In: Prokhorov, D. (ed.) Comput. Intel. in Automotive Applications, ch.7. Springer, Heidelberg (2009)
[40] Puskorius, G.V., Feldkamp, L.A.: Parameter-based Kalman filter training: theory and implementation. In: Haykin, S. (ed.) Kalman filtering and neural networks, ch. 2, pp. 23–67. Wiley, Chichester (2001)

[41] Rakotomamonjy, A., Le Riche, R., Gualandris, D., Harchaoui, Z.: A comparison of statistical learning approaches for engine torque estimation. Control Engineering Practice (2007)
[42] Reed, R.: Pruning algorithms – a survey. IEEE Trans. on Neural Networks 4, 740–747 (1993)
[43] Rychetsky, M., Ortmann, S., Glesner, M.: Support vector approaches for engine knock detection. In: Proc. of the Int. Joint Conf. on Neural Networks (IJCNN), Washington, DC, USA, vol. 2, pp. 969–974 (1999)
[44] Schölkopf, B., Smola, A.J.: Learning with Kernels: Support Vector Machines, Regularization, Optimization, and Beyond. MIT Press, Cambridge (2001)
[45] Shayler, P.J., Goodman, M.S., Ma, T.: The exploitation of neural networks in automotive engine management systems. Engineering Applications of Artificial Intelligence 13(2), 147–157 (2000)
[46] Sjöberg, J., Ngia, L.S.H.: Neural nets and related model structures for nonlinear system identification. In: Suykens, J.A.K., Vandewalle, J. (eds.) Nonlinear Modeling, Advanced Black-Box Techniques, ch. 1, pp. 1–28. Kluwer Academic Publishers, Dordrecht (1998)
[47] Sjöberg, J., Zhang, Q., Ljung, L., Benveniste, A., Delyon, B., Glorennec, P.Y., Hjalmarsson, H., Juditsky, A.: Nonlinear black-box modeling in system identification: a unified overview. Automatica 31(12), 1691–1724 (1995)
[48] Smola, A.J., Schölkopf, B.: A tutorial on support vector regression. Statistics and Computing 14(3), 199–222 (2004)
[49] Smola, A.J., Schölkopf, B., Rätsch, G.: Linear programs for automatic accuracy control in regression. In: Proc. of the 9th Int. Conf. on Artificial Neural Networks, Edinburgh, UK, vol. 2, pp. 575–580 (1999)
[50] Stotsky, A., Kolmanovsky, I.: Application of input estimation techniques to charge estimation and control in automotive engines. Control Engineering Practice 10, 1371–1383 (2002)
[51] Thomas, P., Bloch, G.: Robust pruning for multilayer perceptrons. In: Borne, P., Ksouri, M., El Kamel, A. (eds.) Proc. of the IMACS/IEEE Multiconf. on Computational Engineering in Systems Applications, Nabeul-Hammamet, Tunisia, vol. 4, pp. 17–22 (1998)
[52] Vapnik, V.N.: The nature of statistical learning theory. Springer, New York (1995)
[53] Vong, C.-M., Wong, P.-K., Li, Y.-P.: Prediction of automotive engine power and torque using least squares support vector machines and Bayesian inference. Engineering Applications of Artificial Intelligence 19(3), 277–287 (2006)
[54] Zhang, L., Xi, Y.: Nonlinear system identification based on an improved support vector regression estimator. In: Yin, F.-L., Wang, J., Guo, C. (eds.) ISNN 2004, vol. 3173, pp. 586–591. Springer, Heidelberg (2004)

Recurrent Neural Networks for AFR Estimation and Control in Spark Ignition Automotive Engines

Ivan Arsie, Cesare Pianese, and Marco Sorrentino

Department of Mechanical Engineering - University of Salerno – 84084
Fisciano (SA), Italy
{iarsie, pianese, msorrentino}@unisa.it

Introduction

Since 80's continuous government constraints have pushed car manufacturers towards the study of innovative technologies aimed at reducing automotive exhaust emissions and increasing engine fuel economy. As a result of this stringent legislation, the automotive engine technology has experienced continuous improvements in many areas. In the field of Engine Control Systems (ECS) innovative procedures have been proposed and major efforts have been devoted to the study of transient phenomena related to operation and design of engine control strategies. Particular attention has been given to the control of mixture strength excursions, which is a critical task to assure satisfactory efficiency of three-way catalytic converters and thus to meet exhaust emissions regulations. This goal has to be reached in both steady state and transient conditions by estimating the air flow rate at the injector location and delivering the fuel in the right amount and with the appropriate time dependence. Furthermore, the ECS designers have to face with the On Board Diagnostics (OBD) requirements that were introduced in 1996 in California and later in Europe and represent one of the most challenging targets in the field of Automotive Control. OBD requires a continuous monitoring of all powertrain components in order to prevent those faults that could result in a strong increase of exhaust emissions.

Advanced research on both engine control and fault diagnosis mainly relies on model-based techniques that are founded on a mathematical description of the processes to be controlled or monitored. Thus, appropriate identification methodologies have also to be designed to adapt control and diagnostics schemes to different engines and make them robust with respect to aging effects and vehicle-to-vehicle variability.

In the framework of engine control, the fuel metering is performed through the direct measurement by air flow meters or via model-based control techniques. The former approach is intrinsically more expensive and exhibits some limitations due to difficult sensor placement with respect to engine lay-out and transient accuracy. In case of model-based control a key role is played by the Mean Value Engine Models (MVEM's) since they allow describing the main dynamic processes (i.e. air filling and emptying in the intake manifold and fuel film evaporation) as function of the mean values of the most significant engine variables. This approach, originally proposed by Aquino (1981) and Hendricks and Sorenson (1991), is suitable for the on-line AFR control operation since it is based on a mean value scale and allows observing the dynamic processes with a good level of accuracy and a limited computational demand.

Despite their intrinsic accuracy, these models have some limitations due to the approximation made during the design and the parameters identification. Some physical effects are not directly described (e.g. EGR, manifold heat transfer) and "hand-tuned" correction factor maps are usually needed. Furthermore, the non-linear dependence of model parameters (i.e. time constant of fuel evaporation, fuel fraction impinging on the walls) with the engine operation is often accounted for by means of static maps as function of engine speed and load. Moreover, this approach could be time consuming due to the need of an extended experimental data set and it does not guarantee the estimation of the appropriate parameters throughout engine lifetime. In order to overcome these problems, adaptive methodologies have been proposed to estimate the states and tune the parameters making use of real-time measurements (e.g. Kalman filters) (Turin and Geering 1994; Powell et al. 1998; Arsie et al. 2003) or robust control methodologies (e.g. H-infinity control) (Vigild et al. 1999).

Concerning engine fault detection and diagnosis, the model-based approach is still the most widely adopted. It relies on the analytical redundancy analysis performed by processing on-board measurements and model estimates of states and parameters (Isermann 1997; Dinca et al. 1999).

A promising solution for approaching these problems is given by Neural Networks (NN) based black-box models, which have good mapping capabilities and generalization even with a reduced set of identification data (Patterson 1995; Haykin 1999; Norgaard et al. 2000). Some examples of Neural Network models for automotive application have been proposed in the literature for engine control (Shayler et al. 1996; Manzie et al. 2002), diagnostics (Ortmann et al. 1998; Capriglione et al. 2003), pattern recognition (Heimann et al. 2006; St-Pierre and Gingras 2004) and in papers on sensor data fusion (Marko et al. 1989; Malaczynski and Baker 2003), with satisfactory robustness even in presence of noisy data.

Moreover, NN have been proven to be useful for modeling nonlinear dynamic systems introducing feedback connections in a recursive computational structure (Recurrent Neural Networks - RNN). Among the others, of particular interests are the contributions of Dovifaaz et al. (1999), Tan and Saif (2000), Alippi et al. (2003). It is worth to remark that the parameters (i.e. network weights) of both static and dynamic Neural Networks can be found by means of batch or adaptive identification (i.e. training). In the former case the weights are kept constant throughout system lifetime, while in the latter they are continuously tuned and adapted to the variations of the controlled or simulated system.

The chapter is structured as follows: a general description of the physical process is first presented along with current AFR control state of the art; then the potentialities of RNN to improve both feedback and feedforward control tasks are discussed. Finally, after reviewing the main RNN features together with their related control algorithms, practical development and implementation of RNN for AFR simulation and control in SI engines are presented and commented.

1 Manifold Fuel Film Dynamics

This section is devoted to a general overview of the fuel film dynamics in a Spark Ignition (SI) automotive engine. For both single and multi-point spark ignition engines, a two phases fuel flow occurs in the intake manifold, with a thin film of fuel on the manifold walls and droplets transported by the main stream of air and fuel (Aquino 1981; Heywood 1988); a representation of the air-fuel mixture preparation is shown in Fig. 1.

In order to clarify the physical process, the intake manifold fuel dynamics is described by means of a Mean Value Model (MVM) (i.e., "gray box" modeling approach), which considers the fuel path from the injection location to the engine port, accounting for the fuel puddle on the manifold walls and its later evaporation (Aquino 1981; Hendricks and Sorenson 1991). It is assumed that 1) at any instant uniform conditions exist throughout the intake manifold; 2) a fraction X of the injected fuel is deposited on the wall as liquid film; 3) the evaporation rate, proportional to the mass of fuel film, is modeled considering a first order process with time constant τ. The fuel flow entering the combustion chamber $\dot{m}_{f,e}$ is obtained by considering both the vapor flow rate $\dot{m}_{v,e}$ and the liquid flow rate $\dot{m}_{l,e}$ (Couette flow). Thus, the mass balance for liquid (m_l) and vapor (m_v) can be expressed by the following system of ODE's (Arsie et al. 2003):

$$\dot{m}_l = X \dot{m}_{f,i} - \frac{m_l}{\tau} - \dot{m}_{l,e} \tag{1}$$

$$\dot{m}_v = (1-X) \dot{m}_{f,i} + \frac{m_l}{\tau} - \dot{m}_{a,e}\left(\frac{m_v}{m_a}\right) \tag{2}$$

and the fuel flow entering the combustion chamber is:

$$\dot{m}_{f,e} = \dot{m}_{v,e} + \dot{m}_{l,e} = \left(\frac{m_v}{m_a}\right) \dot{m}_{a,e} + \dot{m}_{l,e} \tag{3}$$

The first term on the right side of eq. (3) represents the vapor flow rate that is function of the fuel mass in the vapor phase and of a characteristic manifold time constant τ_m, given as the ratio between the mass of air in the intake manifold and the air flow rate at the intake port:

$$\tau_m = \frac{m_a}{\dot{m}_{a,e}} = \frac{1}{\dot{m}_{a,e}} \frac{p_m V_m}{R T_m} \tag{4}$$

where R is the gas constant, V_m is the intake manifold volume, p_m and T_m are the mean pressure and temperature in the intake manifold, respectively.

The mass flow rate of liquid fuel through the intake port is considered proportional to the amount of fuel film on the manifold walls by means of the constant L_f, while the process is assumed to be governed by the air mass flow, considering that the liquid fuel is dragged by the air flow (Matthews et al. 1991):

$$\dot{m}_{l,e} = L_f \dot{m}_{a,e} \quad \dot{m}_l = \frac{m_l}{\tau_{L,f}} \tag{5}$$

From this relationship the term $L_f \dot{m}_{a,e}$ corresponds to the inverse of a time delay constant $\tau_{L,f}$. Apart from physical flow parameters (i.e. viscosity and density of air and fuel), Matthews et al. (1991) account for a quadratic dependence of the liquid fuel flow with respect to the mass of fuel film, an inverse quadratic dependence on manifold pressure and a direct dependence with air flow rate to the power 1.75.

By following the relationship (3), the mixture strength of the flow entering the cylinders is given by:

$$AFR = \frac{\dot{m}_{a,e}}{\dot{m}_{f,e}} \tag{6}$$

The described phenomenology has been derived from general physical analysis and can be followed to model any arrangement: continuous or pulsed, throttle-body or port fuel injection systems. However, these systems are characterized by different manifold geometry and model parameters (i.e. τ and X), whose identification is one of the most critical tasks in the framework of fuel film modeling.

In the most popular approach, the model parameters identification is accomplished by fitting the simulated AFR with the measured signal detected by the exhaust oxygen sensor when the injected fuel signal is excited. In previous studies, the authors themselves proposed a classical least square minimization algorithm to estimate the parameters τ and X for different engine operations. The results have been used to build-up steady state maps of τ and X versus engine state variables (i.e. manifold pressure and speed) in order to perform continuous engine transient simulations.

Further studies have been devoted to design an adaptive wall wetting observer based on an extended Kalman filter (EKF) in order to offset the effects of engine age, wear, environmental condition and, referring to practical engine application, also the problems related to engine components differences due to manufacturing tolerances (Arsie et al. 2003). In these applications, the EKF estimates the fuel flowing to the combustion chamber making use of both the feed-forward model estimation plus an innovation term given by the feedback from the direct measurement on the system.

In both off-line (i.e. Least Square technique) and on-line (i.e. Kalman filtering) applications, the procedure must be compensated for the pure time delay due to engine cycle and gas transport from the engine port to the UEGO sensor location. Moreover, it must account for the dynamic effects due to the gas mixing processes and the UEGO sensor response (Onder and Geering 1997). It is worth noting that in case of off-line identification these effects can be compensated by preprocessing input (injected fuel) and output (AFR) signals; while they cannot be taken into account in case of on-line identification, thus affecting the accuracy of the parameters estimation.

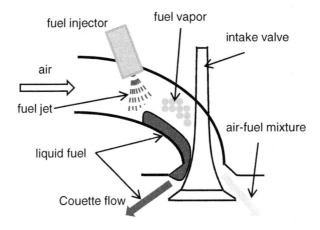

Fig. 1. Schematic representation of the air-fuel mixture preparation in the intake manifold

2 AFR Control

Fig. 2 gives a simplified schematic of the actual fuel control strategies implemented in commercial Engine Control Unit (ECU) to limit mixture strength excursions. The base injection pulse is computed as function of desired AFR, engine speed, manifold pressure, throttle opening and a number of factors to account for changes in ambient conditions, battery voltage and engine thermal state. During transient operation, in order to compensate for the different dynamic response of air and fuel path due to the wall wetting phenomenon, the base injected fuel is compensated by a feedforward controller. This latter is usually based on the MVM approach described above.

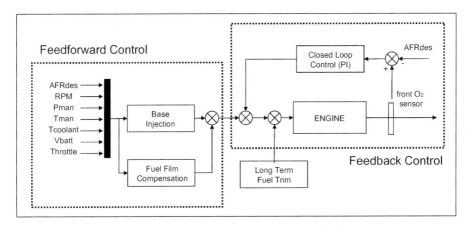

Fig. 2. Simplified schematic of the feedforward and feedback control tasks implemented in commercial ECU

The output of eq. 3 is then further corrected by the closed loop control task and the long term fuel trim. The former is based on a PI controller aimed at keeping the AFR measured by the front oxygen sensor as close as possible to the desired value indicated by the catalyst control strategies (Yasuri et al. 2000). As for the MVM parameters, the PI gains are stored in look-up tables to account for their dependence on engine operating conditions. Long term fuel trim is intended to compensate for imbalance and/or offset in the AFR measurement due to exogenous effects, such as canister purge events, engine wearing, air leaks in the intake manifolds, etc. (Yoo et al. 1999).

The control architecture described above has been well established in the automotive industry for long time, thanks to its good performance during both steady and transient conditions. Nevertheless, the more stringent regulations imposed in the last years by OBD, Low Emission Vehicle, Ultra Low Emission Vehicle, Super Ultra Low Emission Vehicle etc. standards, have pushed car manufacturers towards further developing the actual control strategies implemented on commercial ECU. It is especially important to reduce the calibration efforts required to develop the parameters map and, above all, to improve the performance of the feedback control path in Fig. 2.

Since the nineties, many studies addressed the benefits achievable by replacing the PI controller by advanced control architectures, such as linear observers (Powell et al. 1998), Kalman Filter (Arsie et al. 2003; Fiengo et al. 2002; Kainz and Smith, 1999) and sliding mode (Yasuri et al. 2000, Choi and Hedrick, 1998, Kim et al. 1998). Nevertheless, a significant obstacle to the implementation of such strategies is represented by measuring delay due to the path from injection to O_2 sensor location and the response time of the sensor itself. To overcome these barriers, new sensors have been developed (Nakae et al. 2002) but their practical applicability in the short term is far from being feasible due to high cost and impossibility to remove transport delay. Thus, developing predictive models from experimental data might significantly contribute to promoting the use of advanced closed loop controllers.

2.1 RNN Potential

Recurrent Neural Network, whose modeling features are presented in the following section, have significant potential to face the issues associated with AFR control. The authors themselves (Arsie et al. 2004) and other contributions (e.g. Alippi et al. 2003) showed how an inverse controller made of two RNNs, simulating both forward and inverse intake manifold dynamics, is suitable to perform the feedforward control task. Such architectures could be developed making use of only one highly-informative data-set (Arsie et al. 2006), thus reducing the calibration effort with respect to conventional approaches. Moreover, the opportunity of adaptively modifying network parameters allows accounting for other exogenous effects, such as change in fuel characteristics, construction tolerances and engine wear.

Besides their high potential, when embedded in the framework of pure neural-network controller, RNN AFR estimators are also suitable in virtual sensing applications, such as the prediction of AFR in cold-start phases. RNN training during cold-start can be performed on the test-bench off-line, by pre-heating the lambda sensor before turning on the engine. Moreover, proper post-processing of training data enables to predict AFR excursions without the delay between injection (at intake port) and measuring (in the exhaust) events, thus being suitable in the framework of sliding-mode closed-loop control tasks (Choi and Hedrick 1998; Yasuri et al. 2000). In such an application the feedback provided by a UEGO lambda sensor may be used to adaptively modify the RNN estimator to take into account exogenous effects. Finally, RNN-based estimators are well suited for diagnosis of injection/air intake system and lambda sensor failures (Maloney 2001). In contrast with control applications, in this case the AFR prediction includes the measuring delay.

3 Recurrent Neural Networks

The RNN are derived from the static multilayer perceptron feedforward (MLPFF) networks by considering feedback connections among the neurons. Thus, a dynamic effect is introduced into the computational system by a local memory process. Moreover, by retaining the non-linear mapping features of the MLPFF, the RNN are suitable for black-box nonlinear dynamic modeling (Patterson 1995; Haykin 1999). Depending upon the feedback typology, which can involve either all the neurons or just the output and input ones, the RNN are classified into Local Recurrent Neural Networks and Global or External Recurrent Neural Networks, respectively (Haykin 1999; Nørgaard et al. 2000). This latter kind of network is implemented here and its basic scheme is shown in Figure 3, where, for clarity of representation, only two time delay operators D are introduced. It is worth noting that in the figure a time sequence of external input data is fed to the network (i.e. the input time sequence $\{x(t-1), x(t-2)\}$).

The RNN depicted in Figure 3 is known in the literature as Nonlinear Output Error (NOE) (Nørgaard et al. 2000; Nelles 2000), and its general form is:

$$\hat{y}(t|\theta) = F[\varphi(t|\theta), \theta] \tag{7}$$

where F is the non-linear mapping operation performed by the neural network and $\varphi(t)$ represents the network input vector (i.e. the regression vector):

$$\varphi(t,\theta) = [\hat{y}(t-1|\theta), ..., \hat{y}(t-n|\theta), x(t-1), ..., x(t-m)] \tag{8}$$

where θ is the parameters vector, to be identified during the training process; the indices n and m define the lag space dimensions of external inputs and feedback variables.

Starting from the above relationships and accounting for the calculations performed inside the network, the general form of the NOE RNN can also be written as function of network weights:

$$\hat{y}(t|\theta) = f\left(\sum_{i=1}^{nh} v_i g\left(\sum_{j=1}^{ni} w_{ij} \varphi_j(t)\right)\right) \tag{9}$$

where nh is the number of nodes in the hidden layer and ni the number of input nodes, v_i represents the weight between the i-th node in the hidden layer and the output node, while w_{ij} is the weight connecting the j-th input node and the i-th node in the hidden layer. These weights are the components of the parameters vector and $\varphi_j(t)$ is the j-th element of the input vector (8). The activation function g(.) in the hidden layer nodes is the following non-linear sigmoid function:

$$g(x) = \frac{2}{1+\exp(-2x+b)} - 1 \quad [-1;1] \tag{10}$$

where b is an adjustable bias term; at the output node the linear activation function $f(.)$ is assumed. Then, accounting for all the weights and biases the parameters vector is $\theta = [v_1,...v_{nh}, w_{11},...w_{nh,ni}, b_1,...b_{nh}, b_{no}]$.

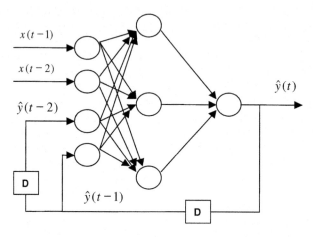

Fig. 3. NOE Recurrent Neural Network Structure with one external input, one output, one hidden layer and two output delays

3.1 Dynamic Network Features

Like the static MLPFF networks, RNN learning and generalization capabilities are the main features to be considered during network development. The former deals with the ability of the learning algorithm to find the set of weights which gives the desired accuracy, according to a criterion function. The generalization expresses the neural network accuracy on a data set different from the training set. A satisfactory generalization can be achieved when the training set contains enough information and the network structure (i.e. number of layers and nodes, feedback connections) has been properly designed. In this context, large networks should be avoided because a high number of parameters can cause, through an overparametrization, the overfitting of the training pattern; whilst under-sized networks are unable to learn from the training pattern; in both cases a loss of generalization occurs. Furthermore, for a redundant network, the initial weights may affect the learning process leading to different final performance. This problem has not been completely overcome yet and some approaches are available in the literature: notably the works of Atiya and Ji (1997) and Sum et al. (1999) give suggestions on the initial values of the network parameters. The authors themselves have faced the problem of network weights uniqueness, making use of clustering methods in the frame of a MonteCarlo parameters identification algorithm (Mastroberti and Pianese 2001), in case of steady state networks. Following recommendations provided by Thimm and Fiesler (1997) and Nørgaard et al. (2000), the weights are initialized randomly in the range [-0.5:0.5] to operate the activation function described by eq. 10 in its linear region. Several studies have also focused on the definition of the proper network size and structure through the implementation of pruning and regularization algorithms. These techniques rearrange the connection levels or remove the redundant nodes and weights as function of their marginal contribution to the network precision gained during consecutive training phases (Haykin 1999; Ripley 2000). For Fully Recurrent Neural Networks the issue of finding the network topology becomes more complex because of the interconnections among parallel nodes. Thus, specific pruning methodologies, such as constructive learning approach, have been proposed in the literature (Giles et al. 1995; Lehtokangas 1999).

Some approaches have been followed to improve the network generalization through the implementation of Active Learning (AL) Techniques (Arsie et al. 2001). These methods, that are derived from the Experimental Design Techniques (EDT), allow reducing the experimental effort and increase the information content of the training data set, thus improving the generalization. For static MLPFF networks the application of information-based techniques for active data selection (MacKay 1992; Fukumizu 2000) addresses the appropriate choice of the experimental data set to be used for model identification by an iterative selection of the most informative data. In the field of Internal Combustion Engine (ICE), these techniques search in the independent variables domain (i.e. the engine operating conditions) for those experimental input-output data that maximize system knowledge (Arsie et al. 2001).

Heuristic rules may be followed to populate the learning data set if AL techniques are awkward to implement. In such a case, the training data set must be composed of experimental data that span the whole operating domain, to guarantee steady-state mapping. Furthermore, in order to account for the dynamic features of the system, the input data must be arranged in a time sequence which contains all the frequencies and amplitudes associated with the system and in such a way that the key features of the dynamics to be modeled will be excited (Nørgaard et al. 2000). As it will be shown in the following section, a pseudo-random sequence of input data has been considered here to meet the proposed requirements.

The RNN described through the relationships (7) and (8) is a discrete time network, and a fixed sampling frequency has to be considered. To build the time dependent input-output data set, the sampling frequency should be high enough to guarantee that the dynamic behaviour of the system is well represented in the input-output sequence. Nevertheless, as reported in Nørgaard et al. (2000), the use of a large sampling frequency may generate numerical problems (i.e. ill-conditioning); moreover, when the dimension of the data sequence becomes too large an increase of the computational burden occurs.

A fundamental feature to be considered for dynamic modeling concerns with the stability of the solution. RNNs behave like dynamic systems and their dynamics can be described through a set of equivalent nonlinear Ordinary Differential Equations (ODEs). Hence, the time evolution of a RNN can exhibit a convergence to a fixed point, periodic oscillations with constant or variable frequencies or a chaotic behavior

(Patterson 1995). In the literature studies on stability of neural networks belong to the field of neurodynamics which provides the mathematical framework to analyze their dynamic behavior (Haykin 1999).

3.2 Recurrent Neural Network Architectures for AFR Control

Neural Network based control systems are classified into two main types: a) Direct Control Systems and b) Indirect Control Systems (Nørgaard et al. 2000) as described in the following sections.

Direct Control System

In the Direct Control Systems (DCS) the neural network acts as a controller which processes the signals coming from the real system. The control signals are evaluated as function of the target value of the controlled variable. In this section two different DCS are presented: the Direct Inverse Model (DIM) and the Internal Model Control (IMC). A description of these two approaches is given below.

Direct Inverse Model

The basic principle of this methodology is to train the Neural Network (i.e. identification of the Network parameters) to simulate the "inverse" dynamics of the real process and then to use the inverse model as a controller. Assuming that the dynamics of the system can be described through an RNN model analogous to the one proposed in eqs. (7) and (8):

$$\hat{y}(t+1) = F[\hat{y}(t),...,\hat{y}(t-n+1), x(t),...,x(t-m+1)] \tag{11}$$

For a generic MIMO the Inverse RNN Model (IRNNM) is obtained by isolating the most recent control input as follows:

$$\hat{u}(t) = F^{-1}[y(t+1), y(t),..., y(t-n+1),...,\hat{u}(t-1),...,\hat{u}(t-m+1)] \tag{12}$$

It is worth remarking that the u notation in eq. (12) replaces 'x' to highlight that in control applications all or some external variables (depending on plant typology e.g. SISO, MISO or MIMO) are control variables too.

Once the network has been trained, it can be used as a controller by substituting the output $\hat{y}(t+1)$ with the desired target value $r(t+1)$. Assuming that the model (12) describes accurately the inverse dynamics of the system, the computed control signal $\hat{u}$ will be able to drive the system output $y(t+1)$ to the desired value $r(t+1)$. Fig. 4 shows the block-diagram of a DIM. It is worth noting that the desired value r is at the time $t+1$ while the control input $\hat{u}$ refers to the time t; hence the controller performs the control action one time step earlier than the desired target.

The advantages of the DIM approach are the simple implementation and the Dead-Beat built-in control properties (i.e. fast response for wide and sudden variations of the state variables) (Nørgaard et al. 2000). Nevertheless, since the controller parameters are identified in off-line, the DIM does not perform as an adaptive control system. In order to develop an adaptive system, on-line training methodologies are required. The suitability of the inverse neural controllers to control complex dynamic

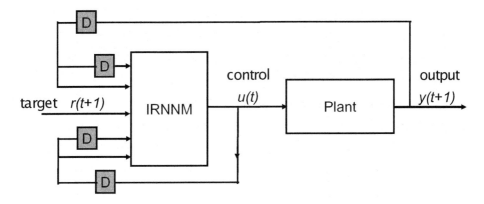

Fig. 4. Direct Inverse Model (DIM) based on a NOE network (Inverse RNNM). The control signal is estimated at the time t as a function of the desired value $r(t+1)$ and the feedbacks. Note that the parameters m and n (see eq. 12) have been assumed equal to 2.

processes is confirmed by several studies, particularly in the field of aerospace control systems (e.g., Gili and Battipede, 2001).

Internal Model Control

The IMC control structure is derived from the DIM, described in the previous section, where a Forward RNN Model (FRNNM) of the system is added to work in parallel with an IRNNM. Fig. 5 shows a block diagram of this control architecture. The output values predicted by the FRNNM are fed back to the IRNNM, that evaluates the control actions as function of the desired output at the next time step $r(t+1)$. The more the FRNNM prediction is accurate, the less the difference between FRNNM and Plant outputs will be. However, the differences between Plant and FRNNM may be used to update the networks online (i.e. both IRNNM and FRNNM), thus providing the neural controller with adaptive features.

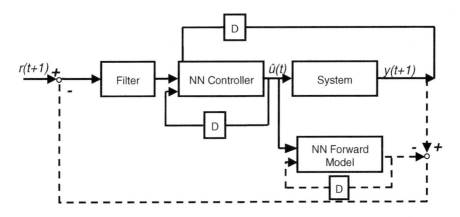

Fig. 5. Neural Networks based Internal Model Control Structure (---- closed loop connections)

The stability of the IMC scheme (i.e. the closed-loop system and the compensation for constant noises) is guaranteed if both the real system and the controller are stable (Hunt and Sbarbaro, 1991). Despite their stringent requirements, the IMC approach seems to be really appealing, as shown by several applications in the field of chemical processes control (e.g. Hussain, 1999).

Indirect Control Systems

Following the classification proposed by Nørgaard et al. (2000), a NN based control structure is Indirect if the control signal is determined by processing the informations provided from both the real system and the NN model. The Model Predictive Control (MPC) structure is an example of such control scheme.

Model Predictive Control

The MPC computes the optimal control signals through the minimization of an assigned Cost Function (CF). Thus, in order to optimize the future control actions, the system performance is predicted by a model of the real system. This technique is suitable for those applications where the relationships between performances, operating constraints and states of the system are strongly nonlinear (Hussain, 1999). In the MPCs the use of the Recurrent Neural Networks is particularly appropriate because of their nonlinear mapping capabilities and dynamic features.

In the MPC scheme, a minimization of the error between the model output and the target values is performed (Nørgaard et al. 2000). The time interval on which the performance are optimized starts from the current time t to the time $t+k$. The minimization algorithm accounts also for the control signals sequence through a penalty factor ρ. Thus, the Cost Function is

$$J(t,u(t)) = \sum_{i=N_1}^{N_2}[r(t+i)-\hat{y}(t+i)] + \rho\sum_{i=1}^{N_3}[\Delta u(t+i-1)]^2 \qquad (14)$$

where the interval $[N_1, N_2]$ denotes the prediction horizon, N_3 is the control horizon, r is the target value of the controlled variable, $\hat{y}$ is the model output (i.e. the NN) and $\Delta u(t+i-1)$ are the changes in the control sequences.

In the literature, several examples dealing with the implementation of the MPCs have been proposed. Manzie et al. (2002) have developed a MPC based on Radial Basis Neural Networks for the AFR control. Furthermore, Hussain (1999) reports the benefits achievable through the implementation of NN predictive models for the control of chemical processes.

4 Model Identification

This section deals with practical issues associated with Recurrent Networks design and identification, such as choice of the most suitable learning approach, selection of independent variables (i.e. network external input x) and output feedbacks, search of optimal network size and definition of methods to assess RNN generalization.

4.1 RNN Learning Approach

The parameters identification of a Neural Network is performed through a learning process during which a set of training examples (experimental data) is presented to the network to settle the levels of the connections between the nodes. The most common approach is the error Backpropagation algorithm due to its easy-to-handle implementation. At each iteration the error between the experimental data and the corresponding estimated value is propagated backward from the output to the input layer through the hidden layers. The learning process is stopped when the following cost function (MSE) reaches its minimum:

$$E(\theta) = \frac{1}{2N} \sum_{t=1}^{N} (\hat{y}(t \mid \theta) - y(t))^2 \qquad (15)$$

N is the time horizon on which the training pattern has been gathered from experiments. The Mean Squared Error (MSE) (eq. 15) evaluates how close is the simulated pattern $\hat{y}(t|\theta)\,[t=1,N]$ to the training $y(t)\,[t=1,N]$. The Backpropagation method is a first-order technique and its use for complex networks might cause long training and in some cases a loss of effectiveness of the procedure. In the current work, a second-order method based on the Levenberg-Marquardt optimization algorithm is employed (Patterson 1995; Haykin 1999; Hecht-Nielsen R. 1987; Nørgaard et al. 2000; Ripley 2000). To mitigate overfitting, a regularization term (Nørgaard et al. 2000) has been added to eq. 15, yielding the following new cost function:

$$E^*(\theta) = \frac{1}{2N} \sum_{t=1}^{N} (\hat{y}(t \mid \theta) - y(t))^2 + \frac{1}{2N} \cdot \theta^T \cdot \alpha \cdot \theta \qquad (16)$$

where α is the weight decay.

The above function minimization can be carried out in either a batch or pattern-by-pattern way. The former is usually preferred at the initial development stage, whereas the latter is usually employed in online RNN implementation, as it allows to adapt network weights in response to the exogenous variations of the controlled/simulated system.

The training process aims at determining RNN models with a satisfactory compromise between precision (i.e. small error on the training-set) and generalization (i.e. small error on the test-set). High generalization cannot be guaranteed if the training data-set is not sufficiently rich. This is an issue of particular relevance for dynamic networks such as RNN, since they require the training-set not only to cover most of the system operating domain, but also to provide accurate knowledge of its dynamic behavior. Thus, the input data should include all system frequencies and amplitudes and must be arranged in such a way that the key features of the dynamics to be modeled are excited (Nørgaard et al. 2000). As far as network structure and learning approach are concerned, the precision and generalization goals are often in conflict. The loss of generalization due to parameters redundancy in model structure is addressed in the literature as overfitting (Nelles 2000). This latter may occur in case of a large number of weights, which in principle improves RNN precision but may cause generalization to decrease. A similar effect can occur if network training is stopped after too many epochs. Although this can be beneficial to precision it may negatively impacts generalization capabilities and is known as overtraining. Based on the above observations and to ensure a proper design of RNNs, the following steps should be taken:

  i) generate a training data set extensive enough to guarantee acceptable generalization of the knowledge retained in the training examples,
  ii) select the proper stopping criteria to prevent overtraining, and
  iii) define the network structure with the minimum number of weights.

As far as the impact of point *i)* on the current application, AFR dynamics can be learned well by the RNN estimator once the trajectories of engine state variables (i.e. manifold pressure and engine speed) described in the training set are informative enough. This means that the training experiments on the test-bench should be performed in such a way to cover most of the engine working domain. Furthermore a proper description of both low- and high-frequency dynamics is necessary, thus the experimental profile should be obtained by alternating steady operations of the engine with both smooth and sharp acceleration/deceleration maneuvers.

Point *ii)* is usually addressed by employing the early stopping method. This technique consists of interrupting the training process, once the MSE computed on a data set different from the training one stops decreasing. Therefore, when the early stopping is used, network training and test require at least three data sets (Haykin 1999): training-set (set A), early stopping test-set (set B) and generalization test-set (set C).

Finally, point *iii)* is addressed by referring to a previous paper (Arsie et al. 2006), in which a trial-and-error analysis was performed to select the optimal network architecture in terms of hidden nodes and lag space. Although various approaches on choosing MLPFF network sizing are proposed in the specific literature, finding the best architecture for recurrent neural network is a challenging task due to the presence of feedback connections and (sometimes) past input values (Nørgaard et al. 2000). In the current work the trial-and-error approach is used to address *iii)*.

4.2 Input Variables and RNNs Formulation

Based on experimental tests, the instantaneous port air mass flow is known to be mainly dependent on both manifold pressure and engine speed (i.e. engine state variables) (Heywood, 1988). On the other hand, the actual fuel flow rate results from the dynamic processes occurring into the inlet manifold (see Fig. 1 and eqs. 1, 2), therefore the manifold can be studied as a Multi Input Single Output (MISO) system. In case of FRNNM for AFR prediction, the output, control and external input variables are $\hat{y} = AFR$, $u = t_{inj}$, $x = [rpm, p_m]$, respectively and the formulation resulting from eqs. (7) and (8) will be:

$$A\hat{F}R(t,\theta) = F\big[A\hat{F}R(t-1 \mid \theta_1),..., A\hat{F}R(t-n \mid \theta_1), t_{inj}(t-1),...,t_{inj}(t-m), \\ ...rpm(t-1),..., rpm(t-m), p_m(t-1),..., p_m(t-m)\big] \quad (17)$$

It is worth noting that the output feedback in the regressors vector (i.e. $A\hat{F}R$ in Eq. 17) is simulated by the network itself, thus the FRNNM does not require any AFR measurement as feedback to perform the online estimation. Despite such a choice could reduce network accuracy (Nørgaard et al., 2000), it allows to properly address the AFR measurement delay issue due to mass transport, exhaust gas mixing and lambda sensor response (Onder and Geering, 1997). This NOE feature is also very appealing because it enables AFR virtual sensing even when the oxygen sensor does not guarantee an accurate measurement, as it happens during cold start phases.

Regarding AFR control, for the current application a DIM structure (see Fig. 4) is considered. Hence the IRNNM can be obtained by inverting Eq. (7), yielding the following RNN expression:

$$\hat{t}_{inj}(t-1|\theta_2) = G\Big[A\hat{F}R(t|\theta_1),...,A\hat{F}R(t-n|\theta_1),\hat{t}_{inj}(t-2|\theta_2),...,\hat{t}_{inj}(t-m|\theta_2) \\ rpm(t-1),...,rpm(t-m), p_m(t-1),..., p_m(t-m)\Big] \quad (18)$$

where $A\hat{F}R$ is the estimate provided by FRNNM and θ_2 is the IRNNM parameters vector.

Afterwards, substituting $A\hat{F}R(t+1|\theta_1)$ by the future target value AFR_{des}, the control action can be computed as:

$$\hat{t}_{inj}(t-1|\theta_2) = G\Big[AFR_{des}(t),...,A\hat{F}R(t-n|\theta_1),\hat{t}_{inj}(t-2|\theta_2),...,\hat{t}_{inj}(t-m|\theta_2) \\ rpm(t-1),...,rpm(t-m), p_m(t-1),..., p_m(t-m)\Big] \quad (19)$$

5 Experimental Set-Up

The RNN AFR estimator and controller were trained and tested vs. transient data sets measured on the engine test bench at the University of Salerno. The experiments were carried out on a commercial engine, 4 cylinders, 1.2 liters, with Multi-Point injection. It is worth noting that due to the absence of a camshaft sensor, the injection is not synchronized with the intake stroke, therefore it takes place twice a cycle for each cylinder. The test bench is equipped with a Borghi & Saveri FE-200S eddy current dynamometer. A data acquisition system, based on National Instruments cards PCI MIO 16E-1 and Sample & Hold Amplifier SC-2040, is used to measure engine variables with a sampling frequency up to 10 kHz. An AVL gravimetric balance is used to measure fuel consumption in steady-state conditions to calibrate the injector flow rate. The engine control system is replaced with a dSPACE© MicroAutobox equipment and a power conditioning unit. Such a system allows to control all the engine tasks and to customize the control laws. To guarantee the controllability and reproducibility of the transient maneuvers, both throttle valve and engine speed are controlled through an AVL PUMA engine automation tool (see Fig. 6).

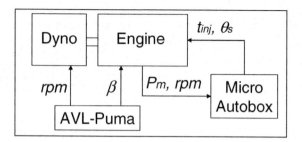

Fig. 6. Lay-out of the experimental plant. β = throttle opening, θ_s = spark advance, t_{inj} = injection time.

Fig. 7. Location of the UEGO sensor (dotted oval)

The exhaust AFR is sensed by an ETAS Lambda Meter LA4, equipped with a Bosch LSU 4.2 UEGO sensor. This is placed right after the exhaust valve of the first cylinder (see Fig. 7) to investigate the air-fuel mixing process in one cylinder only. This choice allows to remove the dynamic effects induced by gas transport and mixing phenomena occurring in the exhaust pipes. Also non predictable effects generated by cylinder-to-cylinder unbalance due to uneven processes such as air breathing, thermal state and fuel injection can be neglected. Therefore, the time shift between injection timing and oxygen sensor measurement mostly accounts for pure engine cycle and lack of synchronization between injection and intake valve timing. This latter term can be neglected in case of synchronized injection. As mentioned before, the time delay could represent a significant problem for control applications (Powell et al. 1998; Choi and Hedrick, 1998; Manzie et al. 2002; Inagaki et al. 1990) and the accomplishment of this task is described in Section 5.1.

5.1 Training and Test Data

The training and test sets were generated by running the engine on the test bench in transient conditions. In order to span most of the engine operating region, perturbations on throttle and load torque were imposed during the transients. Fast throttle actions, with large opening-closing profiles and variable engine speed set points, were generated off-line and assigned through the bench controller to the engine and the dyno, respectively (see Fig. 6). Regarding fuel injection, only the base injection task (see Fig. 2) was executed, thus allowing to investigate the dynamic behavior of the uncontrolled plant, without being affected by neither feedforward nor feedback compensation Furthermore, in order to excite the wall wetting process independently from the air breathing dynamics, a uniform random perturbation was added to the injection base time, limiting the gain in the range +/-15 % of the nominal fuel injection time (i.e. injector opening time). Such an approach protects the RNN from unacceptable accuracy of predictions in the case of injection time deviation at constant engine load and speed as it is the case when a different AFR target is imposed. This matter is extensively described in our previous work (Arsie et al. 2006).

In Fig. 8 the measured signals of throttle opening (a), engine speed (b), manifold pressure (c), injection time (d) and AFR (e) used as training-data (Set A) are shown. The throttle opening transient shown in Fig. 8 (a) allows exciting the filling-emptying dynamics of the intake manifold and the engine speed dynamics, as a consequence of both engine breathing and energy balance between engine and load torque. Fig. 8 (b), (c) indicate that the transient spans most of the engine operating domain with engine speed and manifold pressure ranging from 1000 to 4000 rpm and from low to high load, respectively. The variation of manifold pressure and engine speed affects the intake air flow rate and consequently the amount of fuel to be injected to meet the target AFR. It is worth noting that the manifold pressure signal was filtered in order to cancel the process noise that has negligible effects on the AFR dynamic response. This enhances the FRNNM in learning the main manifold dynamics without being perturbed by second order phenomena (e.g. intake manifold pressure fluctuation).

The injection time trajectory (Fig. 8 d), commanded by the ECS, excites the wall wetting dynamics, which in turn influences the in-cylinder AFR in a broad frequency range. It is also worth to mention that engine speed and throttle opening are decoupled to generate more realistic engine speed-pressure transients. Furthermore, the speed dynamics is more critical as compared to what occur on a real vehicle, due to the lower inertia of the engine test-bench.

Fig. 9 shows the time histories of throttle opening, engine speed, manifold pressure and injected fuel measured for the test-set (SET B), SET B was obtained imposing square wave throttle maneuvers (Fig. 9 a) to excite the highest frequencies of the air dynamics, while keeping the engine speed constant (Fig. 9 b) and removing the fuel pulse random perturbation. Fig. 9 (d) and (e) evidence that the resulting step variations of injected fuel generate wide lean/rich spikes of AFR, due to uncompensated effects of wall wetting dynamics during abrupt throttle opening/closing transients. Such features make SET B suitable as test data-set since RNN accuracy in predicting high frequency dynamic response is demanded. Moreover, the step variations of injected fuel are also appropriate to estimate the AFR delay. Thus, SET B is also very suitable to assess the ability of RNN to cope with the pure time delay.

It is worth noting that the early stopping was not applied directly, in that the maximum number of training epochs was set to 50 following the indications from a previous study where an early stopping data set was used (Arsie et al., 2007).

FRNNM - AFR Prediction

The identification task of the FRNMM was performed on a NOE structure consisting of 12 hidden nodes, with lag spaces n = 2 and m = 5 (see eq. 17) (Arsie et al. 2006). SET A was used to train the RNN with a number of training epochs set to 50. This value was selected according to the results shown in previous work (Arsie et al. 2007) where the early stopping criterion was utilized to maximize RNN generalization. The

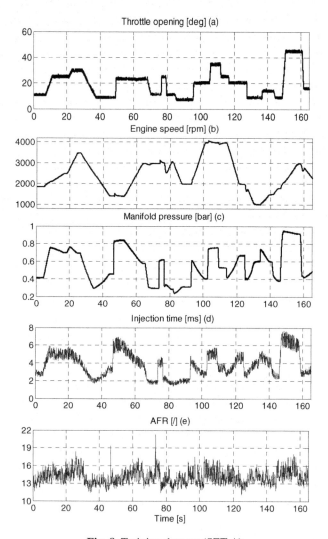

Fig. 8. Training data set (SET A)

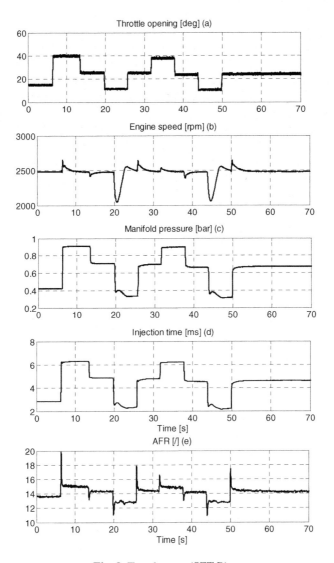

Fig. 9. Test data set (SET B)

Table 1. Delay scenarios analyzed for RNN training

	Case 1	Case 2	Case 3	Case 4
Δt_{AFR} (s)	0	$3\pi/\omega$	$5\pi/\omega$	$6\pi/\omega$

training procedure was applied four times to investigate the influence of the time delay Δt_{AFR} between AFR measurement and triggering signal (i.e. injection timing) on RNN accuracy. Table I lists the 4 cases analyzed, with the time delay expressed as function of the engine speed (ω).

In Cases 2, 3 and 4, the AFR signal was back-shifted by the corresponding delay. Case 2 accounts for the pure delay due to the engine cycle, that can be approximated as one and a half engine revolution, from middle intake stroke to middle exhaust stroke (i.e. $3\pi/\omega$) as suggested by [38]. In Cases 3 and 4, instead, the delay was assumed longer than $3\pi/\omega$, thus allowing to account for other delays that could occur. They include injection actuation, lack of synchronization between injection and intake valve timing, unsteadiness of gas flowing through the measuring section and mixing in the pipe with residual gas from the previous cycle. The value assumed for the extreme Case 4 is in accordance with the delay proposed by Powell et al. (1998) for an oxygen sensor located in the main exhaust pipe after the junction. In this study, the delay in the set-up of Fig. 6 can be determined by comparing the performance achieved by the RNN estimator on SET A and SET B for the four delay scenarios. Fig. 10 and Fig. 11 summarize the results of this comparison.

The general improvement in accuracy was expected due to the simpler dynamics resulting from delay removal. In this way the RNN is not forced to learn the delay as part of the process dynamics, thus enhancing the training task. More specifically, the analysis of the errors indicates that the best result with respect to SET B occurs for Case 3. Fig. 10 and Fig. 11 show the comparison of AFR traces in both rich and lean transients. In Fig. 10 it is evident how the Case 3 RNN performs the best estimation of the rich spike and, moreover, simulates it with the correct dynamics. Fig. 11 confirms this behavior for lean spike reproduction as well.

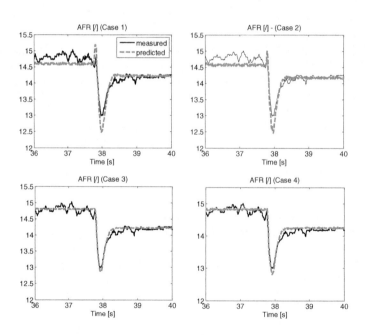

Fig. 10. Comparison between measured and predicted AFR in presence of rich excursions (SET B)

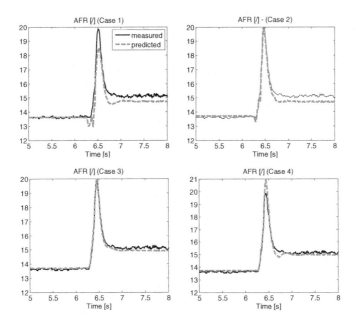

Fig. 11. Comparison between measured and predicted AFR in presence of lean excursions (SET B)

The above observations led to select Case 3 as the best case and, as a consequence, the AFR trace was back-shifted by a time delay corresponding to $5\pi/\omega$ crank-shaft interval. The assumed time interval accounts for the engine cycle delay ($3\pi/\omega$) plus the effect due to the lack of synchronization between injection and intake stroke; this latter can be approximated, in the average, with a delay of $2\pi/\omega$. Furthermore, the delay of $5\pi/\omega$ is also in accordance with the value proposed by Inagaki et al (1990) for similar engine configuration and lambda sensor location.

IRNNM - Real-Time AFR Control

A DIM-based neural controller (see Fig. 4) was set-up to implement the AFR control law detailed above (i.e. eq. (19)). In this configuration, sketched in Fig. 12, the IRNNM is coupled to the FRNNM, which replaces the plant in providing the AFR feedback to the IRNNM without measuring delay, as discussed in the previous section. This FRNNM feature is particularly beneficial because it allows to simplify the IRNNM structure (Norgaard et al., 2000), in that no delay has to be included in the corresponding regression vector. It is also worth remarking that RNN structures can only include constant delay in the time domain, whereas the accounted AFR measurement delay is mainly constant in the frequency domain, thus variable in the time domain (Onder and Geering 1997).

Fig. 12 also shows that the AFR values predicted by the FRNNM are fed as feedbacks to the IRNNM, which evaluates the control actions as function of FRNNM feedbacks, desired AFR at next time step ($t+1$) and current and past external variables provided by engine measurements (i.e. *rpm* (*t*-1, *t-m*), p_m (*t*-1, *t-m*)). The more the FRNNM prediction is accurate, the less the difference between FRNNM and Plant outputs will be. However, the differences between Plant and FRNNM may be used to update the networks online (i.e. both IRNNM and FRNNM), thus providing the neural controller with adaptive features.

According to the selected neural controller architecture, the IRNNM was trained by means of the identification procedure sketched in Fig. 13, on the same SET A assumed for the FRNNM training. Following the formulation described in Eq. 18, the AFR fed to the network is nothing but the prediction performed by the FRNNM,

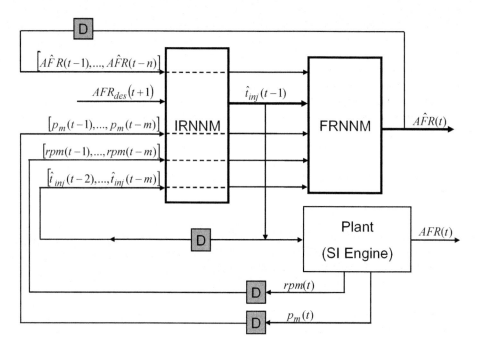

Fig. 12. DIM neural controller based on the integration of DIM and FRNNM. The control signal t_{inj} is estimated as function of the desired AFR value at time *t*+1, the feedbacks from both IRNNM and FRNNM and the external variable values provided by plant measurements. Note that the dashed lines in the IRNNM box mean that the corresponding inputs are passed to the FRNNM too.

while engine speed and manifold pressure values correspond to the measured signals shown in Fig. 8 [set A]. The training algorithm was then aimed at minimizing the residual between simulated and experimental injected fuel time histories.

A preliminary test on the inverse model was accomplished off-line by comparing simulated and experimental injected fuel signals in case of data sets A and B.

Afterwards, a real-time application was carried out by implementing the neural controller architecture (see Fig. 12) in the engine control system on the test bench. The neural controller was employed in Matlab©/Simulink environment and then uploaded to the dSPACE© MicroAutobox equipment. According to the formulation presented in Eq. 19, the controller is intended to provide the actual injection time by processing actual and earlier measurements of engine speed and manifold pressure, and earlier prediction of AFR performed by the FRNMM. Furthermore, unlike the preliminary test, the target AFR has also to be imposed and it was set to the stoichiometric value for the current application.

As mentioned above, due to the UEGO sensor location, the controller was tested on the first cylinder only, while the nominal map based injection strategy was used for the three remaining cylinders. The results of the controller performance in tracking the target AFR along the engine test transient (SET B) are discussed in the next section.

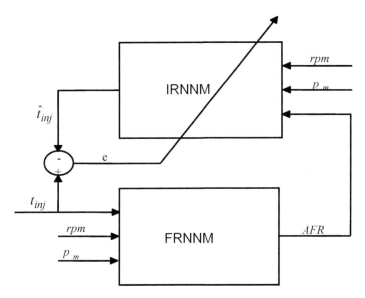

Fig. 13. Training structure adopted for the IRNNM. Note that feedback connections are omitted for simplicity.

6 Results

As described in the previous section, the training SET A has been used to design the RNNs that simulate both forward and inverse dynamics of the fuel film in the intake manifold. The test transient SET B has been simulated to verify the network generalization features. Furthermore, online implementation of the two networks was carried out to test, again on SET B, the performance of the neural controller (see Fig. 12) proposed in this work to limit AFR excursions from stoichiometry.

6.1 FRNNM – AFR Prediction

The accuracy of the developed FRNNM is demonstrated by the small discrepancies between measured and predicted AFR, as shown in detail in Fig. 14, Fig. 15, Fig. 16 and Fig. 17 for both SET A and SET B. Particularly, Fig. 16 and Fig. 17, which refer

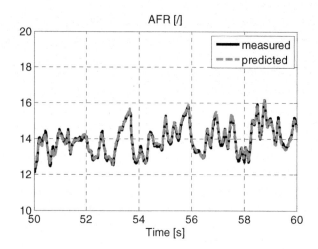

Fig. 14. Trajectories of measured and predicted AFR (SET A, time window [50-60])

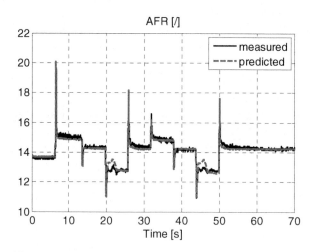

Fig. 15. Trajectories of measured and predicted AFR (SET B)

to abrupt throttle opening/closing maneuvers (see Fig. 9 a), show that the delay removal from the AFR signal enables the FRNNM to capture AFR dynamics in both rich and lean transients very well (Arsie et al. 2007). Fig. 16 and Fig. 17 also indicate how the simulated AFR trajectory is smoother as compared to the more wavy experimental profile.

This behavior is due to the pressure signal filtering, mentioned in section 5.1, and confirms that this procedure allows retaining suitable accuracy in predicting AFR dynamic response during engine transients. Particularly, Fig. 14 shows that the FRNNM correctly follows small-amplitude. AFR oscillations when simulating the training transient (SET A), thus indicating that the signal filtering does not affect FRNNM accuracy in predicting AFR dynamics.

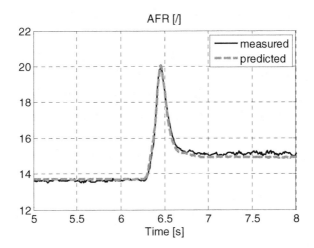

Fig. 16. Trajectories of measured and predicted AFR (SET B, time window [5, 8])

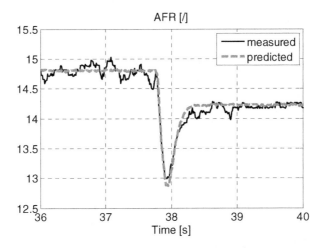

Fig. 17. Trajectories of measured and predicted AFR (SET B, time window [36, 40])

It is worth noting that the inaccurate overshoot predicted by the FRNMM (see Fig. 15 at 20 s and 45 s) is due to the occurrence of a dramatic engine speed drop induced by a tip-out maneuver, resulting in the pressure increase shown on Fig. 9 (c). This experimental behavior is explained considering that the test dynamometer is unable to guarantee a constant engine speed when an abrupt throttle closing is imposed at medium engine speed and low loads (i.e. 2500 rpm, from 25 to 15 deg). Actually, in such conditions, the dynamometer response is not as fast as it should be in counterbalancing the engine torque drop and an over-resistant torque is applied to the engine shaft. It is worth mentioning that such transient is quite unusual with respect to the in-vehicle engine operation.

6.2 IRNNM – AFR Control

As explained in Section 5.1, the IRNNM was trained by feeding as input the AFR predicted by the virtual sensor (i.e. FRNNM). The RNN structure adopted to simulate inverse AFR dynamics was obtained through the inversion of the FRNNM structure (i.e. 12 hidden nodes, n=2, m=5), according to the modeling scheme discussed above (see Eq. 18).

Fig. 18 shows the comparison between IRNNM and experimental injection time for the entire SET B. The trained IRNNM approximates the inverse AFR dynamics as accurately as the FRNNM does for the forward dynamics. Regarding the more wavy profile simulated by the IRNNM, it can be explained considering that this network was trained on t_{inj} profile from Fig. 9 heavily perturbed by the noise as mentioned in Section 5.1. Therefore, it tends to amplify the AFR oscillations in SET B (see Fig. 9 e), although they are of smaller amplitude as compared to the ones measured in the training transient (see Fig. 8). Nevertheless, this effect is well balanced by the benefits provided by the random t_{inj} perturbation, both in terms of IRNNM generalization (Arsie et al. 2006) and direct controller performance, as discussed below.

Online tests of the developed RNNs were performed by integrating the FRNNM and IRNNM in the framework of a dSPACE© MicroAutobox control unit for the adopted neural controller (see Fig. 12). The experiments were conducted reproducing, by means of the automated test bench controller, the same transient used as test-set (i.e. SET B), this time evaluating the injection time with the neural controller. The target value of AFR to be fed to the IRNNM (i.e. Eq. 19), was set to stoichiometry, i.e. 14.67. In order to assess the performance of the proposed controller thoroughly, the resulting AFR trajectory was compared to the trajectory measured from the actions of the reference ECU, as shown in Fig. 19 and Fig. 20. The lines corresponding to AFR_{des} = 14.67, AFR_{des} + 5% and AFR_{des} − 5 % are also plotted in the figures.

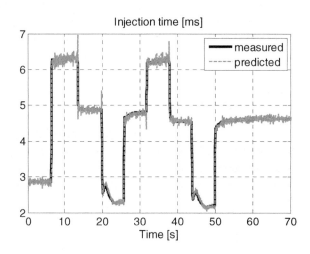

Fig. 18. Trajectories of measured and predicted injection time (SET B)

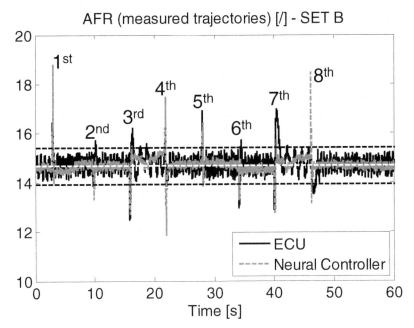

Fig. 19. Comparison between ECU and NN performance in terms of AFR control (SET B)

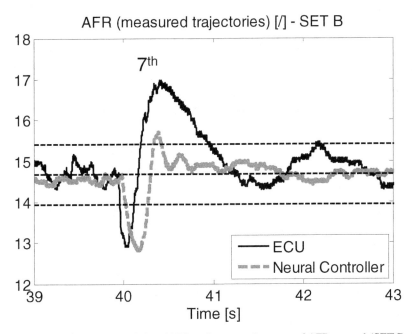

Fig. 20. Comparison between ECU and NN performance in terms of AFR control (SET B, time window [39, 43] of Fig. 19)

Fig. 19 shows how the performance level achieved by the two controllers is very close. It is interesting to note how the neural controller performs better maneuvers 3rd, 5th, 7th or worse maneuvers 1st, 4th, 8th than the ECU in limiting AFR spikes when abrupt throttle opening/closing are performed (see Fig. 9 a). Particularly, in Fig. 20 it can be seen how the feedback action performed by the neural controller on the AFR predicted by the virtual sensor induces a higher-order AFR response as compared to that obtained with ECU. This results in a faster AFR compensation (0.5 s against 1 s) and, particularly, in removing the AFR overshoot observed with the ECU. Of course neither the neural controller nor the ECU can overcome the rich excursion due to the fuel puddle on manifold walls, unless a suitable strategy aimed at increasing the air flow rate is actuated.

Such considerations demonstrate the great potential offered by Neural Networks to improve current AFR control strategies with a significant reduction of experimental burden and calibration effort with respect to conventional approaches.

7 Conclusion

Recurrent Neural Network models have been developed to simulate both forward and inverse AFR dynamics in an SI engine. The main features of RNNs have been reviewed with particular attention to the achievement of reliable models with high degree of generalization for practical engine applications.

Training and test data sets have been derived from experimental measurements on the engine test bench by imposing load and speed perturbations. To enhance RNN generalization, the input variables have been uncorrelated by perturbing the fuel injection time around the base stoichiometric amount. Moreover, the removal of a $5\pi/\omega$ delay from the measured AFR signal was proven to be necessary to ensure accurate prediction of AFR dynamics in correspondence of both rich and lean transients.

The virtual sensor (i.e. FRNNM) has shown satisfactory prediction of the AFR dynamics with an estimation error vs. measured trajectory lower than 2% for most of the test transients, even in presence of wide AFR spikes, thus proving that the RNN dynamic behavior is satisfactorily close to the real system dynamics.

A real-time application of the direct AFR neural controller based on the inverse model (i.e. IRNNM) has been accomplished. The controller, that makes also use of the virtual sensor prediction, has been implemented on the engine control unit and tested over an experimental transient. The comparison with the AFR trajectory resulting from the action of the reference ECU evidences the good performance of the controller. Particularly, the integration with the virtual sensor prediction induces a higher-order response that results in a faster AFR compensation and, particularly, in removing the overshoot observed with the ECU. The results demonstrate the potential offered by neural controllers to improve engine control strategies, particularly considering the significant reduction of experimental burden / calibration efforts vs. actual methodologies. The proposed neural controller is based on an open-loop scheme, therefore adaptive features can be introduced by an on-line training aimed at adjusting IRNNM and FRNNM parameters according to engine aging and/or faults.

References

Alippi, C., de Russis, C., Piuri, V.: Observer-Based Air-Fuel Ratio Control. IEEE Transactions on Systems, Man, and Cybernetics – Part C: Applications and Reviews 33, 259–268 (2003)

Aquino, C.F.: Transient A/F Control Characteristics of the 5 Liter Central Fuel Injection Engine. SAE paper 810494. SAE International, Warrendale (1981)

Arsie, I., Marotta, F., Pianese, C., Rizzo, G.: Information based selection of neural networks training data for S.I. engine mapping. SAE 2001 Transactions - Journal of Engines 110(3), 549–560 (2001)

Arsie, I., Pianese, C., Rizzo, G., Cioffi, V.: An Adaptive Estimator of Fuel Film Dynamics in the Intake Port of a Spark Ignition Engine. Control Engineering Practice 11, 303–309 (2003)

Arsie, I., Pianese, C., Sorrentino, M.: Nonlinear Recurrent Neural Networks for Air Fuel Ratio Control in SI engines. SAE paper 2004-01-1364. SAE International, Warrendale (2004)

Arsie, I., Pianese, C., Sorrentino, M.: A procedure to enhance identification of recurrent neural networks for simulating air-fuel ratio dynamics in SI engines. Engineering Applications of Artificial Intelligence 19, 65–77 (2006)

Arsie, I., Pianese, C., Sorrentino, M.: A Neural Network Air-Fuel Ratio Estimator for Control and Diagnostics in Spark-Ignited Engines. In: Fifth IFAC Symposium on Advances in Automotive Control, Monterey (CA, USA), August 20–22 (2007)

Atiya, A., Ji, C.: How initial conditions affect generalization performance in large networks. IEEE Transactions on Neural Networks 8, 448–454 (1997)

Capriglione, D., Liguori, C., Pianese, C., Pietrosanto, A.: On-line Sensor Fault Detection Isolation, and Accomodation in Automotive Engines. IEEE Transactions on Instrumentation and Measurement 52, 1182–1189 (2003)

Choi, S.B., Hedrick, J.K.: An Observer-Based Controller Design Method for Improving Air/Fuel Characteristics of Spark Ignition Engines. IEEE Transactions on Control Systems Technology 6 (1998)

Dinca, L., Aldemir, T., Rizzoni, G.: A model-based probabilistic approach for fault detection and identification with application to the diagnosis of automotive engines. IEEE Transactions on Automatic Control 44, 2200–2205 (1999)

Dovifaaz, X., Ouladsine, M., Rachid, A.: Optimal Control of a Turbocharged Diesel Engine Using Neural Network. In: 14th IFAC World Congress, IFAC-8b-009, Beijing, China, July 5-9 (1999)

Fiengo, G., Cook, J.A., Grizzle, J.W.: Fore-Aft Oxygen Storage Control. In: American Control Conference, Anchorage, AK, May 8-10 (2002)

Fukumizu, K.: Statistical active learning in multilayer perceptrons. IEEE Transactions on Neural Network 11, 17–26 (2000)

Giles, C.L., Chen, D., Sun, G.Z., Chen, H.H., Lee, Y.C., Goudreau, M.W.: Constructive learning of recurrent neural networks limitations of recurrent cascade correlation and a simple solution. IEEE Transactions on Neural Networks 6, 829–835 (1995)

Gili, P.A., Battipede, M.: Adaptive Neurocontroller for a Nonlinear Combat Aircraft Model. Journal of Guidance, Control and Dynamics 24 (2001)

Haykin, S.: Neural Networks. In: A comprehensive Foundation, 2nd edn., Prentice-Hall, Inc., Englewood Cliffs (1999)

Hecht-Nielsen, R.: Neurocomputing. Addison-Wesley, Reading (1987)

Heimann, B., Bouzid, N., Trabelsi, A.: Road-Wheel Interaction in Vehicles A Mechatronic View of Friction. In: Proceedings of the 2006 IEEE International Conference on Mechatronics, July, 3-5, pp. 137–143 (2006)

Hendricks, E., Sorenson, S.C.: SI Engine Controls and Mean Value Engine Modeling. SAE Paper 910258, SAE International, Warrendale (1991)

Heywood, J.B.: Internal Combustion Engine Fundamental. MC Graw Hill, New York (1988)

Hunt, K.J., Sbarbaro, D.: Neural Networks for nonlinear internal model control. IEEE Procedings-D 138 (1991)

Hussain, M.A.: Review of the applications of neural networks in chemical process control simulation and online implementation. Artificial Intelligence in Engineering 13, 55–68 (1999)

Inagaki, H., Ohata, A., Inoue, T.: An Adaptive Fuel Injection Control with Internal Model in Automotive Engines. In: IECON 1990, 16th Annual Conference of IEEE, pp. 78–83 (1990)

Isermann, R.: Supervision, fault-detection and fault-diagnosis methods—an introduction. Control Engineering Practice 5, 638–652 (1997)

Kainz, J.L., Smith, J.C.: Individual Cylinder Fuel Control with a Switching Oxygen Sensor. SAE Paper 1999-01-0546. SAE International, Warrendale (1999)

Kim, Y.W., Rizzoni, G., Utkin, V.: Automotive Engine Diagnosis and Control via Nonlinear Estimation. IEEE Control Systems Magazine 18, 84–99 (1998)

Lehtokangas, M.: Constructive backpropagation for recurrent networks. Neural Processing Letters 9, 271–278 (1999)

MacKay, D.J.C.: Information—based objective functions for active data selection. Neural Computation 4, 590–604 (1992)

Malaczynski, G.W., Baker, M.E.: Real-Time Digital Signal Processing of Ionization Current for Engine Diagnostic and Control, SAE Paper 2003-01-1119, SAE 2003 World Congress & Exhibition, Detroit, MI, USA (March 2003)

Maloney, P.J.: A Production Wide-Range AFR - Response Diagnostic Algorithm for Direct-Injection Gasoline Application. SAE Paper 2001-01-0558, SAE International, Warrendale (2001)

Manzie, C., Palaniswami, M., Ralph, D., Watson, H., Yi, X.: Observer Model Predictive Control of a Fuel Injection System with a Radial Basis Function Network Observer. ASME Journal of Dynamic Systems, Measurement, and Control 124, 648–658 (2002)

Marko, K.A., James, J., Dosdall, J., Murphy, J.: Automotive control system diagnostics using neural nets for rapid pattern classification of large data sets. In: Proceedings of International Joint Conference on Neural Networks, 1989 IJCNN, Washington, DC, July 18, vol. 2, pp. 13–16 (1989)

Mastroberti, M., Pianese, C.: Identificazione e Validazione diModelli a Rete Neurale per Motori ad Accensione Comandata Mediante Tecniche Monte Carlo. In: 56th National Congress of ATI, Napoli, September 10–14 (2001) (in Italian)

Matthews, R.D., Dongre, S.K., Beaman, J.J.: Intake and ECM sub-model improvements for dynamic SI engine models: examination of tip-in tip-out. SAE Paper No. 910074. SAE International, Warrendale (1991)

Nakae, M., Tsuruta, T., Mori, R., Shinsuke, I.: Development of Planar Air Fuel Ratio Sensor, SAE Paper 2002-01-0474 (2002)

Nelles, O.: Nonlinear System Identification. Springer, Berlin (2000)

Nørgaard, M., Ravn, O., Poulsen, N.L., Hansen, L.K.: Neural Networks for Modelling and Control of Dynamic Systems. Springer, London (2000)

Onder, C.H., Geering, H.P.: Model identification for the A/F path of an SI Engine. SAE Paper No. 970612. SAE International, Warrendale (1997)

Ortmann, S., Glesner, M., Rychetsky, M., Tubetti, P., Morra, G.: Engine Knock Estimation Using Neural Networks based on a Real-World Database. SAE Paper 980513. SAE International, Warrendale (1998)

Patterson, D.W.: Artificial Neural Networks – Theory and Applications. Prentice Hall, Inc., Englewood Cliffs (1995)

Powell, J.D., Fekete, N.P., Chang, C.F.: Observer-Based Air-Fuel Ratio Control. IEEE Transactions on Control Systems 18, 72–83 (1998)

Ripley, B.D.: Pattern Recognition and Neural Networks. Cambridge University Press, Cambridge (2000)

Shayler, P.J., Goodman, M.S., Ma, T.: Transient Air/Fuel Ratio Control of an S.I. Engine Using Neural Networks. SAE Paper 960326. SAE International, Warrendale (1996)

St-Pierre, M., Gingras, D. (2004) Neural Network-Based Data Fusion for Vehicle Positioning in Land Navigation System SAE Paper 2004-01-0752, SAE 2004 World Congress & Exhibition, Detroit, MI, USA (March 2004)

Sum, J., Leung, C., Young, G.H., Kann, W.: On the Kalman filtering method in neural-network training and pruning. IEEE Transactions on Neural Networks 10(1) (1999)

Tan, Y., Saif, M.: Neural-networks-based nonlinear dynamic modeling for automotive engines. Neurocomputing 30, 129–142 (2000)

Thimm, G., Fiesler, E.: Higher order and multilayer perceptron initialization. IEEE Transaction Neural Network 8, 349–359 (1997)

Turin, R.C., Geering, H.P.: Model-Based Adaptive Fuel Control in an SI Engine. SAE Paper 940374. SAE International, Warrendale (1994)

Vigild, C.W., Andersen, K.P.H., Hendricks, E.: Towards Robust H-infinity Control of an SI Engine's Air/Fuel Ratio. SAE Paper 1999-01-0854. SAE International, Warrendale (1999)

Yasuri, Y., Shusuke, A., Ueno, M., Yoshihisa, I.: Secondary O2 Feedback Using Prediction and Identification Type Sliding Mode Control. SAE Paper no. 2000-01-0936. SAE International, Warrendale (2000)

Yoo, I.K., Upadhyay, D., Rizzoni, G.: A Control-Oriented Carbon Canister Model. SAE Paper 1999-01-1103. SAE International, Warrendale (1999)

Intelligent Vehicle Power Management – An Overview

Yi L. Murphey

Department of Electrical and Computer Engineering,
University of Michigan-Dearborn, Dearborn, MI 48128

Abstract. This chapter overviews the progress of vehicle power management technologies that shape the modern automobile. Some of these technologies are still in the research stage. Four in-depth case studies provide readers with different perspectives on the vehicle power management problem and the possibilities that intelligent systems research community can contribute towards this important and challenging problem.

1 Introduction

Automotive industry is facing increased challenges of producing affordable vehicles with increased electrical/electronic components in vehicles to satisfy consumers' needs and, at the same time, with improved fuel economy and reduced emission without sacrificing vehicle performance, safety, and reliability. In order to meet these challenges, it is very important to optimize the architecture and various devices and components of the vehicle system, as well as the energy management strategy that is used to efficiently control the energy flow through a vehicle system [GPI07].

Vehicle power management has been an active research area in the past two decades, and more intensified by the emerging hybrid electric vehicle technologies. Most of these approaches were developed based on mathematical models or human expertise, or knowledge derived from simulation data. The application of optimal control theory to power distribution and management has been the most popular approach, which includes linear programming [TaB00], optimal control [BuF97, BSK02, DLG04], and especially dynamic programming (DP) have been widely studied and applied to a broad range of vehicle models [LPG03, HoD04, AGP04, KKJ05, Pro06]. In general, these techniques do not offer an on-line solution, because they assume that the future driving cycle is entirely known. However these results have been widely used as a benchmark for the performance of power control strategies. In more recently years, various intelligent systems approaches such as neural networks, fuzzy logic, genetic algorithms, etc. have been applied to vehicle power management [JWR00, WLE02, PEB01, SGB04, BaK04, KKJ05, SKC05, WoL05, MOD06, Pro07, SFY07, CMM08, MCK08]. Research has shown that driving style and environment has strong influence over fuel consumption and emissions [Eri00, Eri01]. In this chapter we give an overview on the intelligent systems approaches applied to optimizing power management at the vehicle level in both conventional and hybrid vehicles. We present four in-depth case studies, a conventional vehicle power controller, three different approaches for a parallel HEV power controller, one is a system of fuzzy rules generated from static efficiency maps of vehicle components, a system of rules generated from optimal operation points from a fixed driving cycles with using Dynamic Programming and neural networks, and a fuzzy power controller that incorporates intelligent predictions of driving environment as well as driving patterns. We will also introduce the intelligent system research that can be applied to predicting driving environment and driving patterns, which have strong influence in vehicle emission and fuel consumption.

2 Intelligent Power Management in a Conventional Vehicle System

Most road side vehicles today are standard conventional vehicles. Conventional vehicle systems have been going through a steady increase of power consumption over the past twenty years (about 4% per year) [KNL99, KMT00, NiH00]. As we look ahead, automobiles are steadily going through electrification changes: the core mechanical components such as engine valves, chassis suspension systems, steering columns, brake controls, and shifter controls are replaced by electromechanical, mechatronics, and associated safety critical communications and software technologies. These changes place increased (electrical) power demands on the automobile [GPI07].

To keep up with future power demands, automotive industry has increased its research in building more powerful power net such as a new 42-V power net topologies which should extend (or replace) the traditional 14-V power net from present vehicles [KHM01, EEM03], and energy efficiency components, and vehicle level power management strategies that minimize power loss [Pro07]. In this section, we introduce an intelligent power management approach that is built upon an energy management strategy proposed by Koot, et al [KKJ05]. Inspired by the research in HEVs, Koot et al proposed to use an advanced alternator controlled by power and directly coupled to the engine's crankshaft. So by controlling the output power of alternator, the operating point of the combustion engine can be controlled, thus the control of the fuel use of the vehicle.

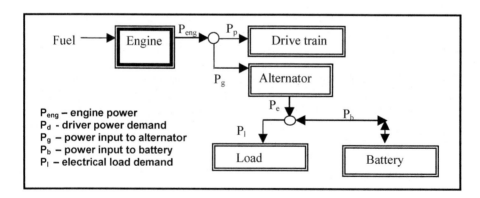

Fig. 1. Power flow in a conventional vehicle system

Figure 1 is a schematic drawing of power flow in a conventional vehicle system. The drive train block contains the components such as clutch, gears, wheels, and inertia. The alternator is connected to the engine with a fixed gear ratio. The power flow in the vehicle starts with fuel that goes into the internal combustion engine. The mapping from fuel consumed to P_{eng} is a nonlinear function of P_{eng} and engine crank speed ω, denoted as fuel rate = $F(P_{eng}, \omega)$, which is often represented through an engine efficiency map (Figure 2(a)) that describes the relation between fuel consumption, engine speed, and engine power.

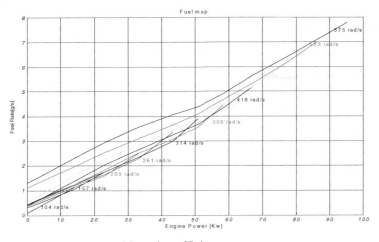

(a) engine efficiency map

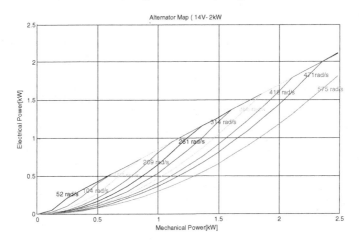

(b) alternator efficiency map

Fig. 2. Static efficiency maps of engine and alternator

The mechanical power that comes out of the engine, P_{eng}, splits up into two components: i.e. $P_{eng} = P_p + P_g$, where P_p goes to the mechanical drive train for vehicle propulsion, whereas P_g goes to the alternator. The alternator converts mechanical power P_g to electric power P_e and tries to maintain a fixed voltage level on the power net. The alternator can be modeled as a nonlinear function of the electric power and engine crank speed, i.e. $P_g = G(P_e, \omega)$, which is a static nonlinear map (see Figure 2 (b)). The alternator provides electric power for the electric loads, P_l, and P_b, power for charging the battery, i.e. $P_e = P_l + P_b$. In the end, the power becomes available for vehicle propulsion and for electric loads connected to the power net. The power flow through the battery, P_b, can be positive (in charge state) or negative (in discharge state), and the power input to the battery, P_b, is more than the actual power stored into the battery, P_s, i.e. there is a power loss during charge and discharge process.

A traditional lead-acid battery is often used in a conventional vehicle system for supplying key-off loads and for making the power net more robust against peak-power demands. Although the battery offers freedom to the alternator in deciding when to generate power, this freedom is generally not yet used in the current practice, which is currently explored by the research community to minimize power loss. Let P_Loss$_{bat}$ represents the power losses function of the battery. P_Loss$_{bat}$ is a function of P_s, E_s and T, where P_s is the power to be stored to or discharged from the battery, E_s is the Energy level of the battery and T is the temperature. To simply the problem, the influence of E_s and T are often ignored in modeling the battery power loss, then P_b can be modeled as a quadratic function of P_s, i.e. $P_b \approx P_s + \beta P_s^2$ [KKJ05]. The optimization of power control is driven by the attempt to minimize power loss during the power generation by the internal combustion engine, power conversion by the alternator, and battery charge/discharge.

Based on the above discussion, we are able to model fuel consumption as a function of ω, P_p, P_l, P_s. In order to keep driver requests fulfilled, the engine speed ω, propulsion power P_p, and electric load P_l are set based on driver's command. Therefore the fuel consumption function can be written as a nonlinear function of only one variable P_s: $\gamma(P_s)$.

One approach to intelligent power control is to derive control strategies from the analysis of global optimization solution. To find the global optimal solution, quadratic and dynamic programming (DP) have been extensively studied in vehicle power management. In general, these techniques do not offer an on-line solution, because they assume that the future driving cycle is entirely known. Nevertheless, their results can be used as a benchmark for the performance of other strategies, or to derive rules for a rule-based strategy. In particular if the short-term future state is predictable based on present and past vehicle states of the same driving cycle, the knowledge can be used in combination with the optimization solution to find effective operating points of the individual components.

The cost function for the optimization is the fuel used during an entire driving cycle: $\int_0^{t_e} \gamma(P_s)dt$ where [0, t_e] is the time interval for the driving cycle. When the complete driving cycle is known *a priori*, the global optimization of the cost function can be solved using either DP or QP with constraints imposed on P_s. But, for an online controller, it has no knowledge about the future of the present driving cycle. Koot et al proposed an online solution by using Model Predict Control strategy based on QP optimization [KKJ05].

The cost function γ(P$_s$) can be approximated by a convex quadratic function:

$$\gamma(P_s) \approx \varphi_2 P_s^2 + \varphi_1 P_s + \varphi_0, \quad \varphi_2 > 0. \tag{2.1}$$

The optimization problem thus can be model as a multistep decision problem with N steps:

$$\underset{\overline{P_s}}{Min} J = \sum_{k=1}^{N} \min_{P_s} \gamma(P_s(k),k) \approx \sum_{k=1}^{N} \min_{P_s} \frac{1}{2}\varphi_2 P_s^2(k) + \varphi_1(k) P_s(k) + \varphi_0 \tag{2.2}$$

where $\overline{P}_s$ contains the optimal setting of P$_s$(k), for k = 0, ..., n, n is the number of time intervals in a given driving cycle has. The quadratic function of the fuel rate is solved by minimizing the following Lagrange function of with respect to P$_s$ and λ.

$$L(Ps(1),...,Ps(N),\lambda) = \sum_{k=1}^{N}\{\varphi_2(k)Ps(k)^2 + \varphi_1(k)Ps(k)\} + \varphi_0 - \lambda \sum_{k=1}^{N} Ps(k) \tag{2.3}$$

The optimization problem is solved by taking the partial derivatives of Lagrange function L with respect to P$_s$ (k), k=1,.. to N and λ respectively and setting both equations to 0. This gives us the optimal setting points

$$P_s^o(k) = \frac{\lambda - \varphi_1(k)}{2\varphi_2(k)}, \tag{2.4}$$

$$\lambda = \sum_{k=1}^{N} \frac{\varphi_1(k)}{2\varphi_2(k)} \bigg/ \sum_{k=1}^{N} \frac{1}{2\varphi_2(k)}, \tag{2.5}$$

for k =1,..., N (driving time span).

The above equations show that $P_s^o(k)$ depends on the Quadratic coefficients at the current time k, which can be obtained online; however, λ requires the knowledge of φ_1 and φ_2 over the entire driving cycle, which is not available to an online controller. To solve this problem, Koot et al proposed to estimate λ dynamically using the PI-type controller as follows [KKJ05]:

$$\lambda(k+1) = \lambda_0 + Kp(Es(0) - Es(k)) + K_I \sum_{i=1}^{k}(Es(0) - Es(i))\Delta t \tag{2.6}$$

where λ_0 is an initial estimate. If we write the equation in an adaptive form, we have

$$\lambda(k+1) = \lambda_0 + K_p(E_s(0) - E_s(k-1) + E_s(k-1) - E_s(k)) + K_I \sum_{i=1}^{k}(E_s(0) - E_s(i))\Delta t \tag{2.7}$$
$$= \lambda(k) + K_p(E_s(k-1) - E_s(k)) + K_I(E_s(0) - E_s(k))\Delta t$$

By incorporating $E_s(k)$, the current energy storage in the battery, into λ dynamically, we are able to avoid draining or overcharging the battery during the driving cycle. The dynamically changed λ reflects the change of the stored energy during the last step of the driving cycle, and the change of stored energy between current and the beginning of the driving cycle. If the stored energy increased (or decreased) in comparison to its value the last step and the initial state, the λ (k+1) will be much smaller(greater) than λ (k).

Koot [Koo06] suggested the following method to tune the PI controller in (6). λ_0 should be obtained from the global QP optimization and is electric load dependant. $\lambda_0 = 2.5$ was suggested. K_P and K_I were tuned such that for average values of $\varphi_1(t)$ and $\varphi_2(t)$ (6) becomes a critically damped second-order system. For $\tilde{\varphi}_2 = 1.67*10^{-4}$, $K_p = 6.7*10^{-7}$, $K_I = 3.3 * 10^{-10}$.

Based on the above discussion, the online control strategy proposed by Koot can be summarized as follows. During an online driving cycle at step k, the controller performs the following three major computations

(1) adapt the Lagrange multiplier,

$$\lambda(k+1) = \lambda_0 + K_p(E_s(0) - E_s(k-1) + E_s(k-1) - E_s(k)) + K_I \sum_{i=1}^{k}(E_s(0) - E_s(i))\Delta t,$$

where λ_0, K_p, K_I are tuned to constants as we discussed above, $E_s(i)$ is the energy level contained in the battery at step i, i = 0, 1, ..., k, and for i=0, it is the battery energy level at the beginning of the driving cycle. All $E_s(i)$ are available from the battery sensor.

(2) calculate the optimal Ps(k) using the following either one of the two formulas

$$P_s^o(k) = \arg\min_{P_s(k)}\{\varphi_2(k)P_s^2(k) + \varphi_1(k)P_s(k) + \varphi_0(k) - \lambda(k+1)P_s(k), \quad (2.8)$$

or

$$P_s^o(k) = \arg\min_{P_s(k)}\{\gamma(P_s(k)) - \lambda(k+1)P_s(k)\} \quad (2.9)$$

Both methods search for the optimal $P_s(k)$ within its valid range at step k [KKJ05], which can be solved using DP with a horizon length of 1 on a dense grid. This step can be interpreted as follows. At each time instant the actual incremental cost for storing energy is compared with the average incremental cost. Energy is stored when generating now is more beneficial than average, whereas it is retrieved when it is less beneficial.

(3) Calculate the optimal set point of engine power

The optimal set point of engine power can be obtained through the following steps.

$$P_{eng}^o = P_g^o + P_p, \text{ where } P_g^o = G(P_e^o, \omega), P_e^o = P\_Loss_{bat}(P_s^o) + P_l.$$

Koot et al implemented their online controllers in a simulation environment in which a conventional vehicle model with the following components was used: a 100-kW 2.0-L SI engine, a manual transmission with five gears, A 42-V 5-kW alternator and a 36-V 30-Ah lead-acid battery make up the alternator and storage components of the 42-V power net. Their simulations show that a fuel reduction of 2% can be obtained by their controllers, while at the same time reducing the emissions. The more promising aspect is that the controller presented above can be extended to a more intelligent power control scheme derived from the knowledge about road type and traffic congestions and driving patterns, which are to be discussed in section C.

3 Intelligent Power Management in Hybrid Vehicle Systems

Growing environmental concerns coupled with the complex issue of global crude oil supplies drive automobile industry towards the development of fuel-efficient vehicles. Advanced diesel engines, fuel cells, and hybrid powertrains have been actively studied as potential technologies for future ground vehicles because of their potential to significantly improve fuel economy and reduce emissions of ground vehicles. Due to the multiple-power-source nature and the complex configuration and operation modes, the control strategy of a hybrid vehicle is more complicated than that of a conventional vehicle. The power management involves the design of the high-level control algorithm that determines the proper power split between the motor and the engine to minimize fuel consumption and emissions, while satisfying constraints such as drivability, sustaining and component reliability [LFL04]. It is well recognized that the energy management strategy of a hybrid vehicle has high influences over vehicle performances.

In this section we focus on the hybrid vehicle systems that use a combination of an internal combustion engine (ICE) and electric motor (EM). There are three different types of such hybrid systems:

- Series Hybrid: In this configuration, an ICE-generator combination is used for providing electrical power to the EM and the battery.
- Parallel Hybrid: The ICE in this scheme is mechanically connected to the wheels, and can therefore directly supply mechanical power to the wheels. The EM is added to the drivetrain in parallel to the ICE, so that it can supplement the ICE torque.
- Series-Parallel Combined System and others such as Toyota Hybrid System (THS).

Most of power management research in HEV has been in the category of parallel HEVs. Therefore this is also the focus of this paper. The design of a HEV power controller involves two major principles:

- meet the driver's power demand while achieving satisfactory fuel consumption and emissions.
- maintain the battery state of charge (SOC) at a satisfactory level to enable effective delivery of power to the vehicle over a wide range of driving situations.

Intelligent systems technologies have been actively explored in power management in HEV's. The most popular methods are to generate rules of conventional or fuzzy logic, based on

- heuristic knowledge on the efficient operation region of an engine to use the battery as a load-leveling component[SSK02].
- knowledge generated by optimization methods about the proper split between the two energy sources determined by minimizing the total equivalent consumption cost [KSL02, LPG03, LPG04]. The optimization methods are typically Dynamic Programming (deterministic or stochastic)
- Driving situation dependent vehicle power optimization based on prediction of driving environment using neural networks and fuzzy logic [LaW05, WoL05, SFY07].

Three case studies will be presented in the following subsections, one from each of the above three categories.

3.1 A Fuzzy Logic Controller Based on the Analysis of Vehicle Efficiency Maps

Schouten, Salman and Kheir presented a fuzzy controller in [SSK02] that is built based on the driver command, the state of charge of the energy storage, and the motor/generator speed. Fuzzy rules were developed for the fuzzy controller to effectively determine the split between the two powerplants: electric motor and internal combustion engine. The underlying theme of the fuzzy rules is to optimize the operational efficiency of three major components, ICE (Internal Combustion Engine), EM (Electric Motor) and Battery.

The fuzzy control strategy was derived based on five different ways of power flow in a parallel HEV: 1) provide power to the wheels with only the engine; 2) only the EM; or 3) both the engine and the EM simultaneously; 4) charge the battery, using part of the engine power to drive the EM as a generator (the other part of ENGINE power is used to drive the wheels); 5) slow down the vehicle by letting the wheels drive the EM as a generator that provides power to the battery (regenerative braking).

A set of 9 fuzzy rules was derived from the analysis of static engine efficiency map and motor efficiency map with input of vehicle current state such as SOC and driver's command. There are three control variables, SOC (battery state of charge), P_{driver} (driver power command), and ω_{EM} (EM speed) and two solution variables, P_{gen} (generator power), scale factor, SF.

The driver inputs from the brake and accelerator pedals were converted to a driver power command. The signals from the pedals are normalized to a value between zero and one (zero: pedal is not pressed, one: pedal fully pressed). The braking pedal signal is then subtracted from the accelerating pedal signal, so that the driver input takes a value between -1 and +1. The negative part of the driver input is sent to a separate brake controller that will compute the regenerative braking and the friction braking power required to decelerate the vehicle. The controller will always maximize the regenerative braking power, but it can never exceed 65% of the total braking power required, because regenerative braking can only be used for the front wheels.

The positive part of the driver input is multiplied by the maximum available power at the current vehicle speed. This way all power is available to the driver at all times [SSK02]. The maximum available power is computed by adding the maximum available engine and EM power. The maximum available EM and engine power depends on EM/engine speed and EM/engine temperature, and is computed using a two-dimensional look-up table with speed and temperature as inputs. However, for a given vehicle speed, the engine speed has one out of five possible values (one for each gear number of the transmission). To obtain the maximum engine power, first the maximum engine power levels for those five speeds are computed, and then the maximum of these values is selected.

Once the driver power command is calculated, the fuzzy logic controller computes the optimal generator power for the EM, P_{gen}, in case it is used for charging the battery and a scaling factor, SF, for the EM in case it is used as a motor. This scaling factor SF is (close to) zero when the SOC of the battery is too low. In that case the EM should not be used to drive the wheels, in order to prevent battery damage. When the SOC is high enough, the scaling factor equals one.

The fuzzy control variable P_{drive} has two fuzzy terms, normal and high. The power range between 0 and 50kw is for "normal", the one between 30kw to the maximum is for "high", the power range for the transition between normal and high, i.e. 30kw ~ 50 kW, is the optimal range for the engine. The fuzzy control variable SOC has four fuzzy terms, too low, low, normal and too high. The fuzzy set for "too low" ranges from 0 to 0.6, "low" from 0.5 to 0.75, "normal" from 0.7 to 0.9, "too high" from 0.85 to 1.

The fuzzy control variable ω_{EM} (EM speed) has three fuzzy sets, "low," "optimal," and "high". The fuzzy set "low" ranges from 0 to 320rad/s, "optimal" ranges from 300 to 470rad/s, "high" from 430 through 1000 rad/s. Fuzzy set "optimal" represents the optimal speed range which gives membership function to 1 at the range of 320rad/s through 430 rad/s. The nine fuzzy rules are shown in Table 1.

Table 1. Rule base of the fuzzy logic controller

1	If SOC is too high then P_{gen} is 0 kw
2	If SOC is normal and P_{drive} is normal and ω_{EM} is optimal then P_{gen} is 10 kw
3	If SOC is normal and ω_{EM} is NOT optimal then P_{gen} is 0 kw
4	If SOC is low and P_{drive} is normal and ω_{EM} is low then P_{gen} is 5 kw
5	If SOC is low and P_{drive} is normal and ω_{EM} is NOT low then P_{gen} is 15 kw
6.	If SOC is too low then P_{gen} is $P_{gen, max}$
7	If SOC is too low then SF is 0
8	If SOC is NOT too low and P_{drive} is high then P_{gen} is 0 kw
9	If SOC is NOT too low then SF is 1

Rule 1 states that if the SOC is too high the desired generator power will be zero, to prevent overcharging the battery. If the SOC is normal (rules 2 and 3), the battery will only be charged when both the EM speed is optimal and the driver power is normal. If the SOC drops to low, the battery will be charged at a higher power level. This will result in a relatively fast return of the SOC to normal. If the SOC drops to too low (rules 6 and 7), the SOC has to be increased as fast as possible to prevent battery damage. To achieve this, the desired generator power is the maximum available generator power and the scaling factor is decreased from one to zero. Rule 8 prevents battery charging when the driver power demand is high and the SOC is not too low. Charging in this situation will shift the engine power level outside the optimum range (30–50 kW). Finally, when the SOC is not too low (rule 9), the scaling factor is one.

The engine power, P_{eng}, and EM power, P_{EM}, are calculated as follows

$$P_{eng} = P_{driver} + P_{gen}, P_{EM} = - P_{gen}$$

except for the following cases:

(1) If $P_{driver} + P_{EM, gen}$ is smaller than the threshold value SF*6 kw then $P_{eng} = 0$ and $P_{EM} = P_{driver}$.
(2) If $P_{driver} + P_{EM, gen}$ is larger than the maximum engine power at current speed ($P_{eng, max@speed}$) then

$P_{eng} = P_{eng, max@speed}$ and $P_{EM} = P_{driver} - P_{eng, max@speed}$
(3) If P_{EM} is positive (EM used as motor), $P_{EM} = P_{EM} * SF$

The desired engine power level is used by the gear shifting controller to compute the optimum gear number of the automated manual transmission. First, the optimal speed-torque curve is used to compute the optimal engine speed and torque for the desired engine power level. The optimal engine speed is then divided by the vehicle speed to obtain the desired gear ratio. Finally, the gear number closest to the desired gear ratio is chosen.

The power controller has been implemented and simulated with PSAT using the driving cycles described in the SAE J1711 standard. The operating points of the engine, EM, and battery were either close to the optimal curve or in the optimal range [SSK02].

3.2 An Intelligent Controller Built Using DP Optimization and Neural Networks

Traditional rule-based algorithms such as the one discussed in the B.1 are popular because they are easy to understand. However, when the control system is multi-variable and/or multi-objective, as often the case in HEV control, it is usually difficult to come up with rules that capture all the important trade-offs among multiple performance variables. Optimization algorithms such as Dynamic Programming (DP) can help us understand the deficiency of the rules, and subsequently serve as a 'role-model' to construct improved and more complicated rules [LFL04, Pro06]. As Lin et al pointed out that using a rule-base algorithm which mimics the optimal actions from the DP approach gives us three distinctive benefits: (i) optimal performance is known from the DP solutions; (ii) the rule-based algorithm is tuned to obtain near-optimal solution, under the pre-determined rule structure and number of free parameters; and

(iii) the design procedure is re-useable, for other hybrid vehicles, or other performance objectives [LFL04].

Lin et al designed a power controller for a parallel HEV that uses deterministic dynamic programming (DP) to find the optimal solution and then extracts implementable rules to form the control strategy [LPG03, LFL04]. Figure 3 gives the overview of the control strategy. The rules are extracted from the optimization results generated by two runs of DP, one is running with regeneration on, and the other with regeneration off. Both require the input of a HEV model and a driving cycle. The DP running with regeneration on generates results from which rules for gear shift logic and power split strategy are extracted, the DP running with regeneration off generates results for rules for charge-sustaining strategy.

When used online, the rule-based controller starts by interpreting the driver pedal motion as a power demand, P_d. When P_d is negative (brake pedal pressed), the motor is used as a generator to recover vehicle kinetic energy. If the vehicle needs to decelerate harder than possible with the 'electric brake', the friction brake will be used. When positive power ($P_d > 0$) is requested (gas pedal pressed), either a Power Split Strategy or a Charge-Sustaining Strategy will be applied, depending on the battery state of charge (SOC). Under normal driving conditions, the Power Split Strategy determines the power flow in the hybrid powertrain. When the SOC drops below the lower limit, the controller will switch to the Charge-Sustaining Strategy until the SOC reaches a pre-determined upper limit, and then the Power Split Strategy will resume.

The DP optimization problem is formulated as follows. Let x(k) represents three state variables, vehicle speed, SOC and gear number, at time step k, and u(k) are the

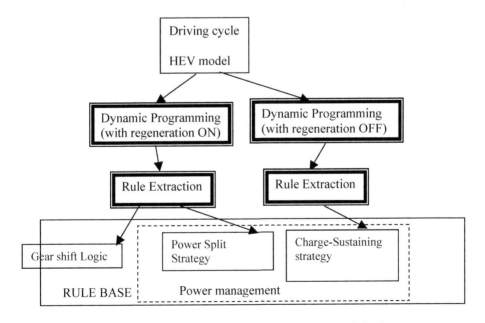

Fig. 3. A rule based system developed based on DP optimization

input signals such as engine fuel rate, transmission shift to the vehicle at time step k. The cost function for fuel consumption is defined as:

$$J = fuel = \sum_{k=1}^{N-1} L(x(k), u(k)) \text{ (kg)}$$

where L is the instantaneous fuel consumption rate, and N is the time length of the driving cycle. Since the problem formulated above does not impose any penalty on battery energy, the optimization algorithm tends to first deplete the battery in order to achieve minimal fuel consumption. This charge depletion behavior will continue until a lower battery SOC is reached. Hence, a final state constraint on SOC needs to be imposed to maintain the energy of the battery and to achieve a fair comparison of fuel economy. A soft terminal constraint on SOC (quadratic penalty function) is added to the cost function as follows:

$$J = \sum_{k=1}^{N-1} L(x(k), u(k)) + G(x(N))$$

where $G(x(N)) = \alpha(SOC(N) - SOC_f)^2$ represents the penalty associated with the error in the terminal SOC; SOC_f is the desired SOC at the final time, α is a weighting factor. For a given driving cycle, D_C, DP produces an optimal, time-varying, state-feedback control policy that is stored in a table for each of the quantized states and time stages, i.e. u*(x(k), k); this function is then used as a state feedback controller in the simulations. In addition, DP creates a family of optimal paths for all possible initial conditions. In our case, once the initial SOC is given, the DP algorithm will find an optimal way to bring the final SOC back to the terminal value (SOC_f) while achieving the minimal fuel consumption.

Note that the DP algorithm uses future information throughout the whole driving cycle, D_C, to determine the optimal strategy, it is only optimal for that particular driving cycle, and cannot be implemented as a control law for general, unknown driving conditions. However, it provides good benchmark to learn from, as long as relevant and simple features can be extracted. Lin et al proposed the following implementable rule-based control strategy incorporating the knowledge extracted from DP results [LFL04]. The driving cycle used by both DP programs is EPA Urban Dynamometer Driving Schedule for Heavy-Duty Vehicles (UDDSHDV) from the ADVISOR drive-cycle library. The HEV model is a medium-duty hybrid electric truck, a 4 × 2 Class VI truck constructed using the hybrid electric vehicle simulation tool (HEVESIM) developed at the Automotive Research Center of the University of Michigan [LFL04]. It is a parallel HEV with a permanent magnet DC brushless motor positioned after the transmission. The engine is connected to the torque converter (TC), the output shaft of which is then coupled to the transmission (Trns). The electric motor is linked to the propeller shaft (PS), differential (D) and two driveshafts (DS). The motor can be run reversely as a generator, by drawing power from regenerative braking or from the engine. The detail of this HEV model can be found in [LFL04, LPG03].

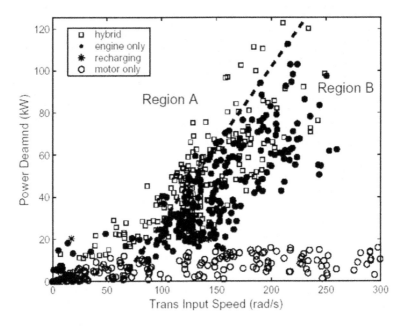

Fig. 4. Optimal operating points generated by DP over UDDSHDV cycle when $P_d > 0$

The DP program that ran with regeneration turned on produced power split graph shown in Figure 4. The graph shows the four possible operating modes in the Power Split Strategy: motor only mode (blue circles), engine only mode (red disks), hybrid mode (both the engine and motor provide power, shown in blue squares), and recharge mode (the engine provides additional power to charge the battery, shown in green diamonds). Note during this driving cycle, recharging rarely happened. The rare occurrence of recharging events implies that, under the current vehicle configuration and driving cycle, it is not efficient to use engine power to charge the battery, even when increasing the engine's power would move its operation to a more efficient region. As a result, we assume there is no recharging during the power split control, other than regeneration, and thus recharge by the engine will only occur when SOC is too low. The following power split rules were generated based on the analysis of the DP results.

Nnet$_1$ is a neural network trained to predict the optimal motor power in a split mode. Since optimal motor power may depend on many variables such as wheel speed, engine speed, power demand, SOC, gear ratio, etc [LFL04], Lin et al first used a regression-based program to select the most dominant variables in determining the motor power. Three variables were selected, power demand, engine speed, and transmission input speed as input to the neural network. The neural network has two hidden layers with 3 and 1 neurons respectively. After the training, the prediction results generated by the neural network are stored in a "look-up table" for real-time online control.

Table 1: Power split rules
If $P_d \leq 15$ kw, use Motor only, i.e. $P_m = P_d$, $P_e = 0$.
Else
If region A, operate in power split mode: $P_m = Nnet_1(P_d, \omega_{trans}, \omega_{eng})$, $P_e = P_d - P_m$.
If region B, use engine only, i.e. $P_m = 0$, $P_e = P_d$.
If $P_e > P_{e\_max}$, $P_e = P_{e\_max}$, $P_m = P_d - P_e$.
Note: P_m is the motor power, P_e is the engine power, and P_d is driver's power demand.

The efficiency operation of the internal combustion engine also depends on transmission shift logic. Lin et al used the DP solution chooses the gear position to improve fuel economy. From the optimization results, the gear operation points are expressed on the engine power demand vs. wheel speed plot shown in Figure 5. The optimal gear positions are separated into four regions, and the boundary between two adjacent regions seems to represent better gear shifting thresholds. Lin et al use a hysteresis function to generate the shifting thresholds. They also pointed out that the optimal gear shift map for minimum fuel consumption can also be constructed through static optimization. Given an engine power and wheel speed, the best gear position for minimum fuel consumption can be chosen based on the steady-state engine fuel consumption map. They found that the steady-state gear map nearly coincides with Figure 5. However for a pre-transmission hybrid configuration, it will be harder to obtain optimal shift map using traditional methods.

Since the Power Split Strategy described above does not check whether the battery SOC is within the desired operating range, an additional rule for charging the battery with the engine was developed by Lin et al to prevent battery from depletion. A traditional practice is to use a thermostat-like charge sustaining strategy, which turns on the recharging mode only if the battery SOC falls below a threshold and the charge continues until the SOC reaches a predetermined level. Although this is an easy to implement strategy, it is not the most efficient way to recharge the battery. In order to improve the overall fuel efficiency further, the questions 'when to recharge' and 'at what rate' need to be answered. Lin et al ran the DP routine with the regenerative braking function was turned off to make sure that all the braking power was supplied by the friction braking and hence there was no 'free' energy available from the regenerative braking. They set the initial SOC at 0.52 for the purpose of simulating the situation that SOC is too low and the battery needs to be recharged. Their simulation result is shown in Figure 6.

Note that the results obtained represent the optimal policy under the condition that the battery SOC has to be recharged from 0.52 to 0.57 using engine power. Note also that negative motor power now represents the recharging power supplied by the engine since there is no regenerative braking. A threshold line is drawn to divide the plot into two regions C and D. A neural network $Nnet_2$ was trained to find the optimal amount of charging power. The basic logic of this recharging control is summarized in Table 2. The rules in Table 1 and Table 2 together provide complete power management of the hybrid propulsion system under any conditions. Figure 7 shows the

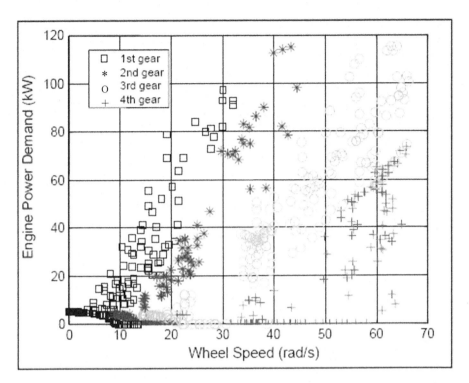

Fig. 5. Transmission gear selection generated by DP algorithm when $P_d > 0$ over UDDSHDV cycle

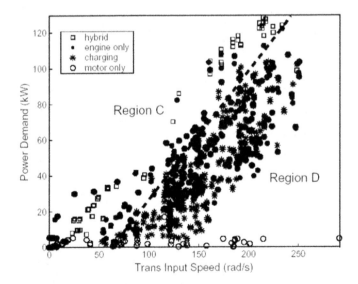

Fig. 6. Optimal operating points from DP (engine recharging scenario) over UDDSHDV cycle

flow charts of the power controller. T_line1 is the function representing the threshold shown in Figure 7 (a), and T_line2 is the function representing the threshold shown in Figure 7 (b).

Table 2. Charge-sustaining rules
If $P_d \leq 8$ kw, use Motor only, i.e. $P_m = P_d$, $P_e = 0$.
Else
If region C, or $\omega_{wheel} < 10$, $P_e = P_d$, $P_m = 0$.
If region D, $P_m = -P_{ch}$, $P_e = P_d + P_{ch}$, $P_{ch} = -Nnet_2(P_d, \omega_{trans}, \omega_{eng})$,
If $P_e > P_{e\_max}$, $P_e = P_{e\_max}$, $P_m = P_d - P_e$.

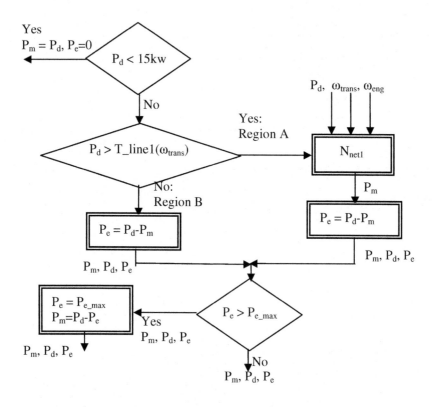

(a) Power Split Strategy

Fig. 7. Power management of the hybrid propulsion system

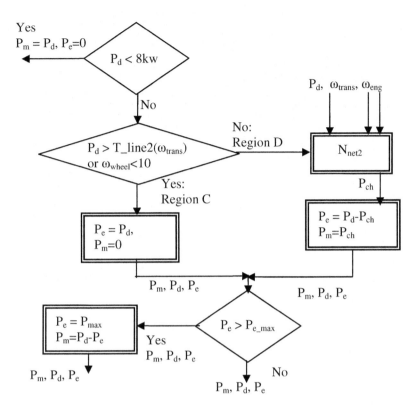

(b) Charge-sustaining strategy

Fig. 7. (*continued*)

Lin et al used the Urban Dynamometer Driving Schedule for Heavy-Duty Vehicles (UDDSHDV) to evaluate the fuel economy of their power management strategy. Their results were obtained for the charge sustaining strategy, with the SOC at the end of the cycle being the same as it was at the beginning. They showed 28% of the fuel economy improvement (over the conventional truck) by the DP algorithm and 24% by their DP-trained rule-based algorithm, which is quite close to the optimal results generated by the DP algorithm.

3.3 Intelligent Vehicle Power Management Incorporating Knowledge about Driving Situations

The power management strategies introduced in the previous sections do not incorporate the driving situation and/or the driving style of the driver into their power management strategies. One step further is to incorporate the optimization into a control strategy that has the capability of predicting upcoming events. In this section we take introduce an intelligent controller, IEMA (intelligent energy management agent) proposed by Langari and Won [LaW05-1, LaW05-2]. IEMA incorporates true drive cycle analysis within an overall framework for energy management in HEVs. Figure 8 shows the two major components in the architecture of IEMA, where T_e is the current engine torque, $T_{ec,\ FTD}$ is the increment of engine torque for propulsion produced by FTD, and $T_{ec,\ TD}$ is the increment engine torque compensating for the effect of variations of driver style. The primary function of IEMA is to distribute the required torque between the electric motor and the IC (internal combustion) engine. In order to accomplish this, IEMA utilizes information on roadway type and traffic congestion level, driver style etc., which are produced by intelligent systems discussed in section C. The FTD is a fuzzy controller that has fuzzy rule sets for each roadway type and traffic congestion level. The SCC, constructed based on expert knowledge about charge sustaining properties in different operating modes, guarantees that the level of electric energy available through the electric energy storage is maintained within a prescribed range throughout the entire driving. Its output, $T_{ec,\ SOC}$, is the increment of engine torque for charging. T_{ec} is engine torque command. The relationship among these variables is characterized as follows:

$$T_{ec} + T_{mc} = T_e + T_{ec,\ FTD} + T_{ec,\ SOC} + T_{mc} = T_c \qquad (3.3.1)$$

where T_{mc} is motor torque command and T_c is the driver's torque command.

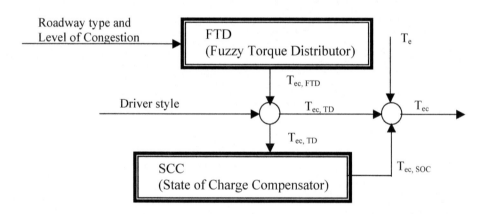

Fig. 8. Architecture view of IEMA proposed by Langari and Wong

The function of FTD is to determine the effective distribution of torque between the motor and the engine. FTD is a fuzzy rule based system that incorporates the knowledge about the driving situation into energy optimization strategies. Fuzzy rules are generated for each of the 9 facility driving cycles defined by Sierra Research [Eri02]. These knowledge bases are indexed as RT1 through RT9.

Six fuzzy membership functions are generated for assessing driving trends, driving modes and the SOC, and for the output variable of FTD, $T_{ec,\ FTD}$. These fuzzy membership functions are generated by the expert knowledge about the systems' characteristics and the responses of the powertrain components. Fuzzy rules are generated based on the postulate that fuel economy can be achieved by operating the ICE at the efficient region of the engine and by avoiding transient operations that would occur in typical driving situations such as abrupt acceleration or deceleration, frequent stop-and-go. The rational for each fuzzy rule set is given below.

(1) Low-speed Cruise Trend. This speed range is defined as below 36.66 ft/s (25mph) with small acceleration/deceleration rates (within ± 0.5 ft/s^2). When the level of the SOC is high, the electric mother (EM) is used to provide the propulsive power to meet the driver's torque demand (T_{dc}). When the SOC is low, ICE is used to generate propulsive power even if it means (temporary) high fuel consumption, since priority is given to maintaining the SOC at certain levels. For low speed region of ICE under low SOC, no additional engine operation for propulsion is made to avoid the ICE operation at inefficient regions of the region. For high engine speed under low SOC, ICE together with EM are used to generate propulsive power. This strategy is applied to all facility-specific drive cycles whenever this driving trend is present.

(2) High-speed Cruise Trend. This speed range is defined as over 58.65 ft/s (40 mph) with small acceleration/deceleration rates (within ± 0.5 ft/s^2). In this speed range, ICE is sued to provide propulsive power. For speeds over about 55 mph, fuel consumption rate increases with the increase of vehicle speed. Therefore, EM is used in this region to easy the overall fuel usage. Note, the continued use of EM will result in making SCC to act to recover the SOC of the battery. This rule is applied to all facility-specific drive cycles whenever this driving trend is present. Depending on the gear ratio during driving, the engine speed is determined according to the speed of the vehicle. Given the speed of the vehicle, the engine speed will be high or low depending on the gear ratio. For the high-speed region of the engine, ICE is to be operated for generating power according to the level of the SOC. The EM is used for low speed region of the engine.

(3) Acceleration/Deceleration Trend. In this acceleration/deceleration, fuzzy rules are devised based on the characteristic features of each drive cycle (i.e., each of the nine facility-specific drive cycles), and is derived by comparing with the characteristics of the neighboring drive cycles. The fuzzy rules were derived based on the speed-fuel consumption analysis, and observations of speed profiles for all 9 Sierra driving cycles.

The IEMA also incorporates the prediction of driver style into the torque compensation. Langari and Won characterizes the driving style of the driver by using a driving style factor α_{DSI}, which is used to compensate the output of the FTD through the following formula:

$$T_{ec, TD} = T_{ec, FTD} * (1 + \text{sgn}(T_{ec, FTD}) * \alpha_{DSI}) \qquad (3.3.2)$$

where $T_{ec, TD}$ is the increment of engine torque compensating for the effect of driver variability, which is represented by α_{DSI}. Combining with B.3.1 with B.3.2 we have

$$T_{ec} + T_{mc} =$$
$$T_e + T_{ec, FTD} * (1 + \text{sgn}(T_{ec, FTD}) * \alpha_{DSI}) + T_{ec, SOC} + T_{mc} = T_c . \qquad (3.3.3)$$

This compensation was devised based on the following assumption. The transient operation of the engine yields higher fuel consumption than steady operation does. Therefore the driver's behavior is predicted and its effect on the engine operation is adjusted. For example, for the aggressive driver, the less use of ICE is implemented to avoid fuel consumption that is due to the transient operation of the engine by the driver. In [WoL05], a maximum 10% of the increment of engine torque was considered for calm drivers, normal 0% and aggressive -10%.

In order to keep the level of electric energy within a prescribed range throughout driving, SCC, the State-of-Charge Compensator was used in IEMA. The SCC detects the current SOC and compares it with the target SOC, and commands additional engine torque command, $T_{ec,\ SOC}$. As shown in (3.3.3), the increment of engine torque from SCC is added to (or subtracted from) the current engine torque for the charge (or discharge) operation.

The charge (or discharge) operations are determined by the instantaneous operating mode, which is characterized into start-up, acceleration, cruise, deceleration (braking), and stationary.

Under the charge sustenance concept, the function of the electric motor can be switched to that of a generator to charge the battery for the next use if surplus power from the engine is available. In certain driving modes, however, particularly in acceleration and cruise modes, battery charge by operating ICE is generally not recommended because this may cause the overall performance to deteriorate and/or the battery to be overcharged. Selective battery charge operation may be required, however, for the operation of HEVs in these modes. In the stop (idle) mode, the charge sustaining operation can be accomplished in an efficient region of the engine while maximizing fuel efficiency if applicable or required. While not considered in this study, external charge operation can be accomplished in the stationary (parking) mode of the vehicle. Langari and Won derived functions for $T_{ec,\ SOC}$ for both hybrid (acceleration, cruise, and deceleration) and stop modes.

In the hybrid mode $T_{ec,\ SOC}$ is determined based on the analysis of the engine-motor torque plane, driver's torque demand, the (engine) torque margin for the charge operation (TMC) and the discharge operation (TMD). The increments of engine torque were obtained by introducing an appropriate function that relates to the current SOC, TMC, and TMD.

The $T_{ec,\ SOC}$ is further adjusted based on the vehicle's mode of operation. The adjustment was implemented by a fuzzy variable β_{hybrid} and then the incremental of engine torque for the charge operation becomes:

$$T_{ec,\ SOC,\ hybrid} = \beta_{hybrid} \times T_{ec,\ SOC},$$

where β_{hybrid} is the output of a mode-based fuzzy inference system that generates a weighted value of [0, 1] to represent the degree of charge according to the vehicle modes. For instance, if the vehicle experiences high acceleration, additional battery charge is prohibited to avoid deteriorating the vehicle performance even in low level of the SOC in the battery. The value of β_{hybrid} is set to zero whenever the level of the SOC is high in all modes. In the cruise or the deceleration mode, battery charge operation is performed according to the engine speed under low SOC level. In the acceleration mode, battery charge operation is dependent on the magnitude of power demand under low SOC level.

The charge sustaining operation in the stop mode is accomplished based on the analysis of efficient region of the engine while maximizing fuel economy and the best point (or region) of operation of the engine and the gear ratio of the transmission device.

Langari and Won evaluated their IEMA system through simulation on the facility-specific drive cycles [CaA97] and the EPA UDDS [EPA-online]. For the simulation study, a typical parallel drivetrain with a continuous variable transmission was used. The vehicle has a total mass of 1655 kg which is the sum of the curb weight of 1467 kg and the battery weight. An internal combustion engine with a displacement of 0.77 L and peak power of 25 kW was chosen. The electric motor is chosen to meet the acceleration performance (zero to 60 mph in less than 15 s.) To satisfy the requirement for acceleration, a motor with a power of 45 kW is selected. The battery capacity is 6 kW h (21.6 MJ) with a weight of 188 kg and is chosen on the basis of estimated values of the lead acid battery type used in a conventional car. Langari and Won gave a detailed performance analysis based on whether any of the intelligent predictions are used: roadway type, driving trend, driving style and operating mode. Their experiments showed their IEMA gave significant improvements when any and all of these four types knowledge were used in the power management system.

4 Intelligent Systems for Predicting Driving Patterns

Driving pattern exhibited in real driving is the product of the instantaneous decisions of the driver to cope with the (physical) driving environment. It plays an important roll in power distribution and needs to be predicted during real time operation. Driving patterns are generally defined in terms of the speed profile of the vehicle between time interval [t- Δw, t], where t is the current time instance during a driving experience, and $\Delta w > 0$ is the window size that characterizes the length of the speed profile that should be used to explore driving patterns. Various research work have suggested that road type and traffic condition, trend and style, and vehicle operation modes have various degrees of impacts on vehicle fuel consumptions [CaA97, BaY94, CHL99, LSM95, Eri00, Eri01, VKK00]. However most of the existing vehicle power control approaches do not incorporate the knowledge about driving patterns into their vehicle power management strategies. Only recently research community in intelligent vehicle power control has begun to explore the ways to incorporate the knowledge about online driving pattern into control strategies [JJP02, KSL02, LaW05, WoL05].

This section discusses the research issues related to the prediction of roadway type and traffic congestions, the driving style of the driver, current driving mode and driving trend.

4.1 Features Characterizing Driving Patterns

Driving patterns can be observed generally in the speed profile of the vehicle in a particular environment. The statistics used to characterize driving patterns include 16 groups of 62 parameters [Eri01], and parameters in 9 out of these 16 groups critically affect fuel usage and emissions [BaY94, Eri00]. However it may not necessary to use all these features for predicting a specific driving pattern, and on the other hand, additional new features may be explored as well. For example in[LaW05], Langari and Won used only 40 of the 62 parameters and then added seven new parameters, trip time; trip distance; maximum speed; maximum acceleration; maximum deceleration; number of stops; idle time, i.e., percent of time at speed 0 kph. To develop the intelligent systems for each of the four prediction problems addressed in this section, namely roadway type and traffic congestion level, driver style, operation mode, driving trend, the selection of a good set of features is one of the most important steps. It has been shown in pattern recognition that too many features may degrade system performances. Furthermore, more features imply higher hardware cost in onboard implementation and more computational time. Our own research shows that a small subset of the features used by Langari and Won [LaW05] can give just as good or better roadway predictions [MCK08].

The problem of selecting a subset of optimal features from a given set of features is a classic research topic in pattern recognition and a NP problem. Because the feature selection problem is computationally expensive, research has focused on finding a quasi optimal subset of features, where *quasi optimal* implies good classification performance, but not necessarily the best classification performance. Interesting feature selection techniques can be found in [FPH94, MuG00, CGM03].

4.2 A Multi-class Intelligent System for Predicting Roadway Types

In general a driving cycle can be described as a composition of different types of roadways such as local, freeway, arterial/collector, etc, and different levels of traffic congestions. Under a contract with the Environmental Protection Agency (EPA), Sierra Research Inc. [CaA97] has developed a set of 11 standard driving cycles, called facility-specific (FS) cycles, to represent passenger car and light truck operations over a range of facilities and congestion levels in urban areas. The 11 drive cycles can be divided into four categories, freeway, freeway ramp, arterial, and local. More recently they have updated the data to reflect the speed limit changes in certain freeways [Sie03]. The two categories, freeway and arterial are further divided into subcategories based on a qualitative measure called level of service (LOS) that describe operational conditions within a traffic stream based on speed and travel time, freedom to maneuver, traffic interruptions, comfort, and convenience. Six types of LOS are defined with labels, A through F, with LOS A representing the best operating conditions and LOS F the worst. Each level of service represents a range of operating conditions and the driver's perception of those conditions; however safety is not included in the measures that establish service levels [TRB00, Sie03]. According to the most recent results [Sie03], the freeway category has been divided into six LOS: freeway-LOS A, through freeway-LOS F; the arterial category into three LOS: Arterial LOS A-B, Arterial LOS C-D, and Arterial LOS E-F. The speed profiles of the 11 Sierra new driving cycles are shown in Figure 9. The statistical features of these 11 drive cycles are listed in Table 3.

One potential use of facility specific driving cycles in vehicle power management is to learn knowledge about optimal fuel consumption and emissions in each of 11 facility specific driving cycles and apply the knowledge to online control. From the discussion presented in section B we can see that most vehicle power management strategies are generally based on a fixed drive cycle, and as such do not deal with the variability in the driving situation. A promising approach is to formulate a driving cycle dependent optimization approach that selects the optimal operation points according to the characteristic features of the drive cycle. The driving cycle specific knowledge can be extracted from all 11 Sierra FS driving cycles through machine learning of the optimal operation points. During the online power control, the power controller needs to predict the current road type and LOS at the current time. Based

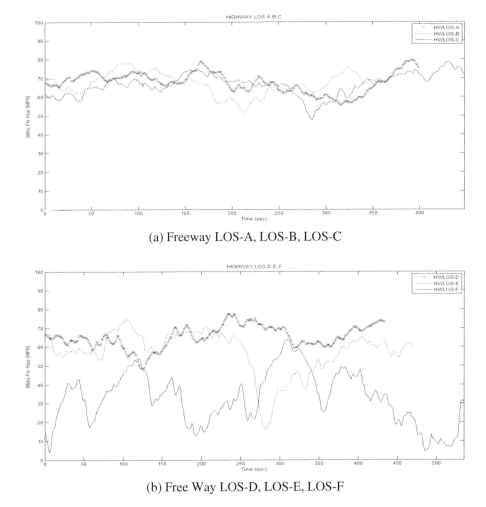

(a) Freeway LOS-A, LOS-B, LOS-C

(b) Free Way LOS-D, LOS-E, LOS-F

Fig. 9. New facility specific driving cycles defined by Sierra Research. The speed is in meters per second.

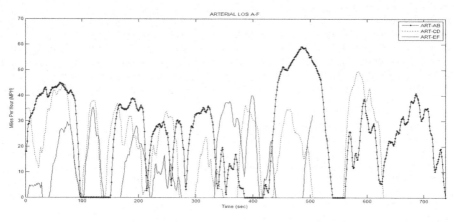

(c) Arterial LOS A-B, LOS C-D, LOS E-F

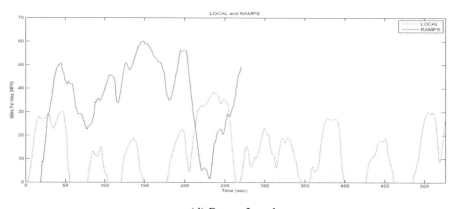

(d) Ramp, Local

Fig. 9. (*continued*)

on the prediction result, the knowledge extracted from the Sierra FS cycle that match the predicted result can be used for the online power controller. The key issue is to develop an intelligent system that can predict the current and a short-term future road type and LOS based on the history of the current driving cycle.

An intelligent system can be developed to classify the current driving environment in terms of roadway type combined with traffic congestion level based on the 11 standard Sierra FS cycles. The intelligent system can be a neural network or decision tree or any other classification techniques. Langari and Won used a learning vector quantization (LVQ) network to classify the current roadway type and congestion level [LaW05]. An LVQ network has a two-stage process. At the first stage, a competitive layer is used to identify the *subclasses* of input vectors. In the second stage, a linear layer is used to combine these subclasses into the appropriate target classes.

The prediction or classification of road type and LOS is on a segment-by-segment basis. Each driving cycle in a training data set is divided into segments of Δw seconds. In [LaW05], Langari and Won used a $\Delta w = 150$ seconds, and adjacent segments

Table 3. Statistics of 11 facility specific driving cycles

Facility Cycles by Sierra Research				
Cycle	V_{avg} (mph)	V_{max} (mph)	A_{max} (mph/s^2)	Length (sec)
Freeway LOS A	67.79	79.52	2.3	399
Freeway LOS B	66.91	78.34	2.9	366
Freeway LOS C	66.54	78.74	3.4	448
Freeway LOS D	65.25	77.56	2.9	433
Freeway LOS E	57.2	74.43	4.0	471
Freeway LOS F	32.63	63.85	4.0	536
Freeway Ramps	34.6	60.2	5.7	266
Arterials LOS A-B	24.8	58.9	5.0	737
Arterials LOS C-D	19.2	49.5	5.7	629
Arterials LOS E-F	11.6	39.9	5.8	504
Local Roadways	12.9	38.3	3.7	525

are overlapped. Features are extracted from each segment for prediction and classification. The features used in [LaW05] consisting of 40 parameters from the 62 parameters defined by Sierra are considered since the information on the engine speed and gear changing behavior are not provided in the drive cycles under their consideration. In addition, seven other characteristic parameters, which they believe can enhance the system performance are: trip time; trip distance; maximum speed; maximum acceleration; maximum deceleration; number of stops; idle time, i.e., percent of time at speed 0 kph.

An intelligent system that is developed for road type prediction will have multiple target classes, e.g. 11 Sierra road type and LOSs. For an intelligent system that has a relative large number of output classes, it is important to select an appropriate method to model the output classes. In general a multi-class pattern classification problem can be modeled in two system architectures, a single system with multiple output classes or a system of multiple classifiers. The pattern classes can be modeled by using either One-Against-One (OAO), One-Against-ALL (OAA) or P-Against-Q, with P>1 and Q>1. A comprehensive discussion on this topic can be found in [OMF04, OuM07].

4.3 Predicting Driving Trend, Operation Mode and Driver Style

Driving trends are referring to the short term or transient effects of the drive cycle such as low speed cruise, high speed cruise, acceleration/deceleration, and so on. These driving trends can be predicted using features such as the magnitudes of average speed, acceleration value, starting and ending speed in the past time segment.

The instantaneous operating mode of the vehicle at every second is the representation of the driver's intention (desire) for the operation of the vehicle, such as start-up, acceleration, cruise, deceleration (braking), and stationary. For each mode, different energy management strategies are required to control the flow of power in the drivetrain and maintain adequate reserves of energy in the electric energy storage device. The operation mode can be determined by examining the torque relations on the drive shaft. According to [LaW05], the operation modes can be characterized by the two torque features, the torque required for maintaining the vehicle speed constant in spite of road load such as rolling resistance, wind drag, and road grade, and the torque required for acceleration or deceleration of the vehicle (driver's command). Langari and Won used the engine speed SP_E is used to infer the road load, which is a function of the road grade and the speed of the vehicle. Under the assumption that mechanical connection between the engine and the wheels through transmission converts the input argument for the speed of the vehicle to the engine speed, and driving occurs on a level road, the road load can be represented by the engine speed.

Driver style or behavior has a strong influence on emissions and fuel consumption [HoN97, VKK00, Vli99, Pre99]. It has been observed that emissions obtained from aggressive driving in urban and rural traffic are much higher than those obtained from normal driving. Similar result is observed in relation to fuel consumption. Drivers style can be categorized in the following three classes [LaW05],

- *Calm driving* implies anticipating other road user's movement, traffic lights, speed limits, and avoiding hard acceleration;
- *Normal driving* implies moderate acceleration and braking;
- *Aggressive driving* implies sudden acceleration and heavy braking.

Acceleration criteria for the classification of the driver's style are based on the acceleration ranges can be found in [VKK00]. In [LaW05], a fuzzy classifier was presented. Two fuzzy variables were used, average acceleration and the ratio of the standard deviation (SD) of acceleration and the average acceleration over a specific driving range were used together to identify the driving style.

5 Conclusion

In this chapter we presented an overview of intelligent systems with application to vehicle power management. The technologies used in vehicle power management have evolved from rule based systems generated from static efficiency maps of vehicle components, driving cycle specific optimization of fuel consumption using Quadratic or Dynamic Programming, Predictive control combined with optimization, to the intelligent vehicle power management based on road type prediction and driving patterns such as driving style, operation mode and driving trend. We have seen more and more research in the use of neural networks, fuzzy logic and other intelligent system approaches for vehicle power management.

During the recent years many new telematic systems have been introduced into road vehicles. For instance, global positioning systems (GPS) and mobile phones have become a de facto standard in premium cars. With the introduction of these systems, the amount of information about the traffic environment available in the vehicle has increased. With the availability of traffic information, predictions of the vehicle propulsion load can be made more accurately and efficiently. Intelligent vehicle power systems will have more active roles in building predictive control of the hybrid powertrain to improve the overall efficiency.

References

[AGP04] Arsie, I., Graziosi, M., Pianese, C., Rizzo, G., Sorrentino, M.: Optimization of supervisory control strategy for parallel hybrid vehicle with provisional load estimate. In: Proc. 7th Int. Symp. Adv. Vehicle Control (AVEC), Arnhem, The Netherlands (August 2004)

[BaK04] Badreddine, B., Kuang, M.L.: Fuzzy Energy Management for Powersplit Hybrid Vehicles. In: Proc. Of Global Powertrain Conference (September 2004)

[BaY94] Bata, R., Yacoub, Y., Wang, W., Lyons, D., Gambino, M., Rideout, G.: Heavy duty testing cycles: survey and comparison. SAE Paper 942 263 (1994)

[BSK02] Back, M., Simons, M., Kirschaum, F., Krebs, V.: Predictive control of drivetrains. In: Proc. IFAC 15th Triennial World Congress, Barcelona, Spain (2002)

[BuF97] Bumby, J., Forster, I.: Optimization and control of a hybrid electric car. Inst. Elect. Eng. Proc. pt. Part D 134(6), 373–387 (1987)

[CaA97] Carlson, T.R., Austin, R.C.: Development of speed correction cycles. Sierra Research, Inc., Sacramento, CA, Report SR97-04-01 (1997)

[CGM03] Crossman, J.A., Guo, H., Murphey, Y.L., Cardillo, J.: Automotive Signal Fault Diagnostics: Part I: signal fault analysis, feature extraction, and quasi optimal signal selection. IEEE Transactions on Vehicular Technology (July 2003)

[CHL99] Cloke, J., Harris, G., Latham, S., Quimby, A., Smith, E., Baughan, C.: Reducing the environmental impact of driving: a review of training and in-vehicle technologies. Transport Res. Lab., Crowthorne, U.K., Report 384 (1999)

[CMM08] Chen, Z., Abul Masrur, M., Murphey, Y.L.: Intelligent Vehicle Power Management using Machine Learning and Fuzzy Logic. In: FUZZ 2008 (2008)

[DLG04] Delprat, S., Lauber, J., Guerra, T.M., Rimaux, J.: Control of a parallel hybrid powertrain: optimal control. IEEE Trans. Veh. Technol. 53(3), 872–881 (2004)

[EEM03] Emadi, A., Ehsani, M., Miller, J.M.: Vehicular Electric Power Systems: Land, Sea, Air, and Space Vehicles. Marcel Dekker, New York (2003)

[EPA-online] http://www.epa.gov/otaq/emisslab/methods/uddscol.txt

[Eri00] Ericsson, E.: Variability in urban driving patterns. Transportation Res. Part D 5, 337–354 (2000)

[Eri01] Ericsson, E.: Independent driving pattern factors and their influence on fuel-use and exhaust emission factors. Transportation Res. Part D 6, 325–341 (2001)

[FPH94] Ferri, F., Pudil, P., Hatef, M., Kittler, J.: Comparative Study of Techniques for Large Scale Feature Selection. In: Gelsema, E., Kanal, L. (eds.) Pattern Recognition in Practice IV, pp. 403–413. Elsevier Science B.V, Amsterdam (1994)

[GPI07] Gharavi, H., Prasad, K.V., Ioannou, P.: Scanning Advanced Automobile Technology. Proceedings of IEEE 95(2) (February 2007)

[HoD04] Hofman, T., van Druten, R.: Energy analysis of hybrid vehicle powertrains. In: Proc. IEEE Int. Symp. Veh. Power Propulsion, Paris, France (October 2004)

[HoN97] Holmén, B.A., Niemeier, D.A.: Characterizing the effects of driver variability on real-world vehicle emissions. Transportation Res. Part D 3, 117–128 (1997)

[JJP02] Jeon, S.-I., Jo, S.-T., Park, Y.-I., Lee, J.-M.: Multi-mode driving control of a parallel hybrid electric vehicle using driving pattern recognition. J. Dyn. Syst., Measure. Contr. 124, 141–149 (2002)

[JWR00] Johnson, V.H., Wipke, K.B., Rausen, D.J.: HEV control strategy for real-time optimization of fuel economy and emissions. SAE Paper-01-1543 (2000)

[KHM01] Ehlers, K., Hartmann, H.D., Meissner, E.: 42 V—An indication for changing requirements on the vehicle electrical system. J. Power Sources 95, 43–57 (2001)

[KKJ05] Koot, M., Kessels, J.T.B.A., de Jager, B., Heemels, W.P.M.H., van den Bosch, P.P.J., Steinbuch, M.: Energy management strategies for vehicular electric power systems. IEEE Transactions on Vehicular Technology 54(3), 771–782 (2005)

[KMT00] Kassakian, J.G., Miller, J.M., Traub, N.: Automotive electronics power up. IEEE Spectrum 37(5), 34–39 (2000)

[KNL99] Kim, C., NamGoong, E., Lee, S., Kim, T., Kim, H.: Fuel economy optimization for parallel hybrid vehicles with CVT. SAE Paper-01-1148 (1999)

[Koo06] Koot, T.M.W.: Energy management for vehicular electric power systems, Ph. D thesis, Library Technische Universiteit Eindhoven(2006) ISBN-10: 90-386-2868-4

[KSL02] Kolmanovsky, I., Siverguina, I., Lygoe, B.: Optimization of powertrain operating policy for feasibility assessment and calibration: stochastic dynamic programming approach. In: Proc. Amer. Contr. Conf., vol. 2, Anchorage, AK, pp. 1425–1430 (May 2002)

[LaW05] Langari, R., Won, J.-S.: Intelligent energy management agent for a parallel hybrid vehicle-part I: system architecture and design of the driving situation identification process. IEEE Transactions on Vehicular Technology 54(3), 925–934 (2005)

[LFL04] Lin, C.-C., Filipi, Z., Louca, L., Peng, H., Assanis, D., Stein, J.: Modelling and control of a medium-duty hybrid electric truck. Int. J. of Heavy Vehicle Systems 11(3/4), 349–370 (2004)

[LPG03] Lin, C.-C., Peng, H., Grizzle, J.W., Kang, J.-M.: Power management strategy for a parallel hybrid electric truck. IEEE Trans. Contr. Syst. Technol. 11(6), 839–849 (2003)

[LPG04] Lin, C.-C., Peng, H., Grizzle, J.W.: A stochastic control strategy for hybrid electric vehicles. In: Proc. Amer. Contr. Conf., Boston, MI, pp. 4710–4715 (June 2004)

[LSM95] LeBlanc, D.C., Saunders, F.M., Meyer, M.D., Guensler, R.: Driving pattern variability and impacts on vehicle carbon monoxide emissions. In: Transportation Res. Rec. Transportation Research Board, National Research Council, pp. 45–52 (1995)

[MCK08] Murphey, Y.L., Chen, Z., Kiliaris, L., Park, J., Kuang, M., Masrur, A., Phillips, A.: Neural Learning of Predicting Driving Environment. In: IJCNN 2008 (2008)

[MOD06] Moreno, J., Ortúzar, M.E., Dixon, J.W.: Energy-Management System for a Hybrid Electric Vehicle, Using Ultracapacitors and Neural Networks. IEEE Transactions On Industrial Electronics 53(2) (April 2006)

[MuG00] Murphey, Y.L., Guo, H.: Automatic Feature Selection – a hybrid statistical approach. In: International Conference on Pattern Recognition, Barcelona, Spain, September 3-8 (2000)

[NiH00] Nicastri, P., Huang, H.: 42V PowerNet: providing the vehicle electric power for the 21st century. In: Proc. SAE Future Transportation Technol. Conf., Costa Mesa, CA, SAE Paper 2000-01-3050 (August 2000)

[OMF04] Ou, G., Murphey, Y.L., Feldkamp, L.: Multiclass Pattern Classification Using Neural Networks. In: International Conference on Pattern Recognition, Cambridge, UK (2004)

[OuM07] Ou, G., Murphey, Y.L.: Multi-class Pattern Classification Using Neural Networks. Journal of Pattern Recognition 40(1), 4–18 (2007)

[PEB01] Paganelli, G., Ercole, G., Brahma, A., Guezennec, Y., Rizzoni, G.: General supervisory control policy for the energy optimization of charge-sustaining hybrid electric vehicles. JSAE Rev. 22(4), 511–518 (2001)

[Pre99] Preben, T.: Positive side effects of an economical driving style: Safety, emissions, noise, costs. In: Proc., ECODRIVE 7th Conf., September 16-17 (1999)

[Pro07] Prokhorov, D.V.: Toyota Prius HEV neurocontrol. In: Proceedings of International Joint Conference on Neural Networks, Orlando, Florida, USA, August 12-17 (2007)

[Pro06] Prokhorov, D.V.: Approximating Optimal Controls with Recurrent Neural Networks for Automotive Systems. In: Proceedings of the 2006 IEEE International Symposium on Intelligent Control Munich, Germany, October 4-6 (2006)

[SFY07] Syed, F.U., Filev, D., Ying, H.: Fuzzy Rule-Based Driver Advisory System for Fuel Economy Improvement in a Hybrid Electric Vehicle. In: Annual Meeting of the NAFIPS, June 24-27, pp. 178–183 (2007)

[SGB04] Sciarretta, A., Guzzella, L., Back, M.: A real-time optimal control strategy for parallel hybrid vehicles with on-board estimation of the control parameters. In: Proc. IFAC Symp. Adv. Automotive Contr., Salerno, Italy, April 19-23 (2004)

[Sie03] Sierra Research, SCF Improvement – Cycle Development, Report SR2003-06-02 (2003)

[SKC05] Syed, F., Kuang, M.L., Czubay, J., Smith, M., Ying, H.: Fuzzy Control to Improve High-Voltage Battery Power and Engine Speed Control in a Hybrid Electric Vehicle. Soft Computing for Real World Applications, NAFIPS, Ann Arbor, MI, June 22-25 (2005)

[SSK02] Schouten, N.J., Salman, M.A., Kheir, N.A.: Fuzzy logic control for parallel hybrid vehicles. IEEE Trans. Contr. Syst. Technol. 10(3), 460–468 (2002)

[TaB00] Tate, E.D., Boyd, S.P.: Finding ultimate limits of performance for hybrid electric vehicles. SAE Paper-01-3099 (2000)

[TRB00] Highway Capacity Manual 2000, Transportation Res. Board, Wash., DC (2000)

[VKK00] De Vlieger, I., De Keukeleere, D., Kretzschmar, J.: Environmental effects of driving behaviors and congestion related to passenger cars. Atmosph. Environ. 34, 4649–4655 (2000)

[Vli99] De Vlieger, I.: Influence of driving behavior on fuel consumption. In: ECODRIVE 7th Conf., September 16-17 (1999)

[WLE02] Won, J.-S., Langari, R., Ehsani, M.: Energy management strategy for a parallel hybrid vehicle. In: Proc. Int. Mechan. Eng. Congress and Exposition (IMECE 2002), New Orleans, LA, IMECE2002–33 460 (November 2002)

[WoL05] Won, J.-S., Langari, R.: Intelligent energy management agent for a parallel hybrid vehicle-part II: torque distribution, charge sustenance strategies, and performance results. IEEE Transactions on Vehicular Technology 54(3), 935–953 (2005)

An Integrated Diagnostic Process for Automotive Systems

Pattipati Krishna[1], Kodali Anuradha[1], Luo Jianhui[3], Choi Kihoon[1], Singh Satnam[1], Sankavaram Chaitanya[1], Mandal Suvasri[1], Donat William[1], Namburu Setu Madhavi[2], Chigusa Shunsuke[2], and Qiao Liu[2]

[1] University of Connecticut, Storrs, CT, 06268
krishna@engr.uconn.edu
[2] Toyota Technical Center USA, 1555 Woodridge Rd., Ann Arbor, MI 48105
[3] Qualtech Systems, Inc., Putnam Park, Suite 603, 100 Great Meadow Road, Wethersfield, CT 06109, USA

1 Introduction

The increased complexity and integration of vehicle systems has resulted in greater difficulty in the identification of malfunction phenomena, especially those related to cross-subsystem failure propagation and thus made system monitoring an inevitable component of future vehicles. Consequently, a continuous monitoring and early warning capability that detects, isolates and estimates size or severity of faults (viz., fault detection and diagnosis), and that relates detected degradations in vehicles to accurate remaining life-time predictions (viz., prognosis) is required to minimize downtime, improve resource management via condition-based maintenance, and minimize operational costs.

The recent advances in sensor technology, remote communication and computational capabilities, and standardized hardware/software interfaces are creating a dramatic shift in the way the health of vehicle systems is monitored and managed. The availability of data (sensor, command, activity and error code logs) collected during nominal and faulty conditions, coupled with intelligent health management techniques, ensure continuous vehicle operation by recognizing anomalies in vehicle behavior, isolating their root causes, and assisting vehicle operators and maintenance personnel in executing appropriate remedial actions to remove the effects of abnormal behavior. There is also an increased trend towards online real-time diagnostic algorithms embedded in the Electronic Control Units (ECUs), with the diagnostic troubleshooting codes (DTCs) that are more elaborate in reducing cross-subsystem fault ambiguities. With the advancements in remote support, the maintenance technician can use an intelligent scanner with optimized and adaptive state-dependent test procedures (e.g., test procedures generated by test sequencing software, e.g. [47],) instead of pre-computed static paper-based decision trees, and detailed maintenance logs ("cases") with diagnostic tests performed, their outcomes, test setups, test times and repair actions can be recorded automatically for adaptive diagnostic knowledge management. If the technician can not isolate the root cause, the history of sensor data and symptoms are transmitted to a technical support center for further refined diagnosis.

The automotive industry has adopted quantitative simulation as a vital tool for a variety of functions, including algorithm design for ECUs, rapid prototyping, programming for hardware-in-the-loop simulations (HILS), production code generation, and process management documentation. Accordingly, fault detection and diagnosis (FDD) and prognosis have mainly evolved upon three major paradigms, viz., model-based, data-driven and knowledge-based approaches.

The model-based approach uses a mathematical representation of the system. This approach is applicable to systems, where satisfactory physics-based models of the system and an adequate number of sensors to observe the state of the system are available. Most applications of model-based diagnostic approach have been on systems with a relatively small number of inputs, outputs, and states. The main advantage of a model-based approach is its ability to incorporate a physical understanding of the process into the process monitoring scheme. However, it is difficult to apply the model-based approach to large-scale systems because it requires detailed analytical models in order to be effective.

A data-driven approach to FDD is preferred when system models are not available, but instead system monitoring data is available. This situation arises frequently when subsystem vendors seek to protect their intellectual property by not providing internal system details to the system or vehicle integrators. In these cases, experimental data

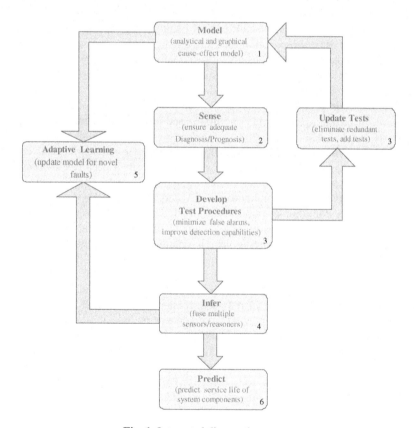

Fig. 1. Integrated diagnostic process

from an operating system or simulated data from a black-box simulator will be the major source of system knowledge for FDD. Neural network and statistical classification methods are illustrative of data-driven techniques. Significant amount of data is needed from monitored variables under nominal and faulty scenarios for data-driven analysis.

The knowledge-based approach uses qualitative models for process monitoring and troubleshooting. The approach is especially well-suited for systems for which detailed mathematical models are not available. Most knowledge-based techniques are based on casual analysis, expert systems, and/or ad hoc rules. Because of the qualitative nature of these models, knowledge-based approaches have been applied to many complex systems. Graphical models such as Petri nets, multi-signal flow graphs and Bayesian networks are applied for diagnostic knowledge representation and inference in automotive systems [34]. Bayesian Networks subsume the deterministic fault diagnosis models embodied in the Petri net and multi-signal models. However, multi-signal models are preferred because they can be applied to large-scale systems with thousands of failure sources and tests, and can include failure probabilities and unreliable tests as part of the inference process in a way which is computationally more efficient than Bayesian networks. Model based, data-driven and knowledge-based approaches provide the "sand box" that test designers can use to experiment with, and systematically select relevant models or combinations thereof to satisfy the requirements on diagnostic accuracy, computational speed, memory, on-line versus off-line diagnosis, and so on. Ironically, no single technique alone can serve as the diagnostic approach for complex automotive applications. Thus, an integrated diagnostic process [41] that naturally employs data-driven techniques, graph-based dependency models and mathematical/physical models is necessary for fault diagnosis, thereby enabling efficient maintenance of these systems.

Integrated diagnostics represents a structured, systems engineering approach and the concomitant information-based architecture for maximizing the economic and functional performance of a system by integrating the individual diagnostic elements of design for testability, on-board diagnostics, automatic testing, manual troubleshooting, training, maintenance aiding, technical information, and adaptation/learning [4][29]. This process, illustrated in Fig. 1, is employed during all stages of a system life cycle, viz., concept, design, development, production, operations, and training. From a design perspective, it has been well-established that a system must be engineered simultaneously with three design goals in mind: performance, ease of maintenance, and reliability [12]. To maximize its impact, these design goals must be considered at all stages of the design: concept to design of subsystems to system integration. Ease of maintenance and reliability are improved by performing testability and reliability analyses at the design stage.

The integrated diagnostic process we advocate contains six major steps: ***model, sense, develop and update test procedures, infer, adaptive learning***, and ***predict***.

(A) Step 1: Model
In this step, models to understand fault-to-error characteristics of system components are developed. This is achieved by a hybrid modeling technique, which combines mathematical models (simulation models), monitored data and graphical cause-effect model (e.g., diagnostic matrix (D-matrix)[34]) in the failure space, through an

understanding of the failure modes and their effects, physical/behavioral models, and statistical and machine learning techniques based on actual failure progression data (e.g., field failure data). The testability analysis tool (e.g., TEAMS [47]) computes percent fault detection and isolation measures, identifies redundant tests and ambiguity groups, and generates updated Failure Modes Effects and Criticality Analysis (FMECA) report [13], and the diagnostic tree [11]. The onboard diagnostic data can also be downloaded to a remote diagnostic server (such as TEAMS - RDS [47]) for interactive diagnosis (by driving interactive electronic technical manuals), diagnostic/maintenance data management, logging and trending. The process can also be integrated with the supply-chain management systems and logistics databases for enterprise-wide vehicle health and asset management.

(B) Step 2: Sense
The sensor suite is typically designed for vehicle control and performance. In this step, the efficacies of these sensors are systematically evaluated and quantified to ensure that adequate diagnosis and prognosis are achievable. If the existing sensors are not adequate for diagnosis/prognosis, use of additional sensors and/or analytical redundancy must be considered without impacting vehicle control and performance. Diagnostic analysis by analysis tools (such as TEAMS [47]) can be used to compare and evaluate alternative sensor placement schemes.

(C) Step 3: Develop and Update Test Procedures
Smart test procedures that detect failures, or onsets thereof, have to be developed. These procedures have to be carefully tuned to minimize false alarms, while improving their detection capability (power of the test and detection delays). The procedures should have the capability to detect trends and degradation, and assess the severity of a failure for early warning.

(D) Step 4: Adaptive Learning
If the observed fault signature does not correspond to faults reflected in the graphical dependency model derived from fault simulation, system identification techniques are invoked to identify new cause-effect relationships to update the model.

(E) Step 5: Infer
An integrated on-board and off-board reasoning system capable of fusing results from multiple sensors/reasoners and driver (or "driver model") to evaluate the health of the vehicle needs to be applied. This reasoning engine and the test procedures have to be compact enough so that they can be embedded in the ECU and/or a diagnostic maintenance computer for real-time maintenance. If on-board diagnostic data is downloaded to a repair station, remote diagnostics is used to provide assistance to repair personnel in rapidly identifying replaceable component(s).

(F) Step 6: Predict (Prognostics)
Algorithms for computing the remaining useful life (RUL) of vehicle components that interface with onboard usage monitoring systems, parts management and supply chain management databases are needed. Model-based prognostic techniques based on singular perturbation methods of control theory, coupled with an interacting multiple model (IMM) estimator [1], provide a systematic method to predict the RUL of system components.

This development process provides a general framework for diagnostic design and implementation for automotive applications. The applications of this process are system specific, and one need not go through all the steps for every system. In this chapter, we focus on fault diagnosis of automotive systems using model-based and data-driven approaches. The above integrated diagnostic process has been successfully applied to automotive diagnosis, including an engine's air intake subsystem (AIS) [35] using model-based techniques and an anti-lock braking system (ABS) [36] using both model-based and data-driven techniques. Data-driven techniques are employed for fault diagnosis on automobile engine data [9][10][38]. The prognostic process is employed to predict the remaining life of an automotive suspension system [37].

2 Model-Based Diagnostic Approach

2.1 Model-Based Diagnostic Techniques

A key assumption of quantitative model-based techniques is that a mathematical model is available to describe the system. Although this approach is complex and needs more computing power, several advantages make it very attractive. The mathematical models are used to estimate the needed variables for analytical (software) redundancy. With the mathematical model, a properly designed detection and diagnostic scheme can be not only robust to unknown system disturbances and noise, but also can estimate the fault size at an early stage. The major techniques for quantitative model-based diagnostic design include parameter estimation, observer-based design and/or parity relations [43][54].

Parity (residual) equations

Parity relations are rearranged forms of the input-output or state-space models of the system [26]. The essential characteristic of this approach is to check for consistency of the inputs and outputs. Under normal operating conditions, the magnitudes of residuals or the values of parity relations are small. To enhance residual-based fault isolation, directional, diagonal and structured residual design schemes are proposed [22]. In the directional residual scheme, the response to each fault is confined to a straight line in the residual space. Directional residuals support fault isolation, if the response directions are independent. In the diagonal scheme, each element of the residual vector responds to only one fault. Diagonal residuals are ideal for the isolation of multiple faults, but they can only handle m faults, where m equals the number of outputs [21]. Structured residuals are designed to respond to different subsets of faults and are insensitive to others not in each subset. Parity equations require less computational effort, but do not provide as much insight into the process as parameter estimation schemes.

Parameter identification approach

The parameter estimation-based method [24][25] not only detects and isolates a fault, but also may estimate its size. A key requirement of this method is that the mathematical model should be identified and validated so that it expresses the physical laws of the system as accurately as possible. If the nominal parameters are not known precisely, they need to be estimated from observed data. Two different parameter identification approaches exist for this purpose.

Equation error method. The parameter estimation approach not only detects and isolates a fault, but also estimate its size, thereby providing FDD as a one-shot process. Equation error methods use the fact that faults in dynamic systems are reflected in the physical parameters, such as the friction, mass, inertia, resistance and so on. Isermann [25] has presented a five-step parameter estimation method for general systems.

(1) Obtain a nominal model of the system relating the measured input and output variables:

$$\underline{y}(t) = f\{\underline{u}(t), \underline{\theta}_0\} \qquad (1)$$

(2) Determine the relationship function $\underline{g}$ between the model parameters $\underline{\theta}$, where underscore notation of the parameters represents a vector, and the physical system coefficients $\underline{p}$:

$$\underline{\theta} = \underline{g}(\underline{p}) \qquad (2)$$

(3) Identify the model parameter vector $\underline{\theta}$ from the measured input and output variables

$$\underline{U}^N = \{\underline{u}(k): 0 \le k \le N\} \text{ and } \underline{Y}^N = \{\underline{y}(k): 0 \le k \le N\} \qquad (3)$$

(4) Calculate the system coefficients (parameters): $\underline{p} = \underline{g}^{-1}(\underline{\theta})$ and deviations from nominal coefficients, $\underline{p}_0 = \underline{g}^{-1}(\underline{\theta}_0)$, viz., $\Delta \underline{p} = \underline{p} - \underline{p}_0$

(5) Diagnose faults by using the relationship between system faults (e.g., short-circuit, open-circuit, performance degradations) and deviations in the coefficients $\Delta \underline{p}$.

Output error (prediction-error) method. For a multiple input-multiple output (MIMO) system, suppose we have collected a batch of data from the system:

$$\underline{Z}^N = [\underline{u}(1), \underline{y}(1), \underline{u}(2), \underline{y}(2), ..., \underline{u}(N), \underline{y}(N)] \qquad (4)$$

Let the output error provided by a certain model parameterized by $\underline{\theta}$ be given by

$$\underline{e}(k, \underline{\theta}) = \underline{y}(k) - \hat{\underline{y}}(k \mid \underline{\theta}) \qquad (5)$$

Let the output-error sequence in Eq. (5) be filtered through a stable filter L and let the filtered output be denoted by $\underline{e}_F(k, \underline{\theta})$. The estimate $\hat{\underline{\theta}}_N$ is then computed by solving the following optimization problem:

$$\hat{\underline{\theta}}_N = \arg \min_{\underline{\theta}} V_N(\underline{\theta}, \underline{Z}^N) \qquad (6)$$

where

$$V_N(\underline{\theta}, \underline{Z}^N) = \frac{1}{N}\sum_{k=1}^{N} \underline{e}_F^T(k,\underline{\theta}) \Sigma^{-1} \underline{e}_F(k,\underline{\theta}) \tag{7}$$

Here Σ is the covariance of error vector. The effect of filter L is akin to frequency weighting [32]. For example, a low-pass filter can suppress high-frequency disturbances. The minimization of Eq. (7) is carried out iteratively. The estimated covariance matrix and the updated parameter estimates at iteration i are

$$\hat{\Sigma}_N^{(i)} = \frac{1}{N-1}\sum_{k=1}^{N} \underline{e}_F(k,\underline{\theta}_N^{(i)}) \underline{e}_F^T(k,\underline{\theta}_N^{(i)})$$

$$\hat{\underline{\theta}}_N^{(i+1)} = \arg\min_{\underline{\theta}} \frac{1}{N}\sum_{k=1}^{N} \underline{e}_F^T(k,\underline{\theta})[\hat{\Sigma}_N^{(i)}]^{-1} \underline{e}_F(k,\underline{\theta}) \tag{8}$$

We can also derive a recursive version for the output-error method. In general, the function $V_N(\underline{\theta}, \underline{Z}^N)$ cannot be minimized by analytical methods; the solution is obtained numerically. The computational effort of this method is substantially higher than the equation error method, and, consequently, on-line real-time implementation may not be achievable.

Observers

The basic idea here is to estimate the states of the system from measured variables. The output estimation error is therefore used as a residual to detect and, possibly, isolate faults. Some examples of the observers are Luenberger observer [52], Kalman filters and Interacting Multiple Models [1], output observers [43][54], nonlinear observers [20][53], to name a few.

In order to introduce the structure of a (generalized) observer, consider a discrete-time, time-invariant, linear dynamic model for the process under consideration in state-space form as follows.

$$\underline{x}(t+1) = A\underline{x}(t) + B\underline{u}(t)$$
$$\underline{y}(t) = C\underline{x}(t) \tag{9}$$

where $\underline{u}(t) \in \Re^r$, $\underline{x}(t) \in \Re^n$ and $\underline{y}(t) \in \Re^m$

Assuming that the system matrices A, B and C are known, an observer is used to reconstruct the system variables based on the measured inputs and outputs $\underline{u}(t)$ and $\underline{y}(t)$:

$$\hat{\underline{x}}(t+1) = A\hat{\underline{x}}(t) + B\underline{u}(t) + H\underline{r}(t)$$
$$\underline{r}(t) = \underline{y}(t) - C\hat{\underline{x}}(t) \tag{10}$$

For the state estimation error $\underline{e}_x(t)$, it follows from Eq. (10) that

$$\underline{e}_x(t) = \underline{x}(t) - \hat{\underline{x}}(t)$$
$$\underline{e}_x(t+1) = (A - HC)\underline{e}_x(t) \tag{11}$$

The state estimation error $\underline{e}_x(t)$, and the residual $\underline{r}(t)) = C\underline{e}_x(t)$ vanish asymptotically

$$\lim_{t \to \infty} \underline{e}_x(t) = \underline{0} \qquad (12)$$

if the observer is stable; this can be achieved by proper design of the observer feedback gain matrix H (provided that the system is detectable). If the process is subjected to parametric faults, such as changes in parameters in $\{A, B\}$, the process behavior becomes

$$\begin{aligned}\underline{x}(t+1) &= (A+\Delta A)\underline{x}(t) + (B+\Delta B)\underline{u}(t) \\ \underline{y}(t) &= C\underline{x}(t)\end{aligned} \qquad (13)$$

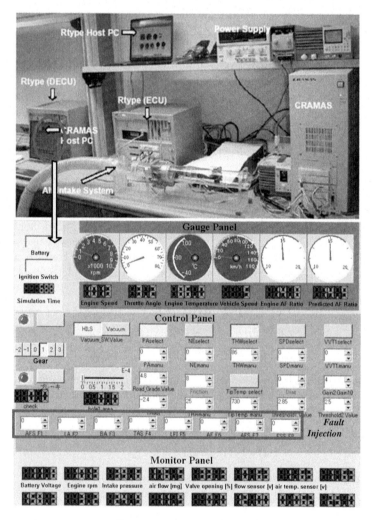

Fig. 2. CRAMAS® Engine Simulation Platform and Operation GUI

Then, the state error $\underline{e}_x(t)$, and the residual $\underline{r}(t)$ are given by

$$\underline{e}_x(t+1) = (A - HC)\underline{e}_x(t) + \Delta A \underline{x}(t) + \Delta B \underline{u}(t)$$
$$\underline{r}(t) = C\underline{e}_x(t) \tag{14}$$

In this case, the changes in residuals depend on the parameter changes, as well as input and state variable changes. The faults are detected and isolated by designing statistical tests on the residuals.

2.2 Application of Model-Based Diagnostics to an Air-Intake System

Experimental set-up: HILS development platform

The hardware for the development platform consists of a custom-built ComputeR Aided Multi-Analysis System (CRAMAS) and two Rapid Prototype ECUs (Rtypes) [19]. The CRAMAS (Fig. 2) is a real-time simulator that enables designers to evaluate the functionality and reliability of their control algorithms installed in ECUs for vehicle sub-systems under simulated conditions, as if they were actually mounted on an automobile. The Rtype is an ECU emulator for experimental research on power train control that achieves extremely high-speed processing and high compatibility with the production ECU [23]. Besides emulating the commercial ECU software, experimental control designs can be carried out in the Rtype host PC using the MATLAB/Simulink environment and compiled through the Real-Time Workshop. Typical model-based techniques include digital filter design to suppress the noise, abrupt change detection techniques (such as the generalized likelihood ratio test (GLRT), cumulative sum (CUSUM), sequential probability ratio test (SPRT)), recursive least squares (RLS) estimation, and output error (nonlinear) estimation for parametric faults, extended Kalman filter (EKF) for parameter and state estimation, Luenberger observer, and the diagnostic inference algorithms (e.g., TEAMS-RT) [2][45][47]). This toolset facilitates validation of model-based diagnostic algorithms.

Combining the Rtype with the CRAMAS, and a HIL Simulator, designers can experiment with different diagnostic techniques, and/or verify their own test designs/

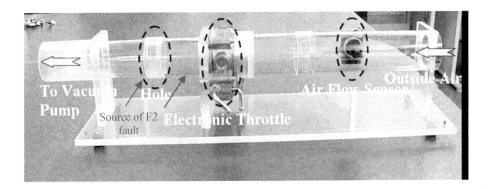

Fig. 3. Photograph of air-intake system

diagnostic inference algorithms, execute simulations, and verify HILS operations. After rough calibration is confirmed, the two Rtypes can also be installed in an actual vehicle, and test drives can be carried out [23]. As a result, it is possible to create high-quality diagnostic algorithms at the initial design stage, thereby significantly shortening the development period ("time-to-market").

The diagnostic experiment employs a prototype air intake subsystem (AIS) as the hardware system in our HILS. The function of AIS is to filter the air, measure the intake air flow, and control the amount of air entering the engine. The reasons for selecting the AIS are its portability and its reasonably accurate physical model. Fig. 3 shows the photograph of our prototype AIS. It consists of a polyvinyl chloride pipe, an air flow sensor, an electronic throttle, and a vacuum pump. It functionally resembles the real AIS for the engine.

The model consists of five primary subsystems: air dynamics, fuel dynamics, torque generation, rotational dynamics, and the exhaust system. We used a mean value model, which captures dynamics on a time-scale spanning over several combustion cycles (without considering in-cycle effects). In the following, we elaborate on the sub-system models. The details of the subsystems and SIMULINK model of air-intake system are available in [35]. Nine faults are considered for this experiment. The air flow sensor fault (F1) is injected by adding 6% of the original sensor measurement. Two physical faults, a leak in the manifold (F2) and a dirty air filter (F3), can be manually injected in the prototype AIS. The leakage fault is injected by adjusting the hole size in the pipe, while the dirty air filter fault is injected by blocking the opening of the pipe located at the right hand side of Fig. 3. The throttle angle sensor fault (F4) is injected by adding 10% of the original sensor measurement. Throttle actuator fault (F5) is injected by adding a pulse to the output of the throttle controller [40]. The pulse lasts for a duration of 3 seconds and the pulse amplitude is 20% of the nominal control signal amplitude. The other faults are modeled using a realistic engine model in CRAMAS, and are injected via the GUI in CRAMAS host PC. Faults F6-F9 injected through the CRAMAS include: less fuel injection (-10%), added

Table 1. Fault list of engine system

F1 [a]	Air flow sensor fault (+6%)
F2 [a]	Leakage in AIS (2cm x 1cm)
F3 [a]	Blockage of air filter
F4 [a]	Throttle angle sensor fault (+10%)
F5 [a]	Throttle actuator fault (+20%)
F6 [b]	Less fuel injection (-10%)
F7 [b]	Added engine friction (+10%)
F8 [b]	AF sensor fault (-10%)
F9 [b]	Engine speed sensor fault (-10%)

[a] real faults in the AIS.
[b] simulated faults in the CRAMAS engine model.

engine friction (+10%), air/fuel sensor fault (-10%), and the engine speed sensor fault (-10%). Here all the % changes are deviations from the nominal values. All the injected faults are listed in Table 1.

Residuals for the engine model

The identified subsystem reference models are used to set up six residuals based on the dynamics of the diagnostic model as follows. The first residual *R1* is based on the difference between the throttle angle sensor reading from the AIS and the predicted throttle angle. The second residual *R2* is based on the difference between air mass flow sensor reading from the AIS and the predicted air flow past the throttle. The third residual *R3* is based on the difference between the engine speed from CRAMAS, and the predicted engine speed. The fourth residual *R4* is based on the difference between the turbine speed from CRAMAS and predicted turbine speed. The fifth residual *R5* is based on the difference between the vehicle speed from CRAMAS, and the predicted vehicle speed. The sixth residual *R6* is obtained as the difference between the air/fuel ratio from CRAMAS, and the predicted air/fuel ratio.

Experimental results

Table 2 shows the diagnostic matrix (D-matrix), which summarizes the test designs for all faults considered in the engine. Each row represents a fault state and columns represent tests. The D-matrix $D = \{ d_{ij} \}$ provides detection information, where d_{ij} is 1 if test *j* detects a fault state *i*. Here, F0 represents "System OK" status, with all the tests passing. Since there are no identical rows in this table, all the faults can be uniquely isolated. For *R2*, there are two tests: *R2_l* and *R2_h*, which correspond to the threshold tests for low (negative) and high (positive) levels. There are also two tests *R6_l* and *R6_h* for *R6* corresponding to low and high levels of the threshold. The other tests have the same name as the residual name (e.g., *R1* is the test for residual *R1*). The goal of using minimum expected cost tests to isolate the faults is achieved by constructing an AND/OR diagnostic tree via an optimal test sequencing algorithm

Table 2. Diagnostic matrix of the engine system

Fault \ Test	R1	R2_h	R2_l	R3	R4	R5	R6_h	R6_l
F0	0	0	0	0	0	0	0	0
F1	0	1	0	0	0	0	0	0
F2	0	0	1	1	0	0	1	0
F3	0	0	1	1	0	0	0	1
F4	0	0	0	1	0	0	0	1
F5	1	0	0	0	0	0	0	0
F6	0	0	0	1	1	1	1	0
F7	0	0	0	1	0	0	0	0
F8	0	0	0	0	0	0	0	1
F9	0	1	0	1	0	0	0	1

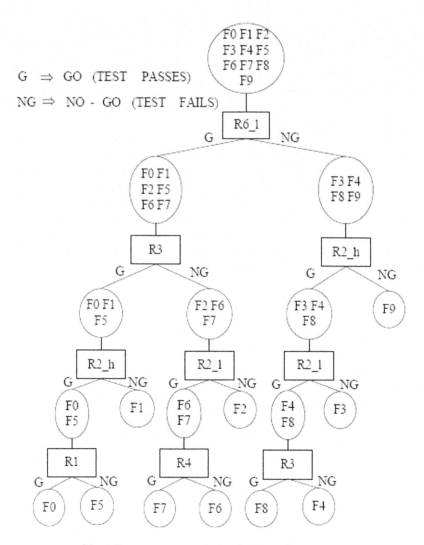

Fig. 4. Test sequence generation for the engine system

[42][48][49]; the tree is shown in Fig. 4, where an AND node represents a test and an OR node denotes the ambiguity group. In this tree, the branch which goes to the left/right below the test means the test passed (G)/failed (NG). It can be seen that tests R5 and R6_h are not shown in the tree, which means that these tests are redundant. This is consistent with the D-matrix, since R4 and R5 have identical columns. One feature of the diagnostic tree is that it shows the set of Go-path tests (R6_l, R3, R2_h and R1) that can respond to any fault. The Go-path tests can be obtained by putting all the tests on the left branches of the tree, which lead to the "system OK" status. Any residual exceeding the threshold(s) will result in a test outcome of 1 (failed). The residuals for throttle position, air mass flow, engine speed, turbine speed, vehicle speed, and air/fuel ratio show the expected behavior. In the fault-free case, the residuals are

almost zero. The results of real-time fault diagnosis are exemplified in Fig. 5 with different faults injected at different times for a particular operating condition of the engine. The noise level of residuals is very low and changes are abrupt under all faulty scenarios. Therefore, simple threshold tests would achieve 100% fault detection and 0% false alarms. Although more advanced tests (such as generalized likelihood ratio test) can be adopted using the fault diagnosis toolset, they provided the same results in this experiment. The diagnostic scheme under other operating conditions (different throttle angles, etc.) was also tested. We found that under nominal conditions, the residuals have the same behavior (near zero).

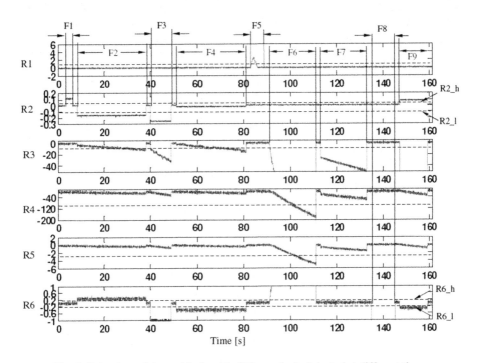

Fig. 5. Behaviors of the residuals with different faults injected at different times

However, under faulty conditions, the deviation of residuals will change as the operating condition changes. Therefore, to obtain the best performance (minimal false alarm and maximum fault detection) of this diagnosis scheme under different operating conditions, a reasonable practice is to use a lookup table for the thresholds on residual tests according to different operating conditions. During the experiment, we found that, for two faults, the engine operating conditions are unstable for large-size faults. The first fault is the air flow sensor fault (F1); the maximum size of the fault is +6%. The second fault is the leakage in air intake manifold; the maximum size of this fault is 4cm × 1cm. These findings are beneficial feedback to control engineers in understanding the limits of their design. If the diagnostic matrix does change under different operating conditions, we can use multi-mode diagnosis techniques to handle this situation [51].

2.3 Model-Based Prognostics

Conventional maintenance strategies, such as corrective and preventive maintenance, are not adequate to fulfill the needs of expensive and high availability transportation and industrial systems. A new strategy based on forecasting of system degradation through a prognostic process is required. The recent advances in model-based design technology facilitate the integration of model-based diagnosis and prognosis of systems, leading to condition-based maintenance to potentially increase the availability of systems. The advantage of model-based prognostics is that, in many situations, the changes in feature vector are closely related to model parameters [7]. Therefore, it can also establish a functional mapping between the drifting parameters and the selected prognostic features. Moreover, as understanding of the system degradation improves, the model can be adapted to increase its accuracy and to address subtle performance problems.

Prognostic methods use residuals as features. The premise is that the residuals are large in the presence of malfunctions, and small in the presence of normal disturbances, noise and modeling errors.

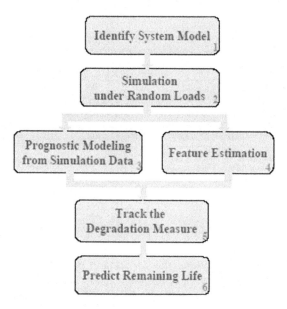

Fig. 6. Model-based prognostic process

Statistical techniques are used to define the thresholds on residuals to detect the presence of faults.

The model-based prognostic process is illustrated in Fig. 6. This process consists of six steps for predicting the remaining life of a system. These are briefly described below.

(A) Step 1: Identify system model
The degradation model of a system is considered to be of the following singular perturbation form:

$$\dot{x} = f(\underline{x}, \lambda(\theta), \underline{u})$$
$$\dot{\theta} = \varepsilon\, g(\underline{x}, \theta) \qquad (15)$$
$$\underline{y} = C\underline{x} + D\underline{u} + \underline{v}$$

Here $\underline{x} \in R^n$ is the set of state variables associated with the fast dynamic behavior of the system; θ is a scalar, slow dynamic variable related to system damage (degradation); $\underline{u} \in R^r$ is the input vector; the drifting parameter λ is a function of θ; the rate constant $0 < \varepsilon \ll 1$ defines the time-scale separation between the fast system state dynamics and the slow drift due to damage [6]; $\underline{y} \in R^m$ is the output vector and $\underline{v} \in R^m$ is the measurement noise. Since ε is very small, Eq. (15) can be considered as a two-time scale system with a slowly drifting parameter.

The modified prognostic model, after scaling the damage variable $\theta\,(\theta_0 \leq \theta \leq \theta_M)$ to a normalized degradation measure $\xi\,(\xi_0 \leq \xi \leq 1)$ [58] and relating the degradation measure from load cycle *(i-1)* to load cycle *i* is:

$$\dot{x} = f(\underline{x}, \lambda(\xi), \underline{u})$$
$$\xi_i = \eta\phi_1(\xi_{i-1})\phi_2(\rho_i) + \xi_{i-1} \qquad (16)$$
$$\underline{y} = C\underline{x} + D\underline{u} + \underline{v}$$

Here ρ_i is the random load parameter during cycle *i* (e.g., stress/strain amplitude). The function $\lambda(\xi)$, which maps the degradation measure to a system parameter, is often assumed to be a polynomial [58]:

$$\lambda(\xi) = \sum_{k=0}^{K} \alpha_k \xi^k \qquad (17)$$

where *K* is the total number of stress levels.

(B) Step 2: Simulation under random loads
The prognostic model is generally nonlinear. Consequently, the evolution of system dynamics (including fast and slow-time) is typically obtained through Monte-Carlo simulations. Since the parameter *p* is a stochastic process in many applications [57], simulation of Eq. (16) for ξ requires the update of degradation parameter ξ for every cycle based on the random load parameter *p* in that cycle, where *p* can be represented as a function of $\underline{x}$. The cycle number n_i and randomly realized load parameter

$\{p_i^j\}_{j=1}^{n_i}$ can be obtained through cycle counting method, viz., the rainflow cycle. This method catches both slow and rapid variations of load by forming cycles that pair high maxima with low minima, even if they are separated by intermediate extremes [28].

Consequently, the updated degradation measure is obtained as

$$\xi_i = \eta \phi_1(\xi_{i-1}) \sum_{j=1}^{n_i} \phi_2(p_i^j) + \xi_{i-1} \tag{18}$$

The initial degradation measure ξ_0 is assumed to be a Gaussian random variable $N(\mu, \sigma^2)$, where μ and σ represent the mean and the standard deviation, respectively.

(C) Step 3: Prognostic modeling
Prognostic modeling is concerned with the dynamics of the degradation measure. Consider a system excited under L different random load conditions as being in modes 1,2,...,L. Assume that M Monte-Carlo simulations are performed for each random load condition. Then the L models can be constructed, one for each mode. The dynamic evolution of degradation measure under mode m is given by:

$$\xi_m(k+1) = \beta_m(\xi_m(k)) + v_m(k) \qquad k = 0,1,... \tag{19}$$

where m is the mode number, β_m is a function of previous state $\xi_m(k)$ and $v_m(k)$ is a zero mean white Gaussian noise with variance $\hat{Q}_m(k)$. The functional form of β_m is obtained from historical/simulated data. The state prediction (function β_m) is obtained in IMM [37] for mode m via:

$$\hat{\xi}_m(k+1/k) = \beta_m(\hat{\xi}_m(k/k)) \qquad k = 0,1,..... \tag{20}$$

Here $\hat{\xi}_m(k/k)$ is the updated (corrected) estimate of ξ_m at time k and $\hat{\xi}_m(k+1/k)$ denotes propagated (predicted) estimate of ξ_m at time $k+1$ based on the measurements up to and including time k.

(D) Step 4: Feature parameter estimation
Since the hidden variable ξ is unobserved, it needs to be estimated from the input/output data $\{\underline{u}(t), \underline{y}(t)\}_{t=0}^{T}$. One way to estimate ξ is to use the update equation for ξ in Eq. (16), where $\phi_2(\rho_i)$ is a function of measurement $\underline{y}$. Since the initial value of ξ_0 is not known, it will produce biased estimates.

Another method is based on estimation of the drifting parameters λ of the fast time process in Eq. (16). Two parameter estimation techniques, equation error method and output error method, can be employed to estimate λ from a time history of measurements $\{\underline{u}(t), \underline{y}(t)\}_{t=0}^{T}$ [33]. Here, the equation error method is employed to estimate λ.

In the equation-error method, the governing equation for estimating λ is the residual equation. The residual equation $r(\underline{u}, \underline{y}, \lambda)$ is the rearranged form of the input-output or state-space model of the system. Suppose N data points of the fast-time process are acquired in an intermediate time interval $[t, t+\alpha T]$, then the optimal parameter estimate is given by:

$$\lambda^* = \arg\min_{\lambda^*} \sum_{i=1}^{N} \|r(u_i, y_i, \lambda)\|^2 \quad (21)$$

Based on the internal structure of the residual equation, two optimization algorithms - linear least squares and nonlinear least squares – can be implemented. If prior knowledge on the range of λ is available, the problem can be solved via constrained optimization. In any case, the measurement equation is constructed as:

$$z(k) = \lambda(\xi(k)) + \vartheta(k) \quad k = 0, 1, 2, ... \quad (22)$$

where $z(k) = \lambda^*$, $\lambda(\xi)$ is typically a polynomial function as in Eq. (17) and $\vartheta(k)$ is a zero mean Gaussian noise with variance $\hat{S}(k)$. The variance is obtained as a by-product of the parameter estimation method.

(E) Step 5: Track the degradation measure
To track the degradation measure, an interacting multiple model (IMM) estimator [1][46] is implemented for online estimation of the damage variable. For a system with L operational modes, there will be L models in the IMM, one for each mode. Each mode will have its own dynamic equation and the measurement equation is of the form in Eq. (22).

(F) Step 6: Predict the remaining life
The remaining life depends on the current damage state $\xi(k)$, as well as the future usage of the system. If the future operation of a system is known *a priori*, the remaining life can be estimated using the knowledge of future usage. Typically, one can consider three types of prior knowledge on future usage.

(1) Deterministic operational sequence: In this case, the system is assumed to be operated according to a known sequence of mode changes and mode durations. Define a sequence $S = \{m_i, T_{si}, T_{ei}\}_{i=1}^{Q}$, where T_{si} and T_{ei} represent the start time and the end time under mode m_i, such that $T_{s1} = 0, T_{ei} = T_{si+1}$ and T_{sQ} is the time at which $\xi = 1$. Suppose M Monte-Carlo simulations are performed for this operational sequence. Then, the M remaining life estimates can be obtained based on $\hat{\xi}(k/k)$, the updated

damage estimate at time instant k. The mean remaining life estimate and its variance are obtained from these M estimates.

(2) Probabilistic operational sequences: In this case, the system is assumed to operate under J operational sequences $S_j = \{m_i^j, T_{si}^j, T_{ei}^j\}_{j=1}^J$, where s_j is assumed to occur with a known probability p_j. If $\hat{r}_j(k)$ is the estimate of residual life based on sequence s_j, then the remaining life estimate $\hat{r}(k)$, and its variance $P(k)$, are given by:

$$\hat{r}(k) = \sum_{j=1}^L p_j(k)\hat{r}_j(k)$$
$$P(k) = \sum_{j=1}^L p_j(k)\left\{P_j(k) + [\hat{r}_j(k) - \hat{r}(k)]^2\right\} \quad (23)$$

(3) On-line sequence estimation: This method estimates the operational sequence based on measured data via IMM mode probabilities. Here, the future operation of the system is assumed to follow the observed history and the dynamics of mode changes. For the ith Monte-Carlo run in mode m, the time to failure until $\xi = 1$ can be calculated as [61]:

$$t_m^i(end) = (\psi_m^i)^{-1} \quad (24)$$

The remaining life estimate from i^{th} Monte-Carlo run for mode m is:

$$\hat{r}_m^i(k) = t_m^i(end) - t_m^i(k) \quad (25)$$

Then, the remaining life estimate and its variance for mode m are:

$$\hat{r}_m(k) = \frac{1}{M}\sum_{i=1}^M \hat{r}_m^i(k)$$
$$P_m(k) = \frac{1}{M-1}\sum_{i=1}^M [\hat{r}_m(k) - \hat{r}_m^i(k)]^2 \quad (26)$$

The above calculation can be performed off-line based on simulated (or historical) data. To reflect the operational history, the mode probabilities from IMM can be used to estimate the remaining life.

2.4 Prognostics of Suspension System

To demonstrate the prognostic algorithms, a simulation study is conducted on an automotive suspension system [35]. The demonstration to estimate the degradation measure will follow the prognostic process discussed above for a half-car two degree of freedom model [64]. Singular perturbation methods of control theory, coupled with

An Integrated Diagnostic Process for Automotive Systems 271

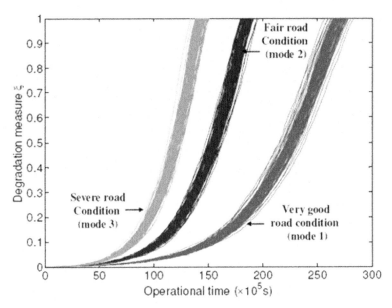

Fig. 7. 100 Monte-Carlo simulations for three random loads

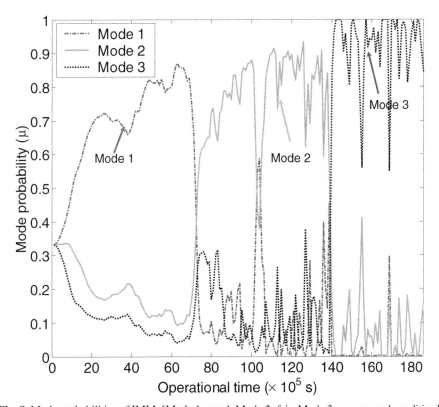

Fig. 8. Mode probabilities of IMM [Mode 1: good, Mode 2: fair, Mode 3: severe road condition]

dynamic state estimation techniques, are employed. An IMM filter is implemented to estimate the degradation measure. The time-averaged mode probabilities are used to predict the remaining life.

The details of the system model are given in [35]. The results of 100 Monte-Carlo simulations for the system under three different road conditions viz., very good, fair, and severe road conditions are presented in Fig. 7. Compared to the severe road condition, the increases in the life times for the fair and very good roads are about 35% and 80%, respectively. If we assume a 10% calendar time usage of the automobile (2.4 hours a day), the expected life of suspension system will be 4.5, 6 and 8 years, respectively, for the three road conditions. Since the suspension system has three random road conditions, the number of modes in the degradation model is 3. IMM may be viewed as a software sensor. It tracks the road condition very well based on noisy data. For IMM implementation, the following transition matrix is used in this scenario:

$$\Phi = \begin{bmatrix} 0.9 & 0.05 & 0.05 \\ 0.05 & 0.9 & 0.05 \\ 0.05 & 0.05 & 0.9 \end{bmatrix}$$

where $\phi_{ij} = P$ (mode j in effect at time $k+1$| mode i in effect at time k). The system mode changes are expected to be as follows. Mode 1 is from 0 to 70×10^5 s, Mode 2 is

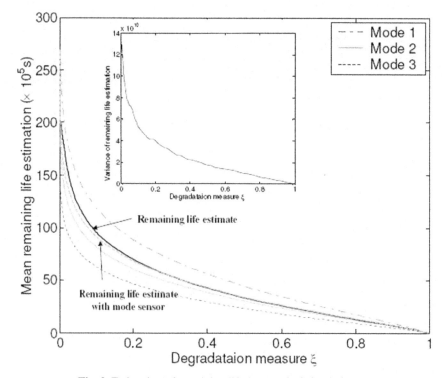

Fig. 9. Estimation of remaining life for a typical simulation run

from 70×10^5 s to 140×10^5 s, Mode 3 is from 140×10^5 s to t_{end}, where t_{end} is the time at which degradation measure $\xi = 1$. Fig. 8 shows the plot of mode probabilities of the IMM. Fig. 9 presents the estimate of remaining life (solid bold line) and its variance for a single run of the scenario considered using the IMM mode probabilities. We can see that the remaining life estimate moves at first to the estimate in between Modes 1 and 2, then gradually approaches the estimate for Mode 2, which is what one would expect. The dashed bold line represents the remaining life estimate assuming that the road surface condition can be measured accurately via a sensor (e.g., an infrared sensor).

In this case, the mode is known, and we can evaluate the accuracy of the remaining life estimate. The dashed bold line represents the remaining life estimate with the mode sensor. In Fig. 9, we can see the IMM produces remaining life estimate close to the estimate of the mode sensor. The difference between these two estimates is relatively high (about 6%) at the beginning ($\xi < 0.1$), and they become virtually identical as degradation measure ξ increases.

3 Data-Driven Diagnostic Approach

Data-driven FDD techniques seek to categorize the input-output data into normal or faulty classes based on training information. Efficient data reduction techniques are employed to handle high-dimensional and/or large volumes of data. Typically, one experiments with a number of classification algorithms and fusion architectures, along with the data reduction techniques, to achieve high diagnostic accuracy. The fusion architectures minimize variability in diagnostic accuracy, and ensure better collective reliability and efficiency when compared with a single classifier.

3.1 Data-Driven Techniques

Data preprocessing and reduction

The sensor data obtained from a system is typically noisy and often incomplete. The data may be continuous or discrete (categorical). A linear trend, a signal mean, or noise in the raw data, outliers and drift can cause errors in the FDD analysis. Hence, it is important to preprocess the data. Data preprocessing involves filtering the data to isolate the signal components, de-trending, removing drift and outliers, smart fill in of missing values, pre-filtering and auto-scaling to adjust scale differences among variables to obtain normalized data (typically zero mean and unit variance), to name a few. In addition, traditional methods of data collection and storage capabilities become untenable mainly because of the increase in the number of variables associated with each observation ("dimension of the data") [16]. Data reduction, an intelligent preprocessing technique, synthesizes a smaller number of features to overcome the "curse of dimensionality". One of the issues with high-dimensional datasets (caused by multiple modes of system operation and sensor data over time) is that all the measurements are not salient for understanding the essential phenomena of interest. The salient features extracted using the data reduction techniques enable real-time implementation of data-driven diagnostic algorithms via compact memory footprint, improved computational efficiency and generally enhanced diagnostic accuracy.

In the data reduction process, the entire data is projected onto a low-dimensional space, and the reduced space often gives information about the important structure of the high-dimensional data space. Feature extraction for data reduction involves signal processing methods, such as wavelets, fast Fourier transforms (FFT) and statistical techniques to extract relevant information for diagnosing faults. Statistical data reduction techniques, such as multi-way partial least squares (MPLS) and multi-way principal component analysis (MPCA), are among the widely investigated techniques. The MPCA is used to reduce the dimensionality of data, and produces a representation that preserves the correlation structures among the monitored variables. The PCA is optimal in terms of capturing the variation in data [8]. The MPLS is another dimensionality reduction technique that considers both pattern (independent data) and class (response) information. MPLS technique is widely used for its ability to enhance classification accuracy on high-dimensional datasets, and its computational efficiency [5]. The reduced data can be processed with classifiers for categorizing the various fault classes.

Multi-way partial least squares (MPLS). In an MPLS technique, the dimensionality of the input and output spaces are transformed to find latent variables, which are most highly correlated with the output, i.e., those that not only explain the variation in the input tensor $X \in R^{I \times J \times K}$, but which are most predictive of output matrix $Y \in R^{I \times M}$. The input tensor X (data samples x sensors x time steps) is decomposed into one set of score vectors (latent variables) $\{\underline{t}_f \in R^I\}_{f=1}^L$, and two sets of weight vectors $\{\underline{w}_f \in R^J\}_{f=1}^L$ and $\{\underline{v}_f \in R^K\}_{f=1}^L$ in the second and third dimensions, respectively [50]. The Y matrix is decomposed into score vectors $\underline{t}$ and loading vectors $\underline{q}$. Formally,

$$x_{ijk} = \sum_{f=1}^L t_{if} w_{jf} v_{kf} + e_{ijk}; 1 \leq i \leq I, 1 \leq j \leq J, 1 \leq k \leq K$$
$$y_{im} = \sum_{f=1}^L t_{if} q_{mf} + u_{im}; 1 \leq i \leq I, 1 \leq m \leq M \qquad (27)$$

where L is the number of factors, J is number of sensor readings, K is time variations, and e_{ijk} and u_{ijk} are residuals. The problem of finding $\underline{t}$, $\underline{w}$, $\underline{v}$, and $\underline{q}$ is accomplished by nonlinear iterative partial least squares (NIPALS) algorithm [63]. This reduced space (score matrix) will be applied to the classifiers, discussed in the following section, for fault isolation.

Classifiers

Data-driven techniques for fault diagnosis have a close relationship with pattern recognition, wherein one seeks to categorize the input-output data into normal or one of several fault classes. Many classification algorithms use supervised learning to develop a discriminant function. This function is used to determine the support for a given category or class, and assigns it to one of a set of discrete classes. Once a classifier is constructed from the training data, it is used to classify test patterns. The pattern classifiers used extensively for automobile fault diagnosis are discussed below.

Support vector machine (SVM). Support vector machine transforms the data to a higher dimensional feature space, and finds an optimal hyperplane that maximizes the margin between two classes via quadratic programming [3][15]. There are two distinct advantages of using the SVM for classification. One is that the features are often associated with the physical meaning of data, so that it is easy to interpret. The second advantage is that it requires only a small amount of training data. A kernel function, typically a radial basis function, is used for feature extraction. An optimal hyperplane is found in the feature space to separate the two classes. In the multi-class case, a hyperplane separating each pair of faults (classes) is found, and the final classification decision is made based on a majority vote among the binary classifiers.

Probabilistic Neural Network (PNN). The probabilistic neural network is a supervised method to estimate the probability distribution function of each class. In the recall mode, these functions are used to estimate the likelihood of an input vector being part of a learned category, or class. The learned patterns can also be weighted, with the *a priori* probability, called the relative frequency, of each category and misclassification costs to determine the most likely class for a given input vector. If the relative frequency of the categories is unknown, then all the categories can be assumed to be equally likely and the determination of category is solely based on the closeness of the input vector to the distribution function of a class. The memory requirements of PNN are substantial. Consequently, data reduction methods are essential in real-world applications.

K-Nearest Neighbor (KNN). The *k*-nearest neighbor classifier is a simple non-parametric method for classification. Despite the simplicity of the algorithm, it performs very well, and is an important benchmark method [15]. The KNN classifier requires a metric d and a positive integer k. A new input vector $\underline{x}_{new}$ is classified using a subset of k–feature vectors that are closest to $\underline{x}_{new}$ with respect to the given metric d. The new input vector $\underline{x}_{new}$ is then assigned to the class that appears most frequently within the k–subset. Ties can be broken by choosing an odd number for k (e.g., 1, 3, 5). Mathematically, this can be viewed as computing a posteriori class probabilities $P(c_i | \underline{x}_{new})$ as,

$$P\left(c_i \mid \underline{x}_{new}\right) = \frac{k_i}{k} p\left(c_i\right) \qquad (28)$$

where k_i is the number of vectors belonging to class c_i within the subset of k nearest vectors. A new input vector $\underline{x}_{new}$ is assigned to the class c_i with the highest posterior probability $P(c_i | \underline{x}_{new})$. The KNN classifier needs to store all previously observed cases, and thus data reduction methods should be employed for computational efficiency.

Principal Component Analysis (PCA). Principal component analysis transforms correlated variables into a smaller number of uncorrelated variables, called principal components. PCA calculates the covariance matrix of the training data and the corresponding eigenvalues and eigenvectors. The eigenvalues are then sorted, and the vectors (called scores) with the highest values are selected to represent the data in a reduced space. The number of principal components is determined by cross-validation [27]. The score vectors from different principal components are the coordinates of the original data sample in the reduced space. A classification of a new test pattern (data sample) is made by obtaining its predicted scores and residuals. If the test pattern is similar to a specific class in the trained classifier, the scores will be located near the origin of the reduced space, and the residual should be small. The distance of test pattern from the origin of the reduced space can be measured by Hotelling statistic [39].

Linear Discriminant Analysis (LD) and Quadratic Discriminant Analysis (QD). Discriminant functions can be related to the class-conditional density functions through Bayes' theorem [44]. The decision rule for minimizing the probability of misclassification may be cast in terms of discriminant functions. Linear discriminant function can be written as

$$g_i(\underline{x}) = \underline{w}_i^T \underline{x} + w_{i0} \qquad (29)$$

where $\underline{w}_i$ and w_{i0} are the weight vector and bias for the ith class, respectively. Decision boundaries corresponding to linear discriminant functions are hyper planes. Quadratic discriminant function can be obtained by adding terms corresponding to the covariance matrix with $c(c+1)/2$ coefficients to produce more complicated separating surfaces [15]. The separating surfaces can be hyperquadratic, hyperspheric, hyperellipsoid, hyperhyperboloid, etc.

Classifier Fusion Techniques

Fusion techniques combine classifier outputs, viz., single class labels or decisions, confidence (or probability) estimates, or class rankings, for higher performance and more reliable diagnostic decisions than a single classifier alone. The accuracy of each classifier in a fusion ensemble is not the same. Thus, the classifier's priority or weight needs to be optimized as part of the fusion architecture for improved performance.

Many fusion techniques are explored in the area of fault diagnosis [10][14]. Some of these are discussed below:

Fusion of Classifier Output Labels. If a classifier's final decision on a test pattern is a single class label, an ensemble of R classifiers provides R discrete output labels. The following algorithms use output labels from each classifier and combine them into a final fused decision.

(1) Majority Voting: The simplest type of fusion, majority (plurality) voting counts votes for each class from the classifiers. The class with the most votes is declared the winner. If a tie exists for the most votes, either it can be broken arbitrarily or a "tie class label" can be assigned. This type of fusion does not require any training or optimized architecture.

(2) Weighted Voting: In weighted voting, a weight calculated during training of the fusion architecture is used to calculate the overall score of each class. The higher the classifier accuracy, the more weight that classifier is given. A score is constructed for each class by using a sum of weighted votes for each class c_i [14]. The class with the highest score is declared the winner.

(3) Naïve Bayes: Classifiers often have very different performance across classes. The confusion matrix of a classifier (derived from training data) contains this information. The entries in a confusion matrix $cm_{k,s}^i$ represent the number of times true class c_k is labeled class c_s by the classifier D_i. Support for each class k on pattern x is developed as [31]:

$$\mu_k(x) \propto \frac{1}{N_k^{L-1}} \left\{ \prod_{i=1}^{L} cm_{k,s_i}^i \right\} \quad (30)$$

Here, N_k is the number of training samples from class c_k, L is the number of classifiers. The class with the highest support is declared as the winner.

Fusion of Classifier Output Ranks. The output of classifiers can be a ranking of the preferences over the C possible output classes. Several techniques operating on this type of output are discussed below.

(1) Borda Count: The ranked votes from each classifier are assigned weights according to their rank. The class ranked first is given a weight of C, the second a weight of (C-1) and so on until a weight of 1 is assigned for the class ranked last. The score for each class is computed as the sum of the class weights from each classifier and the winner is the class with the highest total weight [31].

(2) Ranked Pairs: Ranked Pairs is a voting technique where each voter participates by listing his/her preference of the candidates from the most to the least preferred. In a ranked pair election, the majority preference is sought as opposed to the majority vote or the highest weighted score. That is, we combine the outputs of classifiers to maximize the mutual preference among the classifiers. This approach assumes that voters have a tendency to pick the correct winner [31]. This type of fusion, as in majority voting, does not require any training. If a crisp label is required as a final output, the first position in the ranked vector RV is provided as the final decision.

Fusion of Classifier Posterior Probabilities. The output of a classifier can be an array of confidence estimates or posterior probability estimates. These estimates represent the belief that the pattern belongs to each of the classes. The techniques in this section operate on the values in this array to produce a final fusion label.

(1) Bayesian Fusion: Class-specific Bayesian approach to classifier fusion exploits the fact that different classifiers can be good at classifying different fault classes. The most-likely class is chosen given the test pattern and the training data using the total probability theorem. The posterior probabilities of the test pattern along with the associated posterior probabilities of class c_i from each of the R classifiers obtained during training are used to select the class with the highest posterior probability [10].

(2) Joint Optimization of Fusion Center and of Individual Classifiers: In this technique, the fusion center must decide on the correct class based on its own data and the evidence from the R classifiers. A major result of distributed detection theory (e.g., [44][59][60]) is that the decision rules of the individual classifiers and the fusion center are *coupled*. The decisions of individual classifiers are denoted by $\{u_k\}_{k=1}^{L}$ while

the decision of fusion center by u_0: The classification rule of kth classifier is $u_k = \gamma_k(\underline{x}) \in \{1, 2, ..., C\}$ and that of the fusion center is $u_0 = \gamma_0(u_1, u_2, ..., u_L) \in \{1, 2, ..., C\}$. Let $J(u_0, c_j)$ be the cost of decision u_0 by the committee of classifiers when the true class is c_j. The joint committee strategy of the fusion center along with the classifiers is formulated to minimize the expected cost $E\{J(u_0, c_j)\}$. For computational efficiency, an assumption is made to correlate each classifier only with the best classifier during training to avoid the computation of exponentially increasing entries with the number of classifiers in the joint probability [10]. The decision rule can be written as

$$\gamma_k : u_k = \arg \min_{d_k \in \{1, 2, \cdots, C\}} \sum_{j=1}^{C} \sum_{u_0=1}^{C} P_r(c_j | \underline{x}) \hat{J}(u_0, c_j) \quad (31)$$

where

$$\hat{J}(u_0, c_j) = \sum_{u_0=1}^{C} P(u_0 | \underline{x}, u_k = d_k, c_j) J(u_0, c_j)$$

$$\approx \sum_{u_0=1}^{C} P(u_0 | u_k = d_k, c_j) J(u_0, c_j). \quad (32)$$

Dependence Tree Architectures. We can combine classifiers using a variety of fusion architectures to enhance the diagnostic accuracy [44][59][60]. The class-dependent fusion architectures are developed based on the diagnostic accuracies of individual classifiers on the training data for each class. The classifiers are arranged as a dependence tree to maximize the sum of mutual information be

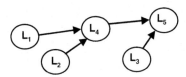

Fig. 10. Generic decision tree architecture

tween all pairs of classifiers [31]. For illustrative purposes, consider Fig. 10, where 5 classifiers are arranged in the form of a tree. Suppose that the classifiers provide class labels $\{L_j\}_{j=1}^{5}$. Then, the support for class c_i is given by:

$$P(\{L_j\}_{j=1}^{5} | c_i) = P(L_5 | c_i) P(L_5 | L_4, c_i) P(L_5 | L_3, c_i) P(L_4 | L_1, c_i) P(L_4 | L_2, c_i) \quad (33)$$

Here, the term $P(L_5 | c_i)$ denotes the probability of label L_5 given the true class c_i from the confusion matrix of classifier 5. The double entries of the form $P(L_k | L_j, c_i)$ represent the output labels of classifiers k and j in the coincidence matrix developed from classifiers k and j on class c_i during training. The final decision corresponds to the class with the highest probability in Eq. (33).

Adaptive Boosting (AdaBoost). AdaBoost [18], short for adaptive boosting, uses the same training set randomly and repeatedly to create an ensemble of classifiers for fusion. This algorithm allows adding weak learners, whose goal is to find a weak hypothesis with small pseudo-loss[1], until a desired low level of training error is achieved. To avoid more complex requirement on the performance of the weak hypothesis, pseudo loss is chosen in place of the prediction error. The pseudo-loss is minimized when correct labels y_i are assigned the value 1, and incorrect labels are assigned the value 0, and it is also calculated with respect to a distribution over all pairs of patterns and incorrect labels. By controlling the distribution, the weak learners can focus on the incorrect labels, thereby hopefully improving the overall performance.

Error-Correcting Output Codes (ECOC). Error-correcting output codes (ECOC) can be used to solve multi-class problems by separating the classes into dichotomies and solving the concomitant binary classification problems, one for each column of the ECOC matrix. The dichotomies are chosen using the principles of orthogonality to ensure maximum separation of rows and columns to enhance the error-correcting properties of the code matrix and to minimize correlated errors of the ensemble, respectively. The maximum number of dichotomies for C classes is $2^{C-1}-1$; however, it is common to use much less than this maximum as in robust design [44]. Each dichotomy is assigned to a binary classifier, which will decide if a pattern belongs to the 0 or 1 group. Three approaches to fuse the dichotomous decisions are discussed below:

(1) Hamming Distance: Using Hamming distance, we compute the number of positions which are different between the row representing a class in the ECOC matrix and the output of the classifier bank. The class which has the minimum distance is declared as the output.

(2) Weighted Voting: Each classifier j detects class i with a different probability. As the multi-class problem is converted into dichotomous classes using ECOC, the weights of each classifier can be expressed in terms of the probability of detection (Pd_j) and the probability of false alarm (Pf_j). These parameters are learned as part of fusion architecture during training. The weighted voting follows the optimum voting rules for binary classifiers [44].

(3) Dynamic fusion: Dynamic fusion architecture, combining ECOC and dynamic inference algorithm for factorial hidden Markov models, accounts for temporal correlations of binary time series data [30][55]. The fusion process involves three steps: the first step transforms the multi-class problem into dichotomies using error correcting output codes (ECOC) and thus solving the concomitant binary classification problems; the second step fuses the outcomes of multiple binary classifiers over time using a sliding-window dynamic fusion method. The dynamic fusion problem is formulated as a maximum *a posteriori* decision problem of inferring the fault sequence based on uncertain binary outcomes of multiple classifiers over time. The resulting problem is solved via a primal-dual optimization framework [56]. The third step optimizes the fusion parameters using a genetic algorithm. The dynamic fusion process is shown in Fig. 11. The probability of detection Pd_j and

---

[1] True loss is non-differentiable and difficult to optimize [3].

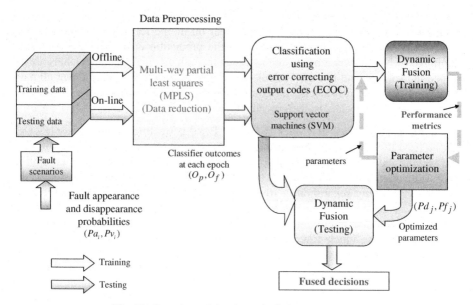

Fig. 11. Overview of the dynamic fusion architecture

false alarm probability Pf_j of each classifier are employed as fusion parameters or the classifier weights; these probabilities are jointly optimized with the dynamic fusion in the fusion architecture, instead of optimizing the parameters of each classifier separately.

This technique allows tradeoff between the size of the sliding window (diagnostic decision delay) and improved accuracy by exploiting the temporal correlations in the data; it is suitable for an on-board application [30]. A special feature of the proposed dynamic fusion architecture is the ability to handle multiple and intermittent faults occurring over time. In addition, the ECOC-based dynamic fusion architecture is an ideal framework to investigate heterogeneous classifier combinations that employ data-driven (e.g., support vector machines, probabilistic neural networks), knowledge-based (e.g., TEAMS-RT [47]), and model-based classifiers (e.g., parity relation-based or observer-based) for the columns of the ECOC matrix.

Fault Severity Estimation

Fault severity estimation is performed by regression techniques, such as the partial least squares (PLS), SVM regression (SVMR), and principal component regression (PCR) in a manner similar to their classification counterparts. After a fault is isolated, we train with the training patterns from the isolated class using the associated severity levels as the targets (Y), i.e., we train the fault severity estimator for each class. Pre-classified test patterns are presented to the corresponding estimator, and the estimated severity levels are obtained [9].

3.2 Application of Data-Driven Techniques

We consider the CRAMAS® engine data considered earlier, but now from a data-driven viewpoint[2]. A 5x2 cross-validation[3] is used to assess the classification performance of various data-driven techniques.

Table 4. Data Driven Classification and Fusion Results on CRAMAS® Engine Data

CRAMAS®	Method	Classification Error ± Std Dev in %					
		SVM	KNN(k=1)	PNN	PCA	LD	QD
Raw Data (25.6MB)	Individual Classification	8.8 ± 2.5	12.9 ± 2.2	14.8 ± 2.1	22.5 ± 2.3	N/A	N/A
Reduced Data via MPLS (12.8 KB)	Individual Classification	8.2 ± 2.5	12.8 ± 2.1	14.1 ± 2.1	21.1 ± 3.7	33.1 ± 3.2	16.3 ± 2.3
	Tandem (Serial) Fusion	15.87 ± 2.49					
	Fusion Center (Parallel)	14.81 ± 3.46					
	Majority Voting	12.06 ± 1.89					
	Naïve Bayes	11.81 ± 1.96					
	ECOC fusion with hamming distance	9.0 ± 2.85					
	Adaboost	7.625 ± 2.14					
	Bayesian Fusion	6.25 ± 2.29					
	Joint Optimization with Majority Voting	5.87 ± 2.04					
	Dynamic Fusion	4.5 ± 1.6					

The diagnostic results, measured in terms of classification errors, with ± representing standard deviations over 5x2 cross validation experiments, are shown in Table 4. We achieved not only smaller fault isolation error, but also significant data reduction (25.6 MB → 12.8 KB for the size of training and testing data). The proposed approaches are mainly evaluated on the reduced data. The Bayesian and dynamic fusion outperformed majority voting, naïve Bayes techniques and serial and parallel fusion approaches. We are able to further improve classification performance

---

[2] The throttle actuator fault F5 is not considered in the data driven approach. HILS data was available only for the remaining eight faults.
[3] A special case of cross validation where the data is divided into two halves, one for training and other for testing. Next time the sets are reversed. This process is repeated for 5 times for a total of 10 training and test sets.

of joint optimization by applying majority voting after getting decisions from the joint optimization algorithm. Posterior probabilities from PNN, KNN (*k*=3), and PCA are fed into the joint optimization algorithm, and then SVM and KNN (*k*=1) are used for majority voting with decisions from the joint optimization algorithm. Majority voting alone provided poor isolation results, which means that the joint optimization approach is definitely a contributor to the increased accuracy. We believe that this is because the joint optimization of fusion center and individual classifiers increases the diversity of the classifier outputs, which is a vital requirement for reducing the diagnostic errors.

For the dynamic fusion approach, we employ SVM as the base classifier for all the columns of the ECOC matrix. This approach achieves low isolation errors as compared to single classifier results. We experimented with two different approaches for *Pd* and *Pf* in dynamic fusion process. The first approach used *Pd* and *Pf* learned from the training data, while coarse optimization is applied to learn *Pd* and *Pf*, and the optimal parameters are *Pd* = 0.5~0.6 and *Pf* = 0~0.02. We found that the dynamic fusion approach involving the parameter optimization reduces diagnostic errors to about 4.5%. Dynamic fusion with parameter optimization is superior to all other approaches considered in this analysis.

Table 5. Comparison of Severity Estimation Performance on Raw and Reduced Data

Fault	Average Error, 100% x (true severity level - its estimate)/true level					
	PLS		SVMR		PCR	
	Raw	Reduced	Raw	Reduced	Raw	Reduced
Air Flow Sensor Fault (F1)	-66.88%	+4.02%	-9.21%	-6.14%	+23.13%	+1.06 %
Leakage in Air Intake System (F2)	-10.11%	+0.76%	-0.20%	-0.72%	-11.22%	+0.75%
Blockage of Air Filter (F3)	-75.55%	+6.42%	+1.37%	+0.75%	-44.20%	+6.38%
Throttle Angle Sensor Fault (F4)	+0.63%	-1.28%	-1.19%	+1.31%	+5.51%	-0.35%
Less Fuel Injection (F6)	-73.42%	-30.92%	+8.04%	+6.77%	-51.36%	-28.60%
Added Engine Friction (F7)	+23.38%	+1.43%	+4.84%	+6.97%	+27.20%	+1.73%
Air/Fuel Sensor Fault (F8)	+36.32%	+0.40%	-2.01%	-2.90%	-26.28%	-0.16%
Engine Speed Sensor Fault (F9)	-7.14%	+10.46%	-25.19%	-26.23%	-1.55%	-3.08%
Overall % of Error	-21.60%	-1.09%	-2.94%	-2.52%	-9.85%	-2.78%

The severity estimation results for raw data and reduced data are shown in Table 5. For training and testing, we randomly selected 60% for training (24 levels for each class) and 40% for testing (16 levels for each class). Relative errors in % are averaged for 16 severity levels in Table 5. We have applied three different estimators, PLS, SVMR, and PCR. Large errors with the raw data can be attributed to ill-conditioning of the parameter estimation problem due to collinearity of data when compared to the reduced data. It is evident that faults 1, 3, and 6 provided poor estimation performance on raw data due to difficulties in estimating low severity levels. However, significant performance improvement can be observed when the estimators are applied to the reduced data. PLS is slightly better than SVMR and PCR in terms of severity estimation performance and provides good estimation results for high severity levels, although estimating low severity levels remains a problem. In all cases, SVMR and PCR are comparable to the PLS in terms of fault severity estimation performance. It is also observed that our techniques perform better on the reduced dataset in terms of severity estimation accuracy.

In addition to individual classifiers, such as the SVM, PNN, KNN, and PCA for fault isolation, posterior probabilities from these classifiers can be fused by the novel Bayesian fusion, joint optimization of fusion and individual classifiers, and dynamic fusion approaches. Our results confirm that fusing individual classifiers can increase the diagnostic performance substantially and that fusion reduces variability in diagnostic classifier performance. In addition, regression techniques such as the PLS, SVMR and PCR estimate the severity of the isolated faults very well when the data is transformed into a low-dimensional space to reduce noise effects.

4 Hybrid Model-Based and Data-Driven Diagnosis

Due to the very diverse nature of faults and modeling uncertainty, no single approach is perfect on all problems (no-free-lunch theorem). Consequently, a hybrid approach that combines model-based and data-driven techniques may be necessary to obtain the required diagnostic performance in complex automotive applications. Here, we present an application involving fault diagnosis in an anti-lock braking system (ABS) [36], where we integrated model and data-driven diagnostic schemes. Specifically, we combined parity equations, nonlinear observer, and SVM to diagnose faults in an ABS. This integrated approach is necessary since neither model-based nor data-driven strategy could adequately solve the entire ABS diagnosis problem, i.e., isolate faults with sufficient accuracy.

4.1 Application of Hybrid Diagnosis Process

We consider longitudinal braking with no steering, and neglect the effects of pitch and roll. The model considers the wheel speed and vehicle speed as measured variables, and the force applied to the brake pedal as the input. The wheel speed is directly measured and vehicle speed can be calculated by integrating the measured acceleration signals, as in [62]. Further details of the model are found in [36]. One commonly occurring sensor fault and four parametric faults are considered for diagnosis in the ABS system. In the case of a wheel speed sensor fault, the sensor systematically misses the detection of teeth in the wheel due to incorrect wheel speed sensor gap caused by loose wheel bearings or worn parts. In order to model the wheel speed sensor fault (F1), we consider two fault severity cases: greater than 0 but less than 5% reduction in the nominal wheel speed (F1.1), and greater than 5% reduction in the nominal wheel speed (F1.2). The four parametric faults (F2-F5) are changes in radius of the wheel (R_w), torque gain (K_f), rotating inertia of the wheel (I_w) and the time constant of the Master Cylinder (τ_m). Fault F2 is the tire pressure fault, F3 and F5 correspond to cylinder faults, while F4 is related to vehicle body. Faults corresponding to more than 2% decrease in R_w are considered. We distinguish among two R_w

faults: greater than 2% but less than 20% (F2.1) decrease in R_w, and greater than 20% decrease in R_w (F2.2). The severities or sizes for K_f and I_w faults considered are as follows: ±2%, ± 3%, ..., ±10%. The size for τ_m fault corresponds to a more than 15% increase in the time constant. Table 6 shows the list of considered faults. The minimum fault magnitude is selected such that changes in the residual signals can not be detected if we choose fault magnitude less than this minimum. The measurement variables for vehicle and wheel speed are corrupted by the zero mean white noise with

Table 6. Simulated fault list of ABS system

F1.1	sensor fault (< 5% decrease)
F1.2	sensor fault (≥ 5% decrease)
F2.1	R_w fault (< 20% decrease)
F2.2	R_w fault (≥ 20% decrease)
F3	K_f fault (±2% ~ ±10%)
F4	I_w fault (±2% ~ ±10%)
F5	τ_m fault (≥ 15% increase)

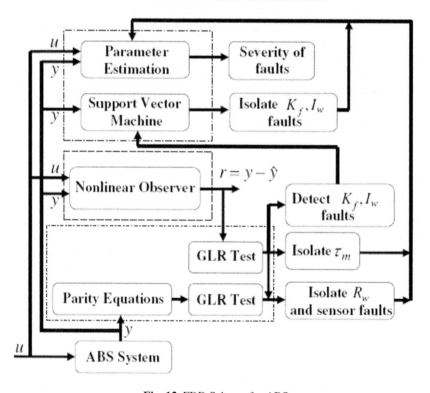

Fig. 12. FDD Scheme for ABS

variances of 0.004 each. The process noise variables are also white with variance of 0.5% of the mean square values of the corresponding states (which corresponds to a signal-to-noise ratio of +23db).

A small amount of process noise is added based on the fact that these states are driven by disturbances from combustion processes in the engine (un-modeled dynamics of wheel and vehicle speeds), and non-linear effects in the ABS actuator (for brake torque and oil pressure).

Fig. 12 shows the block diagram of our proposed FDD scheme for the ABS. The parity equations and GLRT test (G_P_1) are used to detect severe Rw ($\geq$ 20%) and wheel speed sensor ($\geq$ 5%) faults. Then, a nonlinear observer [17][36] is used to generate two additional residuals. The GLRTs based on these two residuals (G_O_1 and G_O_2) and their time dependent GLRT test ($G\_O\_T_1$ and $G\_O\_T_2$) are used to isolate the τ_m fault, less severe (small) Rw and sensor faults. They are also used to detect K_f and I_w faults. Finally, we use the SVM to isolate the K_f and I_w faults. After training, a total of 35 patterns are misclassified in the test data, which results in an error rate of 4.7%. We designed two tests S_K_f and S_I_w using the SVM, which assigns $S\_K_f = 1$ when the data is classified as the K_f fault or assigns $S\_I_w = 1$ when the data is classified as the I_w fault. The diagnostic matrix of the ABS system is shown in Table 7. With the subset of tests, all the faults considered here can be detected. Subsequently, a parameter estimation technique is used after fault isolation to estimate the severity of the fault. After parametric faults are isolated, an output error method is used to estimate the severity of isolated faults. In the ABS, the nonlinear output error parameter estimation method produces biased estimates when all the parameters are estimated as a block. Therefore, the subset parameter estimation techniques are well suited for our application. The subset of parameters to be estimated is chosen by the detection and isolation of the parametric fault using the GLRT and SVM. When a parametric fault is isolated, this parameter is estimated via the nonlinear output error method. Table 8 compares the accuracies of parameter estimation averaged over 20 runs via the two methods: estimating all the parameters versus reduced (one-at-a-time) parameter estimation after fault detection and isolation. The parameters *err* and *std* shows the mean relative errors and standard deviations of the estimated parameters, respectively, normalized by their "true" values (in %).

Table 7. Diagnostic matrix for ABS test design

Fault \ Test	G_P_1	G_O_1	G_O_2	$G\_O\_T_1$	$G\_O\_T_2$	S_K_f	S_I_w
F0	0	0	0	0	0	0	0
F1.1	0	1	0	0	0	0	0
F1.2	0	0	1	1	0	0	1
F2.1	0	0	1	1	0	0	0
F2.2	0	0	0	1	0	0	0
F3	1	0	0	0	0	0	0
F4	0	0	0	1	1	1	1
F5	0	0	0	1	0	0	0

Table 8. Mean relative errors and normalized standard deviations in parameter estimation

		\multicolumn{4}{c}{Block Estimation}	Subset Parameter
		R_w	K_f	I_w	τ_m	Estimation
K_f	err	3.2	5.0	6.0	25.0	1.05
K_f	std	1.2	3.5	6.8	22.2	0.12
I_w	err	2.0	4.5	4.0	19.0	0.52
I_w	std	1.6	4.8	7.2	39.3	0.35
τ_m	err	3.5	7.8	10.3	27.5	2.0
τ_m	std	2.4	5.2	5.6	46.5	0.80
R_w	err	0.39	0.33	2.98	279.33	0.004
R_w	std	0.25	0.12	1.48	33.4	0.014

$$err = \frac{\text{mean relative error}}{\text{"true" value}} \times 100\%$$

$$std = \frac{\text{standard deviation of estimated parameters}}{\text{"true" value}} \times 100\%$$

From Table 8, it is evident that subset parameter estimation provides much more precise estimates than the method which estimates all four parameters as a block. This is especially significant with single parameter faults.

5 Summary and Future Research

This chapter addressed an integrated diagnostic development process for automotive systems. This process can be employed during all stages of a system life cycle, viz., concept, design, development, production, operations, and training of technicians to ensure ease of maintenance and high reliability of vehicle systems by performing testability and reliability analyses at the design stage. The diagnostic design process employs both model-based and data-driven diagnostic techniques. The test designers can experiment with a combination of these techniques that are appropriate for a given system, and trade-off several performance evaluation criteria: detection speed, detection and isolation accuracy, computational efficiency, on-line/off-line implementation, repair strategies, time-based versus preventive versus condition-based maintenance of vehicle components, and so on. The use of condition-based maintenance, on-line system health monitoring and smart diagnostics and reconfiguration/self-healing/repair strategies will help minimize downtime, improve resource management, and minimize operational costs. The integrated diagnostics process promises a major economic impact, especially when implemented effectively across an enterprise.

In addition to extensive applications of the integrated diagnostics process to real-world systems, there are a number of research areas that deserve further attention.

These include: dynamic tracking of the evolution of degraded system states (the so-called "gray-scale diagnosis"), developing rigorous analytical framework for combining model-based and data-driven approaches for adaptive knowledge bases, adaptive inference, agent-based architectures for distributed diagnostics and prognostics, use of diagnostic information for reconfigurable control, and linking the integrated diagnostic process to supply chain management processes for effective parts management.

References

1. Bar-Shalom, Y., Li, X.R., Kirubarajan, T.: Estimation with applications to tracking and navigation. John Wiley and Sons, Inc., Chichester (2001)
2. Basseville, M., Nikiforov, IV.: Detection of abrupt changes. Prentice-Hall Inc., New Jersey (1993)
3. Bishop, C.M.: Pattern Recognition and Machine Learning. Springer, Heidelberg (2006)
4. Bohr, J.: Open Systems Approach - Integrated Diagnostics Demonstration Program. In: NDIA Systems Engineering and Supportability Conference and Workshop (1998), http://www.dtic.mil/ndia/support/bohr.pdf
5. Bro, R.: Multiway Calibration. Multilinear PLS. Journal of Chemometrics 10, 47–61 (1996)
6. Chelidze, D.: Multimode damage tracking and failure prognosis in electro mechanical system. In: SPIE Conference Proceedings, pp. 1–12 (2002)
7. Chelidze, D., Cusumano, J.P., Chatterjee, A.: Dynamical systems approach to damage evolution tracking, part I: The experimental method. Journal of Vibration and Acoustics 124, 250–257 (2002)
8. Chen, J., Liu, K.: On-line batch process monitoring using dynamic PCA and dynamic PLS models. Chemical Engineering Science 57, 63–75 (2002)
9. Choi, K., Luo, J., Pattipati, K., Namburu, M., Qiao, L., Chigusa, S.: Data reduction techniques for intelligent fault diagnosis in automotive systems. In: Proc. IEEE AUTOTESTCON, Anaheim, CA, pp. 66–72 (2006)
10. Choi, K., Singh, S., Kodali, A., Pattipati, K., Namburu, M., Chigusa, S., Qiao L.: A novel Bayesian approach to classifier fusion for fault diagnosis in automotive systems. In: Proc. IEEE AUTOTESTCON, Baltimore, MD, pp. 260–269 (2007)
11. Deb, S., Pattipati, K., Raghavan, V., Shakeri, M., Shrestha, R.: Multi-signal Flow Graphs: A Novel Approach for System Testability Analysis and Fault Diagnosis. In: IEEE Aerospace and Electronics Magazine, pp. 14–25 (1995)
12. Deb, S., Pattipati, K., Shrestha, R.: QSI's Integrate Diagnostics Toolset. In: Proc. of the IEEE AUTOTESTCON, Anaheim, CA, pp. 408–421 (1997)
13. Deb, S., Ghoshal, S., Mathur, A., Pattipati, K.: Multi-signal Modeling for Diagnosis, FMECA and Reliability. In: IEEE Systems, Man, and Cybernetics Conference, San Diego, CA (1998)
14. Donat, W.: Data Visualization, Data Reduction, and Classifier Output Fusion for Intelligent Fault Detection and Diagnosis. M.S Thesis, University of Connecticut (2007)
15. Duda, R.O., Hart, P.E., Stork, D.G.: Pattern Classification, 2nd edn. John Wiley and Sons, Chichester (2001)
16. Fodor, K.: A survey of dimension reduction techniques, http://www.llnl.gov/CASC/sapphire/pubs/148494.pdf
17. Frank, P.M.: On-line fault detection in uncertain nonlinear systems using diagnostic observers: a survey. International Journal of System Science 25, 2129–2154 (1994)

18. Freund, Y., Schapire, R.E.: Experiments with a new boosting algorithm. Machine Learning. In: Proc. of the Thirteenth Inter. Conf. (1996)
19. Fukazawa, M.: Development of PC-based HIL simulator CRAMAS 2001. FUJITSU TEN Technical Journal 19, 12–21 (2001)
20. Garcia, E.A., Frank, P.: Deterministic nonlinear observer based approaches to fault diagnosis: a survey. Control Engineering Practice 5, 663–670 (1997)
21. Gertler, J.: Fault detection and isolation using parity relations. Control Eng. Practice. 5, 1385–1392 (1995)
22. Gertler, J., Monajmey, R.: Generating directional residuals with dynamic parity relations. Automatica 33, 627–635 (1995)
23. Higuchi, T., Kanou, K., Imada, S., Kimura, S., Tarumoto, T.: Development of rapid prototype ECU for power train control. FUJITSU TEN Technical Journal 20, 41–46 (2003)
24. Isermann, R.: Process fault detection based on modeling and estimation methods: a survey. Automatica 20, 387–404 (1984)
25. Isermann, R.: Fault diagnosis of machines via parameter estimation and knowledge processing-tutorial paper. Automatica 29, 815–835 (1993)
26. Isermann, R.: Supervision, fault-detection and fault-diagnosis methods – an introduction. Control Eng. Practice 5, 639–652 (1997)
27. Jackson, J.E.: A User's Guide to Principal Components. John Wiley & Sons, New York (1991)
28. Johannesson: Rainflow cycles for switching processes with Markov structure. Probability in the Engineering and Informational Sciences 12, 143–175 (1998)
29. Keiner, W.: A Navy Approach to Integrated Diagnostics. In: Proc. of the IEEE AUTOTESTCON, pp. 443–450 (1990)
30. Kodali, A., Donat, W., Singh, S., Choi, K., Pattipati, K.: Dynamic fusion and parameter optimization of multiple classifier systems. In: Proceedings of GT 2008, Turbo Expo 2008, Berlin, Germany (2008)
31. Kuncheva, L.I.: Combining Pattern Classifiers. John Wiley, Chichester (2004)
32. Ljung, L.: System identification: theory for the user. Prentice-Hall, Englewood Cliffs (1987)
33. Luo, J., Tu, F., Azam, M., Pattipati, K., Qiao, L., Kawamoto, M.: Intelligent model-based diagnostics for vehicle health management. In: Proc. SPIE Conference, Orlando, pp. 13–26 (2003)
34. Luo, J., Tu, H., Pattipati, K., Qiao, L., Chigusa, S.: Graphical models for diagnostic knowledge representation and inference. IEEE Instrument and Measurement Magazine 9, 45–52 (2006)
35. Luo, J., Pattipati, K., Qiao, L., Chigusa, S.: An integrated diagnostic development process for automotive engine control systems. IEEE Transactions on Systems, Man, and Cybernetics: Part C – Applications and Reviews 37, 1163–1173 (2007)
36. Luo, J., Namburu, M., Pattipati, K., Qiao, L., Chigusa, S.: Integrated model-based and data-driven diagnosis of automotive anti-lock braking systems. IEEE System, Man, and Cybernetics – Part A: Systems and Humans (to appear)
37. Luo, J., Pattipati, K., Qiao, L., Chigusa, S.: Model-based Prognostic Techniques Applied to a Suspension System. IEEE Transactions on Systems, Man, and Cybernetics – Part C: Applications and Reviews (to appear)
38. Namburu, M.: Model-based and data-driven techniques and their application to fault detection and diagnosis in engineering systems and information retrieval. M.S Thesis, University of Connecticut (2006)

39. Nomikos, P.: Detection and Diagnosis of Abnormal Batch Operations Based on Multi-way Principal Component Analysis. ISA Transactions 35, 259–266 (1996)
40. Nyberg, M., Nielsen, L.: Model based diagnosis for the air intake system of the SI-engine. SAE Transactions, Journal of Commercial Vehicles 106, 9–20 (1997)
41. Pattipati, K.: Combinatorial Optimization Algorithms for Fault Diagnosis in Complex Systems. In: International Workshop on IT-enabled Manufacturing, Logistics and Supply Chain Management, Bangalore, India (2003)
42. Pattipati, K., Alexandridis, M.: Application of heuristic search and information theory to sequential fault diagnosis. IEEE Transactions on Systems, Man, and Cybernetics–Part A 20, 872–887 (1990)
43. Patton, R.J., Frank, P.M., Clark, R.N.: Issues of fault diagnosis for dynamic systems. Springer, London (2000)
44. Pete, A., Pattipati, K., Kleinman, D.L.: Optimization of Detection Networks with Generalized Event Structures. IEEE Transactions on Automatic Control, 1702–1707 (1994)
45. Phadke, M.S.: Quality Engineering Using Robust Design. Prentice-Hall, Englewood Cliffs (1989)
46. Phelps, E., Willett, P. (2002) Useful lifetime tracking via the IMM. In: SPIE Conference Proceedings, pp. 145–156 (2002)
47. QSI website, http://www.teamsqsi.com
48. Raghavan, V., Shakeri, M., Pattipati, K.: Test sequencing algorithms with unreliable tests. IEEE Transactions on Systems, Man, and Cybernetics–Part A 29, 347–357 (1999)
49. Raghavan, V., Shakeri, M., Pattipati, K.: Optimal and near-optimal test sequencing algorithms with realistic test models. IEEE Transactions on Systems, Man, and Cybernetics–Part A 29, 11–26 (1999)
50. Rasmus, B.: Multiway calibration. Multilinear PLS. Journal of Chemometrics 10, 259–266 (1996)
51. Ruan, S., Tu, F., Pattipati, K., Patterson-Hine, A.: On a multimode test sequencing problem. IEEE Transactions on Systems, Man and Cybernetics–Part B 34, 1490–1499 (2004)
52. Schroder, D.: Intelligent observer and control design for nonlinear systems. Springer, Berlin (2000)
53. Shakeri, M.: Advances in System Fault Modeling and Diagnosis. Ph.D Thesis, University of Connecticut (1998)
54. Simani, S., Fantuzzi, C., Patton, R.J.: Model-based fault diagnosis in dynamic systems using identification techniques. Springer, London (2003)
55. Singh, S., Choi, K., Kodali, A., Pattipati, K., Namburu, M., Chigusa, S., Qiao, L.: Dynamic classifier fusion in automotive systems. In: IEEE SMC Conference, Montreal, Canada (2007)
56. Singh, S., Kodali, A., Choi, K., Pattipati, K., Namburu, M., Chigusa, S., Prokhorov, D.V., Qiao, L.: Dynamic multiple fault diagnosis: mathematical formulations and solution techniques. IEEE Trans. on SMC- Part A (to appear, 2008)
57. Sobczyk, K., Spencer, B.: Random fatigue: from data to theory. Academic Press Inc., San Diego (1993)
58. Sobczyk, K., Trebicki, J.: Stochastic dynamics with fatigue induced stiffness degradation. Probabilistic Engineering Mechanics 15, 91–99 (2000)
59. Tang, Z.B., Pattipati, K., Kleinman, D.L.: Optimization of Detection Networks: Part I - Tandem Structures. IEEE Transactions on Systems, Man, and Cybernetics: Special issue on Distributed Sensor Networks 21, 1045–1059 (1991)

60. Tang, Z.B., Pattipati, K., Kleinman, D.L.: A Distributed M-ary Hypothesis Testing Problem with Correlated Observations. IEEE Transactions on Automatic Control, 1042–1046 (1992)
61. Terry, B., Lee, S.: What is the prognosis on your maintenance program. Engineering and Mining Journal 196, 32 (1995)
62. Unsal, C., Kachroo, P.: Sliding mode measurement feedback control for antilock braking system. IEEE Transactions on Control Systems Technology 7, 271–281 (1999)
63. Wold, S., Geladi, P., Esbensen, K., Ohman, J.: Principal component analysis. Chemometrics and Intell. Lab. Sys. 2, 37–52 (1987)
64. Yoshimura, T., Nakaminami, K., Kurimoto, M., Hino, J.: Active suspension of passenger cars using linear and fuzzy logic controls. Control Engineering Practice 41, 41–47 (1999)

Automotive Manufacturing: Intelligent Resistance Welding*

Mahmoud El-Banna[1,**], Dimitar Filev[2], and Ratna Babu Chinnam[3]

[1] University of Jordan, Amman 11942, Jordan
m.albanna@ju.edu.jo
[2] Ford Motor Company, Dearborn, MI 48121, USA
dfilev@ford.com
[3] Wayne State University, Detroit, MI 48202, USA
r_chinnam@wayne.edu

1 Introduction

Resistance spot welding (RSW) is an important process in the automotive industry. The advantages of spot welding are many: an economical process, adaptable to a wide variety of materials (including low carbon steel, coated steels, stainless steel, aluminum, nickel, titanium, and copper alloys) and thicknesses, a process with short cycle times, and overall, a relatively robust process with some tolerance to fit-up variations. Although used in mass production for several decades, RSW poses several major problems, most notably, large variation in weld quality. Given the variation and uncertainty in weld quality (attributed to factors such as tip wear, sheet metal surface debris, and fluctuations in power supply), it is a common practice in industry to add a significant number of redundant welds to gain confidence in the structural integrity of the welded assembly [1]. In recent years, global competition for improved productivity and reduced non-value added activity, is forcing automotive OEMs and others to eliminate these redundant spot welds. The emphasis on reduction of the redundant welds significantly increases the need for monitoring of weld quality and minimizing weld process variability. Traditionally, destructive and nondestructive tests for weld quality evaluation are predominantly off-line or end-of-line processes. While this test information is useful and valuable for quality and process monitoring, it cannot be utilized in process control because of the significant delays that are associated with the off-line test analysis. In order to minimize the number of spot welds and still satisfy essential factors such as strength and surface integrity, weld quality has to be monitored and controlled in real-time. Advances over the last decade in the area of non-intrusive electronic sensors, signal processing algorithms, and computational intelligence, coupled with drastic reductions in computing and networking hardware costs, have now made it possible to develop non-intrusive intelligent resistance welding systems that overcome the above shortcomings.

---

* A short version of this paper was presented at the 2006 IEEE World Congress on Computational Intelligence, IEEE International Conference on Fuzzy Systems, Vancouver, Canada, July 2006. Portions reprinted, with permission, from Proc. of 2006 IEEE World Congress of Computational Intelligence, 2006 IEEE International Conference on Fuzzy Systems, Vancouver, 1570 – 1577, © 2006 IEEE.
** Dr. Mahmoud El-Banna was with Wayne State University, Detroit, MI 48202, USA.

The importance of weld quality monitoring and process variability reduction is further amplified by the recent changes in the materials used by automotive manufacturers. The demand for improved corrosion resistance has led the automotive industry to increasingly use zinc coated steel in auto body construction. One of the major concerns associated with welding coated steel is the mushrooming effect (the increase in the electrode diameter due to deposition of copper into the spot surface) resulting in reduced current density and undersized welds (cold welds). The most common approach to this problem is based on the use of simple unconditional incremental algorithms (steppers) for preprogrammed current scheduling. The main objective of the weld current steppers is to maintain weld nugget size within acceptable limits while at the same time minimizing electrode growth. Large current steps could lead to an increase in electrode tip growth due to the use of high current levels. This in turn requires even larger increases in current, thereby causing a runaway process of electrode growth. Under these conditions, weld size would deteriorate at a rapid rate. On the other hand, small increases in welding current result in a slow rate of electrode tip growth, which is advantageous in terms of electrode life, provided the small increases in current are sufficient to maintain adequate current density to produce the required weld nugget size. Since the direct measurement of the main process characteristics - weld quality and expulsion rate - is not feasible in an automotive plant environment one reasonable approach is to estimate these variables by virtual or soft (indirect) sensors. A soft sensor for indirect estimation of the weld quality can provide a real time approximate assessment of the weld nugget diameter. Another opportunity for soft sensing in weld process control is determined by the need to predict the impact of the current changes on the expulsion rate of the weld process. The combination of soft sensing with adequate control algorithms can have dramatic impact on reducing variability of the weld process and effectiveness of weld equipment. The final goal is to develop a control algorithm that can be applied in an automotive assembly plant environment with the final objective of improving the weld quality and consistency, in turn, improving overall manufacturing quality and productivity while reducing redundant welds.

In this chapter we discuss two specific topics: 1) Development of accurate in-process non-destructive evaluation (NDE) of nugget quality by using the dynamic resistance (or secondary voltage) profile during the welding process and 2) Design of closed-loop supervisory control algorithm for adapting the weld controller set points for weld quality enhancement and reduction of process variability.

We propose and demonstrate the performance of a Linear Vector Quantization (LVQ) network for on-line nugget quality classification in conjunction with an intelligent algorithm for adjusting the amount of current to compensate for the electrodes degradation. The algorithm works as a fuzzy logic controller using a set of engineering rules with fuzzy predicates that dynamically adapt the secondary current to the state of the weld process. The state is identified by indirectly estimating two of the main process characteristics - weld quality and expulsion rate. A soft sensor for indirect estimation of the weld quality employing an LVQ type classifier is designed to provide a real time approximate assessment of the weld nugget diameter. Another soft sensing algorithm is applied to predict the impact of changes in current on the expulsion rate of the weld process in real time. By maintaining the expulsion rate just below a minimal acceptable level, robust process control performance and satisfactory weld quality are achieved. The Intelligent Constant Current Control for Resistance Spot Welding is implemented and validated on a Medium Frequency Direct Current (MFDC) Constant Current Weld Controller. Results demonstrate a substantial improvement of weld quality and reduction of process variability due to the proposed new control algorithm.

2 Resistance Spot Welding - Background

A schematic diagram for resistance spot welding is illustrated in Fig. 1. It consists of primary (high voltage, low current) and secondary circuits (low voltage, high current). The process employs a combination of pressure and heat to produce a weld between the sheet metal work pieces in the secondary circuit. Resistance heating occurs as electrical welding current flows through the work pieces in the secondary circuit of a transformer. The transformer converts high-voltage, low current commercial power into suitable high current, low voltage welding power.

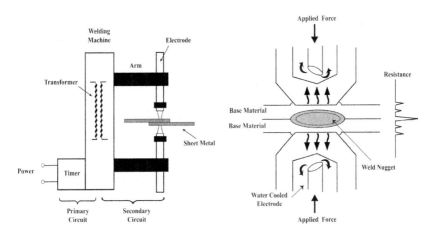

Fig. 1. Schematic diagram for resistance spot welding

The energy required to produce a given resistance weld is determined by several factors. Key among them is the weld area (heated volume), the peak temperature, the specific heat of the work pieces, and the heat loss through the surrounding metal and electrodes. An increase in magnitude of one or more of these factors requires a corresponding increase in energy to produce the weld. A typical spot welding operation is controlled by a weld schedule, whose time steps are controlled by a spot welding controller.

The dynamic resistance technique involves monitoring the resistance in the secondary circuit during the welding process. It is least intrusive, very economical, and seems to provide reasonable and adequate information about the state of the weld process. The word dynamic comes from fact that the resistance changes during the welding cycle. While the electrical resistances of the transformer and the mechanical assembly, R_t and R_m, can be assumed to be reasonably constant during the welding process (see Fig. 2), the sheet metal stack resistance (R_l) varies with nugget formation.

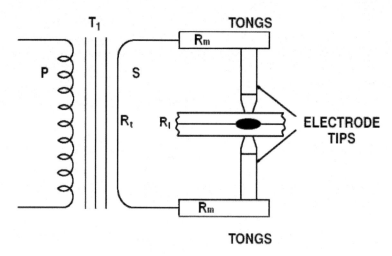

Fig. 2. Dynamic resistances in the secondary circuit

Two of the commonly used types of resistance welding systems (welding machines) in automotive industry are alternating current (AC) type and Medium Frequency Direct Current (MFDC) type. The AC resistance welding machine is inexpensive and its electrodes wear out slowly. However, a disadvantage is that the current supplied to the weld can be controlled only within fairly loose time interval [2]. The major advantage of the MFDC type of welding system is that the current supplied to the weld can be controlled within relatively stringent limits. This is one of the reasons for the increasing share of the MFDC type systems in the automotive assembly plants. In this chapter we pay special attention to the MFDC weld controller, and more specifically to the MFDC controller that is combined with a Constant Current strategy (MFDC-CC). This type of RSW controller is employed to achieve a constant current in each millisecond within the weld but the current can be changed from weld to weld based on a supervisory control algorithm.

3 Online Nugget Quality Evaluation Using Linear Vector Quantization Network

The problem of real-time estimation of the weld quality from process data is one of the key objectives in current weld control systems. The most common techniques can be grouped into four major groups: Ultrasonic technique, Thermal Force technique, Displacement technique, and Dynamic Resistance technique. It should be noted here that some of these techniques tend to be too intrusive and/or expensive for wide-scale deployment (for example, the ultrasonic technique), and in that sense, not compatible for main-stream application in automotive resistance welding. Most of the methods offered in the literature to predict nugget diameter from the process data employ measurements such as voltage and force and are not suitable in an industrial environment for two major reasons: the input signals for prediction model are taken from intrusive sensors (which affect the performance or capability of the welding machine), and the methods often required very large training and testing datasets.

This task can be alleviated if the weld controller is equipped with a voltage sensor in the secondary circuit, facilitating evaluation of dynamic resistance. Further simplification that significantly increases the feasibility of the mission of indirect estimation of weld quality follows from replacing the goal of quantifying the weld quality in terms of button size and integrity by the more modest objective of indirect estimation the class of the weld, e.g. satisfactory (acceptable, "normal" button size), unsatisfactory (under sized, "cold" welds), and defective ("expulsion") - Figure 3. We consider normal the welds within the specifications, i.e. those that have nugget diameter more than the minimum acceptable limit and exhibit no expulsion. Those welds that do not meet the specification are characterized as cold welds. Additionally, we count as expulsion welds the welds that indicate ejection of molten metal – an undesirable event that has detrimental effect on weld nugget integrity (the loss of metal from the fusion zone can reduce the weld size and result in weld porosity), which may significantly reduce the strength and durability of the welded joints.

Fig. 3. Examples of normal, cold and expulsion welds ([3])

Given its non-intrusive nature, relatively low cost of implementation, and reasonable performance in many laboratory and industrial settings, we have adopted the dynamic resistance approach to monitor and control the process on-line. The measurements of voltage and current (at primary or secondary side) are used to calculate dynamic resistance.

Given its well-defined physical meaning and the ease of measurement, a number of studies on the problem of estimation of weld quality from the secondary dynamic resistance have been performed. Cho and Rhee [4] showed that the process variables, which were monitored in the primary circuit of the welding machine, can be used to obtain the variation of the dynamic resistance across electrodes. They introduced an artificial intelligence algorithm for estimation of the weld quality using the primary dynamic resistance. Cho and Rhee used uncoated steel welding (low carbon cold rolled steel) to verify their model but fall short from discussing the impact of coated steel (the material mainly used in the auto industry). Lee et al [5] proposed a quality assurance technique for resistance spot welding using a neuro-fuzzy inference system. They however used the displacement signal (something impractical in an automotive plant environment) as input to their model. Podrzaj et al [6] proposed an LVQ neural network system to detect expulsion. The results showed that the LVQ neural network was able to detect the expulsion in different materials. However, they identified the welding force signal as the most important signal for classification of the expulsion occurrence. Availability of force signal is limited to certain types of guns, and they are more expensive than other types of sensors. Park and Cho [7] used LVQ as well as a multi-layer perceptron neural network to classify the weld quality (strength and indentation) by using the force signal. All those studies targeted AC weld controller while the MFDC controller was not examined.

In order to overcome these shortcomings, in this section, we propose an algorithm for estimation of weld nugget quality through classification of button size based on a small number of patterns for cold, normal, and expulsion welds. Our approach uses an LVQ neural network for nugget quality classification that employs the easily accessible dynamic resistance profile as input. Our focus is on the Medium Frequency Direct Current Constant Current (MFDC-CC) controller. A more general LVQ based soft sensing algorithm considering also alternating current (AC) weld controllers is presented in [8]. The goal is to develop a method and algorithm for on-line classification between normal welds, cold welds, and expulsion welds that can be applicable for weld process control. It should be mentioned that LVQ classification of the weld status is performed after each weld, not during the welding time.

Figure 4 shows prototypical dynamic resistance profiles for three types of welds; cold, normal, and expulsion, for MFDC–CC controller. It can be seen that these profiles are not easily distinguishable. The cold weld dynamic resistance profile tends to be lower than the other profiles, while the expulsion weld dynamic resistance profile tends to have a sharp drop especially towards the end. In order to classify them we apply an LVQ neural net classifier.

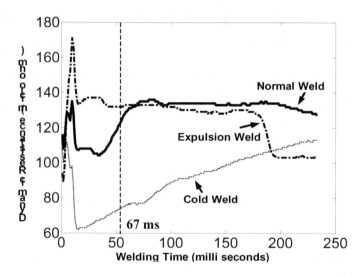

Fig. 4. Dynamic resistance profiles for cold, expulsion and normal welds for MFDC with constant current control

Learning vector quantization (LVQ) [9] is a method for training competitive layers of a neural network in a "supervised" manner. It consists of three layers: an input layer, a competitive layer, and an output layer (Figure 5). The "classes" that the competitive layer finds are dependent only on the distance between input vectors. If two input vectors are very similar, the competitive layer assigns them to the same class. LVQ shows good performance for complex classification problems because of its fast learning nature, reliability, and convenience of use. It particularly performs well with small training sets. This property is significantly important for industrial application, where training data is very limited; take considerable time, cost, or even impractical to get more data.

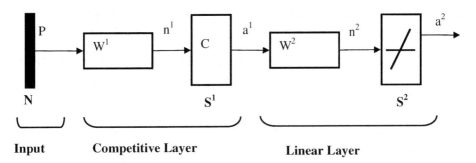

Fig. 5. Learning vector quantization (LVQ) neural network architecture

The network parameters are as follows: **P** is the N dimensional input vector, **W**^{i} is the weight matrix for the *i*th layer, S^i number of neurons in the *i*th layer, **n**^{i} the net input vector of the *i*th layer, and **a**^{i} the output of the *i*th layer. The first (competitive) layer is used to find the prototype vector **W**^1_s, (i.e., a row of the weight matrix **W**^{1}) that points in the direction closest to the input vector, i.e.,

$$\text{Min}_i \left\| \mathbf{P} - \mathbf{W}^1_i \right\|^2 \quad \forall i, \text{ where } i \in (1, 2...S^1)$$

The neurons that possess the least distance between vector weight matrix and input vector are assigned a value of one and the other neurons are assigned a value of zero. Finally, the output layer (linear layer) joins the subclasses (**S**^{1}) from the competitive layer and **W**^{2} weight matrix into target classes (**S**^{2}) through a linear transfer function. Matrix **W**^{2} defines a linear combiner and remains constant while the elements of **W**^{1} change during the training process. The weights of the winning neuron (a row of the input weight matrix) are adjusted using the Kohonen learning rule [20]. For example, supposing that the ith neuron wins the competition, the elements of the ith row of the input weight matrix are adjusted as shown below:

$$w^1(i) = w^1(i-1) + \rho \, (P(i) - w^1(i-1)),$$

where **P**(i) is the input vector of the *i*[th] iteration and ρ is the learning rate.

If just the Kohonen learning rule is employed, the neural network is called LVQ1. LVQ2 is an improved version of LVQ1, with the main difference being that in the latter case, the prototype vectors of two neurons are updated if the input vector **P**(i) is classified incorrectly. The weights of the neuron that wrongly won the competition are also updated as follows:

$$w^1(i) = w^1(i-1) - \rho \, (P(i) - w^1(i-1))$$

The LVQ2 was applied to estimate weld quality by classifying the dynamic resistance vectors corresponding to cold, normal an expulsion welds. The inputs to the network were the vectors of dynamic resistance sampled at 1ms sampling rate.

An experiment was conducted with an MFDC welding machine with capacity of 180 kVA, 680 lb welding force provided by a servo gun, HWPAL25 electrode type with 6.4 mm face diameter, 233 ms welding time, 11.5 kA initial input secondary current, and an incremental stepper of 1 ampere per weld. The nugget diameter was measured for a total of 550 welds: 411 were found to be good welds, 22 were cold welds, and 117 welds with expulsion. In this experiment, LVQ2 network was trained on three, six, and five patterns for cold, normal, and expulsion welds, respectively. Twelve hidden neurons were used with a learning rate $\rho = 0.01$.

The performance of the LVQ-based on-line nugget quality classification algorithm was evaluated in terms of type 1 (α) and type 2 errors (β) for cold, normal, and expulsion welds. Type 1 error (α) (known as false alarm rate) defines the probability of "rejecting" the null hypothesis, while it is true. For example, if the null hypothesis defined the weld as expulsion weld, Type 1 error (α) defines the probability that the weld is misclassified as normal or cold weld, while it really is an expulsion weld. Type 2 error (β) defines the probability of not rejecting the null hypothesis, while it is false. It is important to note that that there is a trade off between Type 1 error and Type 2 error. If the model is too sensitive (i.e., type 2 error is very low), it is normal to have a larger number of false alarms (i.e., type 1 error will be high). Tables 1-3 report type 1 errors (α) and type 2 errors (β) for cold, normal, and expulsion welds when using the entire discretized dynamic resistance profile as an input vector to the LVQ neural network. It can be seen that the percent of false alarms are lowest for the cold weld case at 0%, 11% for normal welds, and 40% for expulsion welds. As for type 2 errors, they are once again lowest for cold welds at 4%, 6% for expulsion welds, and 34% for normal welds.

In order to reduce the dimensionality of the LVQ neural network input vector (dynamic resistance profile), different features were tested as possible candidates to replace the dynamic resistance profile vector as input, i.e. reducing the input of the LVQ network to a feature vector (the first ten models have a single feature input while the last one has a 5-feature input vector):

- Maximum value of the dynamic resistance profile
- Minimum value of the dynamic resistance profile
- Mean value of the dynamic resistance profile
- Standard deviation value of the dynamic resistance profile
- Range value of the dynamic resistance profile
- Root mean square (RMS) value of the dynamic resistance profile
- First region slope (S1) value of the dynamic resistance profile
- Second region slope (S2) value of the dynamic resistance profile
- Third region slope (S3) value of the dynamic resistance profile
- Fourth region slope (S4) value of the dynamic resistance profile
- Binned RMS of dynamic resistance profile: dynamic resistance vector is divided into 5 bins and RMS values are calculated for each bin.

Table 1. Type1 and 2 errors for classification of cold welds when using the entire dynamic resistance profile as input to the LVQ neural network

H_0: Weld is Cold	True State of H_0	
Statistical Decision	H_0 is true	H_0 is false
Reject H_0	$\alpha = 0.00$	$1-\alpha = 1.00$
Don't reject H_0	$1-\beta = 0.96$	$\beta = 0.04$

Table 2. Type 1 and 2 errors for normal welds classification when using the entire dynamic resistance profile as input to the LVQ neural network

H_0: Weld is Normal	True State of H_0	
Statistical Decision	H_0 is true	H_0 is false
Reject H_0	$\alpha = 0.11$	$1-\alpha = 0.89$
Don't reject H_0	$1-\beta = 0.66$	$\beta = 0.34$

The criteria for features selection was based on power of the test (i.e. $1-\beta$) for the cold, normal, and expulsion welds as shown in Table 4. The feature that demonstrates the highest classification performance for the three types of welds was chosen as input for the LVQ network (the first row in Table 4). In order to simplify features selection, we assume that interactions among features are negligible.

Table 3. Type 1 and 2 errors for expulsion welds classification when using the entire dynamic resistance profile as input to the LVQ neural network

H_o: Weld is Expulsion	True State of H_0	
Statistical Decision	H_0 is true	H_0 is false
Reject H_0	$\alpha = 0.40$	$1-\alpha = 0.60$
Don't reject H_0	$1-\beta = 0.94$	$\beta = 0.06$

In our work, we just employed the most promising feature identified by power of the test criteria, the maximum value of the dynamic resistance vector, as input for LVQ neural network. Tables 5 -7 show the type 1 and 2 error results from the network when employing just this feature. It can be seen that both types of errors are reduced by using the maximum resistance feature instead of the entire vector of resistance for normal and expulsion welds. On the other hand, for cold welds, the type 2 error degrades.

LVQ network shows good performance for complex classification problems because of its fast learning nature, reliability, and convenience of use. It particularly performs well with small training sets. This property is especially important for

Table 4. Power of the test (1-β) for different features inputs to the LVQ neural network

Feature	Cold Welds	Normal Welds	Expulsion Welds
Maximum	99.8%	78.6%	83.0%
Minimum	94.6%	13.0%	100.0%
Mean	98.3%	13.7%	100.0%
Standard deviation	74.9%	60.3%	72.2%
Range	100.0%	38.2%	75.0%
Root Mean Square (RMS)	92.1%	14.5%	100.0%
Slope 1	53.6%	80.2%	79.2%
Slope 2	67.7%	100.0%	30.7%
Slope 3	73.9%	90.1%	45.8%
Slope 4	100.0%	37.4%	99.8%
Bin 1	83.6%	31.3%	76.2%
Bin 2	90.7%	16.0%	88.7%
Bin 3	89.6%	14.5%	100.0%
Bin 4	92.1%	100.0%	14.4%
Bin 5	98.1%	20.6%	98.6%

Table 5. Type1 and 2 errors for cold welds classification when using the maximum of dynamic resistance profile as a single input to the LVQ neural network

H_0: Weld is Cold	True State of H_0	
Statistical Decision	H_0 is true	H_0 is false
Reject H_0	α =0.00	1-α =1.00
Don't reject H_0	1-β =0.88	β =0.12

Table 6. Type1 and 2 errors for normal welds classification when using maximum of dynamic resistance profile as a single input to the LVQ neural network

H_0: Weld is Normal	True State of H_0	
Statistical Decision	H_0 is true	H_0 is false
Reject H_0	α =0.29	1-α =0.71
Don't reject H_0	1-β =0.81	β =0.19

automotive manufacturing applications, where the process of obtaining large training data sets may require considerable time and cost. Overall, the results are very promising for developing practical on-line quality monitoring systems for resistance spot-welding machines and complete automation of the welding process.

Table 7. Type 1 and 2 errors for expulsion welds classification when using maximum of dynamic resistance profile as a single input to the LVQ neural network

H_0: Weld is Expulsion	True State of H_0	
Statistical Decision	H_0 is true	H_0 is false
Reject H_0	α =0.23	1-α =0.77
Don't reject H_0	1-β =0.87	β =0.13

4 Intelligent Constant Current Control Algorithm

Most of the conventional weld control systems are based on the concept of "stepper" type preprogrammed scheduling of the primary current. A basis for setting up a current stepper can be developed by determining the pattern of electrode growth obtained in a particular welding cell. Different approaches are used for setting up a weld current stepper, including subjective methods, fixed increments, constant current density, gradient following, and iterative approaches. In a subjective or "best guess" approach, current steps are based on maintaining a slight red glow at the electrode/sheet interface and/or regularly adjusting the current to a level just below the splash or expulsion level. This approach has been found to give significant improvements in electrode life. While acceptable results can be achieved by this means, an extreme skill is required in determining the point at which current is to be increased.

In a fixed (preprogrammed scheduling) increment approach, a current stepper can be based on increasing either the heat control (i.e. phase shift control) or the actual welding current, in fixed increments after performing a predetermined number of welds. Generally, the increment of phase shift can be set between 1% and 5%. It was concluded [10] that a stepper function based on a fixed increment of the heat control or phase shift control was not a viable means of extending electrode life in many instances.

Multiple alternative approaches for adjusting the stepper algorithms based on different criteria have been reported (constant current density [10], gradient following approach [11], fuzzy controlled adaptation of delivered power [12], dynamic resistance profile estimation [13], prediction of weld strength [14], etc.) but have not found strong acceptance in automotive industry for various reasons (sensitivity to the coating type, undesirable rapid growth of electrode diameter, assumption of intrusive (electrode displacement) sensors, lack of robustness with respect to expulsions, etc.

In this section, we present an intelligent control algorithm that addresses the problem of constant current weld control of coated steels in the presence of significant electrode degradation [15]. The algorithm is implemented as a fuzzy logic controller using a set of engineering rules with fuzzy predicates that dynamically adapt the secondary current to the state of the welding process. Since the direct measurement of the main process characteristics - weld quality and expulsion rate - is not feasible in an industrial environment, these variables are estimated by soft (indirect) sensors.

A soft sensor for indirect estimation of the weld quality employing an LVQ type classifier that was described in the previous section provides a real time approximate assessment of the weld nugget diameter. Another soft sensing algorithm that is based on continuous monitoring of the secondary resistance is applied to predict the instantaneous impact of the current changes on the expulsion rate of the weld process. The reason for using the second soft sensor is to monitor the expulsion during the actual welding time (i.e. in each millisecond) so if expulsion is detected during the welding process, current should be turned off or reduced for the remaining welding time. Therefore, the second soft sensor complements the LVQ based soft sensor that was introduced in Section 3 with a real time estimation of potential expulsion conditions, while the LVQ soft sensor provides estimation of weld quality only after the completion of the weld process.

The main objective of the rule set of the fuzzy logic control algorithm is to describe a nonlinear control strategy that adjusts the secondary current to maintain the expulsion rate just below a minimal acceptable level guaranteeing satisfactory weld quality with robust process control performance, and minimize the electrode degradation. The fuzziness of the rules predicates reflects the uncertainty of the indirectly estimated weld quality and expulsion rate variables. The Intelligent Constant Current Control algorithm was implemented and validated on a Medium Frequency Direct Current Constant Current (MFDC-CC) Weld Controller. Results demonstrate a substantial improvement of weld quality and reduction of process variability due to the proposed new control algorithm.

The fuzzy logic control algorithm is implemented in a supervisory control mode (Figure 5) - it replaces the conventional "stepper" type constant current weld control algorithm. The primary current remains unchanged during the weld process but the primary current level for each weld is continuously adjusted based on the estimated state of the weld process during the last p welds (parameter p represents the size of a

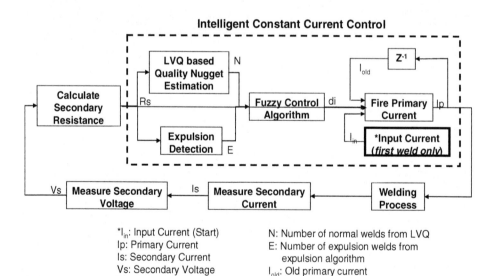

Fig. 6. Intelligent Constant Current Control

moving process window). The adjustment of the primary current results in a consequent adjustment of the secondary current. Two of the main process characteristics that are used as inputs to the fuzzy logic controller – the expulsion rate and the size of the weld nugget – are not directly measured but are derived from the secondary resistance profiles of the last p welds. The dynamic resistance is calculated from the measured secondary voltage and the calculated secondary current (Figure 6).

On the other hand, in order to get the optimum strength for the weld, the input parameters (current, time, force) need to be targeted just below the expulsion level.

The nugget quality estimation algorithm is used to determine the number of normal welds produced during the last process window of p welds based on a LVQ neural network that was discussed in detail in the previous section. We consider the full size input vector, i.e. **P** is a vector of dimension 167 (i.e. N=167), which is equal to the number of millisecond samples in one weld after the pre-heat and cooling phase. The reason for using the vector of dynamic resistance profile rather than a single feature input (the maximum of the profile) is to guarantee robustness of the proposed control algorithm. While an LVQ classifier with a single feature input can be applied for process monitoring for the purpose of supervisory control we consider the full size input vector classifier that contains complete information of the welding process. The number of hidden neurons in the LVQ neural network is 12 while the number of output neurons is 3 corresponding to the three categories of welding status; cold, normal, and expulsion. Consequently, the weight matrices $\mathbf{W}^1$ and $\mathbf{W}^2$ are of size (167X12) and (12X3), respectively.

The LVQ model (Figure 5) was trained on three, six, and five patterns of the secondary resistance vectors for cold, normal, and expulsion welds, respectively. Twelve hidden neurons were trained with a learning rate of 0.01.

Since the number of expulsions over time (expulsion rate) plays very significant role in the proposed control algorithm, we complement the estimation of the expulsion welds with an alternative algorithm for indirect estimation of the expulsion rate. Expulsion is estimated indirectly from the resistance profile. The main indicator for expulsion, as pointed out in [6, 16, 17], is the instantaneous drop in the resistance (Figure 4). In this chapter we use a modified version of the expulsion algorithm from reference [18].

Lets R(k) denote the dynamic resistance value at the current millisecond cycle (the MFDC weld process takes 233 ms), and R(k-1) and R(k-2) the two previous resistance values. The soft sensing expulsion algorithm continuously checks for a resistance drop with respect to a dynamically defined expulsion threshold $E_{level}(k)$ (after the cooling period, i.e. in our experiment after 67 milliseconds) that is represented by the following condition for the resistance:

If Max{R(k-2),R(k-1),R(k)}> Max{R(k-1),R(k)}

Then $E_{level}(k) = \dfrac{Max\{R(k-2),R(k-1),R(k)\} - Max\{R(k-1),R(k)\}}{Max\{R(k-1),R(k)\}} * 100$

Else

$E_{level}(k) = 0$

To determine if there is an expulsion in the examined weld, the following conditions are checked against $E_{level}(k)$:

If $E_{level}(k) \geq A$

Or

If $\{E_{level}(67) + ... + E_{level}(k)\} \geq B$,

where A and B are threshold parameters for expulsion detection (in our experiment A=3, and B=14).

In order to enhance the indirect estimation of the weld status, another soft sensing algorithm (LVQ based quality nugget estimation block in Figure 6) based on quality nugget estimation is introduced. Quality nugget estimation employing an LVQ classifier is designed to provide a real time approximation of the weld nugget status. The primary current for the next window of p welds is calculated by using a fuzzy control algorithm relating the number of expulsion welds and number of normal welds.

Let "E" denote the number of expulsion welds detected from the expulsion algorithm, "N" the number of normal welds detected from LVQ neural network, for the last window of p welds, and dI be the change of current that is inferred by the algorithm. We define the mechanism for adjusting the current gain based on the number of expulsion and normal welds in the last window of p welds through a set of rules with fuzzy predicates (Table 8).

In the rules of the fuzzy logic controller *low*, *medium*, and *high* are fuzzy subsets defined on the [0, p] universe for the number of expulsions "E", and the number of normal welds "N" (Figure 7). $N_g < 0$ and $P_a > 0$ are constants (fuzzy singletons) defining control changes of the current.

Table 8. Fuzzy Logic Controller Rule-Base

	Rule
1	If "E" is *low* AND "N" is *low* THEN $dI = P_a$
2	If "E" is *medium* AND "N" is *low* THEN $dI = N_g/2$
3	If "E" is *high* AND "N" is *low* THEN $dI = N_g$
4	If "E" is *low* AND "N" is *medium* THEN $dI = P_a/2$
5	If "E" is *medium* AND "N" is *medium* THEN $dI = N_g/4$
6	If "E" is *high* AND "N" is *medium* THEN $dI = N_g/2$
7	If "E" is *low* AND "N" is *high* THEN $dI = P_a/4$
8	If "E" is *medium* AND "N" is *high* THEN $dI = N_g/8$
9	If "E" is *high* AND "N" is *high* THEN $dI = N_g/4$

The first three fuzzy rules deal with the case where the number of normal welds "N" in the last window is low. Based on the number of detected expulsions, three alternative strategies for changing current level are considered:

- If the number of expulsions is *low*, it is reasonable to think that the state of the welds is close to the cold welds status. Hence, it is necessary to increase gradually the amount of current, i.e. the current is changed by. dI = Pa
- If the number of detected expulsions is *medium* or *high*, it is reasonable to think that the state of the welds is close to the expulsion state. Hence, it is necessary to decrease the amount of current. This is performed selectively, based on the number of expulsions (*high* vs. *medium*), resulting in negative changes of the current dI = Ng vs. dI = Ng / 2.

When the number of normal welds N in the process window is *medium*, the strategies for adjusting the current level are as follows (rules 4 – 6):

- When we have low expulsion detection rate, the weld state is likely approaching a cold weld. Therefore, the level of current should be increased. This is done by increasing the current level, i.e. dI = Pa / 2. Note that the amount of increase when the number of normal welds "N" is *medium* (dI = Pa / 2) is less than in the case when that number "N" is *low* (dI = Pa).
- The next case deals with medium expulsion rate, i.e. the weld state is close to the expulsion status. This requires a gradual reduction of the current dI. Note that the amount of decrease when "N" is *medium* (dI = Ng / 4) is also less than the case when the "N" is *low* (dI = Ng / 2).
- The last case appears when the expulsion rate is *high*. Since this is an undesirable state, the level of current should be lowered dramatically to minimize the number of expulsions. This is also done by modifying the secondary current dI = Ng / 2 when "N" is *medium* and dI = Ng when "N" is *low*.

The last three fuzzy rules (7 – 9) consider high level of normal welds, i.e. satisfactory weld quality. Their corresponding control strategies are:

- If we have low expulsion detection, the state of the welds will be approaching a cold weld status. Therefore, current level should be increased to prevent potential cold welds. This is done by a minor positive change of the current to $dI = P_a/4$.
- If we have medium expulsion detection, it is reasonable to consider that the state of the welds is close to the expulsion welds status. Therefore the current level should be decreased gradually to dI = Ng / 8).
- In the last case, when the expulsion detection is high, the level of the current should be decreased. The corresponding change of the current is slightly negative (dI = Ng / 4), i.e. significantly less than in the cases when "N" is *medium* (dI = Ng/2) or when "N" is *low* (dI = Ng).

Applying the Simplified Fuzzy Reasoning algorithm [19], we obtain an analytical expression for the change of the current dI depending on the rates of expulsion welds "E" and normal welds "N" as follows:

$$dI = \frac{\sum_{\forall i}\sum_{\forall j} \mu_i(x) v_j(y) \Delta_{i,j}}{\sum_{\forall i}\sum_{\forall j} \mu_i(x) v_j(y)}$$

where:

μ_i : membership function of the linguistic value of the expulsion welds {*low, medium, high*}.

v_j: membership function of the linguistic value of the normal welds {*low, medium, high*}.

x: number of expulsion welds in the process window detected by the expulsion algorithm.

y: number of normal welds in the process window detected by the LVQ soft sensing algorithm.

$\mu_i(x)$: firing level for the expulsion membership function

$v_j(y)$: firing level for the normal membership function

$\Delta_{i,j}$: amount of increment/decrement when the linguistic value of expulsion welds is "i" and the linguistic value of normal welds is "j"(for example, if the linguistic value of the expulsion welds is *high* and the linguistic value of the normal welds is *low* then $\Delta_{high,low} = Ng$, where Ng negative value determines the change of the current *dI*); see Table 8.

Triangular shape membership functions μ_i, v_j are used in the fuzzy control algorithm (Figure 7) to define the linguistic values of the numbers of expulsion and normal welds in the process window. These membership functions depend on the scalar parameters a, b, c as given by:

$$\mu_i, v_j(x, y; a, b, c) = \begin{cases} 0, & x \le a \\ \frac{x-a}{b-a} & a \le x \le b \\ \frac{c-x}{c-b} & b \le x \le c \\ 0, & c \le x \end{cases}$$

The new target current (I_{new}) for the next window of p welds will be:

$$I_{new} = I_{old} + dI\, I_{old}$$

where I_{old} is the current in the previous window of *p* welds and *dI* is the change of the current that is calculated from the fuzzy control algorithm.

Proposed Intelligent Constant Current Control algorithm was implemented in Matlab/Simulink and was experimentally tested in a supervisory control mode in conjunction with an MFDC Constant Current Controller. Four sets of experiments were performed as follows. The first group of tests (with/without sealer) was performed using the Intelligent Constant Current Controller. The second group (with/without sealer) was carried out by using a conventional stepper. The role of the sealer in this test is to simulate a typical set of disturbances that are common for

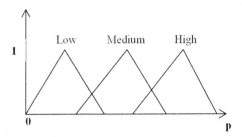

Fig. 7. Membership functions of the number of expulsion welds and the number of normal welds in a process window of the last p welds (p is a fixed parameter). Parameter p defines a universe $[0, p]$ of the all possible expulsion and normal welds within that moving window.

automotive weld processes. Sealer is commonly used to examine the performance of weld controllers and their capability to control process variability.

Each group of tests consists of sixty coupons, i.e. 360 welds (for each test without sealer), and ten coupons, i.e. 60 welds (for each test with sealer) with two metal stacks for each coupon are used for each test. Both tests involved welding 2.00 mm gage hot tip galvanized HSLA steel with 0.85 mm gage electrogalvanized HSLA steel. Thirty six coupons (216 welds) without a sealer between sheet metals and ten coupons (60 welds) with a sealer for each group of tests were examined. Cold and expulsion welds were checked visually in each coupon.

The length of the moving window in the Intelligent Constant Current Controller algorithm was $p = 10$, i.e. the soft sensing of expulsion and normal welds was performed on a sequence of 10 consecutive welds. The negative and positive consequent singleton values in the rule-base of the fuzzy control algorithm were set at Ng = -0.09 and Pa = +0.07.

In the stepper mode test, an increment of one ampere per weld was used as a stepper for this test. The initial input current was set at 11.2 kA for all tests, with no stabilization process to simulate the actual welding setup conditions in the plant after tip dressing.

Intelligent Constant Current Control and Stepper Based Control without Sealer

Figure 8 shows the weld secondary current generated by the Intelligent Constant Current Control algorithm without sealer. It can be seen that at the beginning of the welding process, there were a couple of cold welds, so the fuzzy control scheme increased the current gradually until expulsion began to occur. When expulsion was identified by the soft sensing algorithm, the fuzzy control algorithm began to decrease the current level until expulsion was eliminated and normal welds were estimated again. After that it continued to increase the current until expulsion occurred again and so on.

It can be concluded from the test above that the secondary current in the intelligent control scheme was responding to the weld status; in case of expulsion welds, the secondary current was decreased, and in case of cold welds, the secondary current was increased. Thus, the fuzzy control scheme was able to adapt the secondary current level to weld state estimated by the soft sensing algorithms.

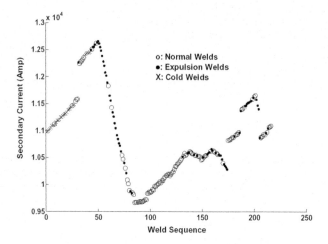

Fig. 8. Secondary current using the intelligent constant current fuzzy control algorithm

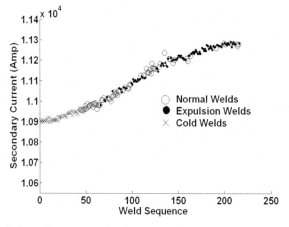

Fig. 9. Secondary current for the stepper based algorithm without sealer

Table 9. Number of expulsion welds for the fuzzy control algorithm and the conventional stepper mode without sealer

Number of expulsion welds using fuzzy controller	Number of expulsion welds using stepper mode
68 / 216 = 31.5%	98 / 216 = 45.4%

Figure 9 shows the secondary current in the case of conventional stepper mode. The weld primary current was set to a constant value at the beginning of the test, and then an increment of one ampere per weld was used as a stepper to compensate for the increase in electrodes diameter (mushrooming of the electrode); observed nonlinearity of the secondary current is a result of the transformer nonlinearities. It can be seen that there were several cold welds at the beginning of the test, followed by some of normal welds, and then expulsion welds were dominant until the end of the test.

Evidently, the secondary current in the stepper mode was too aggressive towards the end of the welding process, resulting in many expulsion welds. On the other hand, at the beginning, the secondary current was not enough, resulting in cold welds. The stepper mode does not really adapt the current to the actual weld state at the beginning or at the end of the welding process.

Tables 9 and 10 show the number of expulsion and cold welds for the Intelligent Constant Current Control algorithm versus the conventional stepper mode implementation. As expected, the number of expulsion welds in the stepper mode (98/216=45.4%) is higher than the number of expulsion welds in the fuzzy control scheme (68/216=31.5%). It can also be seen that the number of cold welds in the fuzzy control scheme test (31/216=14.4%) was less than the number of cold welds in the stepper mode (44/216 =20.4%).

Table 10. Number of cold welds for the fuzzy control algorithm and the conventional stepper mode without sealer

Number of cold welds using fuzzy controller	Number of cold welds using stepper mode
31 / 216 = 14.4%	44 / 216 = 20.4%

Intelligent Constant Current Control and Stepper Based Control with Sealer

It is a common practice in the automotive industry to intentionally introduce sealer material between the two sheet metals to be welded. The purpose of this sealer is to prevent water from collecting between the sheets and in turn reduce any potential

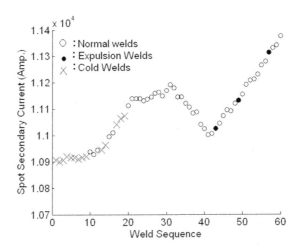

Fig. 10. Spot secondary current for the fuzzy control algorithm with sealer

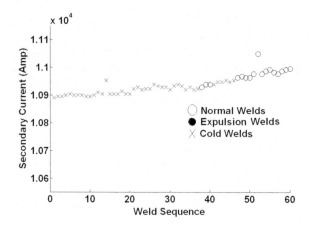

Fig. 11. Spot secondary current for the stepper mode with sealer

corrosion of the inner surface of sheet metals. However, the sealer creates problems for the spot welding process. In particular, the sealer increases the resistance significantly between the two sheet metals to be welded. When the welding process starts, high current will be fired, which is faced by high resistance (because of the sealer) in the desired spot to be welded, that prevents the current from flowing in that direction. The other alternative direction for this current is to flow in the direction of less resistance; this is what is known as shunting effect. Shunting effect produces cold welds, or at least small welds, which will cause a serious problem to the structure.

Figure 10 shows the spot secondary current for the Intelligent Constant Current Control algorithm with sealer. It demonstrates a performance similar to the case with no sealer – increasing/decreasing of the current level to adapt to the estimated cold/expulsion welds.

Figure 11 shows the spot stepper mode secondary current in the presence of sealer. The weld secondary current was set to a constant value at the beginning of the test with subsequent increments of one ampere per weld. It can be seen that the cold welds were dominant until just before the end of the test. There were a couple of normal welds towards the end of the test. No expulsion welds occurred in this test. Apparently, the secondary current was not enough to produce cold welds. Using stepper mode does not adapt the secondary current according to the weld status.

Tables 11 and 12 compare the number of expulsion and cold welds for the Intelligent Constant Current Control algorithm and the conventional stepper mode implementation in the case of welding with sealer. The number of expulsion welds in

Table 11. Number of expulsion welds for the fuzzy control scheme, the stepper, and the no stepper modes with sealer

Number of expulsion welds using fuzzy controller	Number of expulsion welds using stepper mode
3 / 60 = 5%	0 / 60 = 0.0%

Table 12. Number of cold welds for the fuzzy control algorithm, the stepper, and the no stepper modes with sealer

Number of cold welds using fuzzy controller	Number of cold welds using stepper model
14 / 60 = 23.3%	43 / 60 = 71.7%

the fuzzy control scheme test (3/60=5%) is slightly higher than the number of expulsion welds in the stepper mode test (0/60=0.0%). However, the number of cold welds in the case of application of the fuzzy control algorithm (14/60=23.3%) is much less than the number of cold welds in the conventional stepper mode test (43/60=71.7%).

5 Conclusions

The problem of real time estimation of the weld quality from the process data is one of the major issues in the weld quality process improvement. This is particularly the case for resistance spot welding. Most of the models offered in the literature to predict nugget diameter from the process data employ measurements such as ultrasonics, displacement, and thermal force and are not suitable in an industrial environment for two major reasons: the input signals for prediction model are taken from intrusive sensors (which will affect the performance or capability of the welding cell), and, the methods often required very large training and testing datasets.

In order to overcome these shortcomings, we proposed a Linear Vector Quantization (LVQ) neural network for nugget quality classification that employs the easily accessible dynamic resistance profile as input. Instead of estimating the actual weld nugget size the algorithm provides an on-line estimate of the weld quality by classifying the vectors of dynamic resistance profiles into three classes corresponding to normal, cold, and expulsion welds. We also demonstrated that the algorithm can be successfully applied when the dynamic resistance profile vector is replaced by a limited feature set. Based on the results from LVQ, a control algorithm called the Intelligent Constant Current Control for Resistance Spot Welding was proposed for adapting the weld current level to compensate for electrode degradation in resistance spot welding. The algorithm employs a fuzzy logic controller using a set of engineering rules with fuzzy predicates that dynamically adapt the secondary current to the state of the weld process. A soft sensor for indirect estimation of the weld quality employing an LVQ type classifier was implemented in conjunction with the intelligent control algorithm to provide a real time approximate assessment of the weld nugget status. Another soft sensing algorithm was applied to predict the impact of the current changes on the expulsion rate of the weld process. By maintaining the expulsion rate just below a minimal acceptable level, robust process control performance and satisfactory weld quality were achieved. The Intelligent Constant Current Control for Resistance Spot Welding was implemented and experimentally validated on a Medium Frequency Direct Current (MFDC) Constant Current Weld Controller.

Results were verified by benchmarking the proposed algorithm against the conventional stepper mode constant current control. In the case when there was no sealer between sheet metal, it was found that the proposed intelligent control approach reduced the relative number of expulsion welds and the relative number of cold welds by 31% (from absolute 45.4% to 31.5%) and 29% (from absolute 20.4% to 14.4%) respectively, when compared to the stepper mode approach.

In the case when there was a sealer type disturbance, the proposed control algorithm once again demonstrated robust performance by reducing the relative number of cold welds by 67% compared to the stepper mode algorithm (from absolute 71.7% to 23.3%), while increasing the absolute number of expulsion welds by only 5%.

Our Intelligent Constant Current Control Algorithm is capable of successfully adapting the secondary current level according to weld state and to maintain a robust performance.

Our focus in this chapter was on the Medium Frequency Direct Current (MFDC) weld controllers. An alternative version of the Intelligent Constant Current Control Algorithm that is applicable to the problem of alternating current (AC) weld control in conjunction with the Constant Heat Control Algorithm [8] is under development.

References

[1] Jou, M.: Real time monitoring weld quality of resistance spot welding for the fabrication of sheet metal assemblies. Journal of Materials Processing Technology 132, 102–113 (2003)
[2] Nishiwaki, Y., Endo, Y.: Resistance welding Controller. W. T. Corporation, Ed. USA, April 10 (2001)
[3] http://www.romanmfg.com/
[4] Cho, Y., Rhee, S.: New Technology for measuring dynamic resistance and estimating strength in resistance spot welding. Measurement Science & Technology 11, 1173–1178 (2000)
[5] Lee, S.R., Choo, Y.J., Lee, T.Y., Kim, M.H., Choi, S.K.: A quality assurance technique for resistance spot welding using a neuro-fuzzy algorithm. Journal of Manufacturing Systems 20, 320–328 (2001)
[6] Podrzaj, P., Polajnar, I., Diaci, J., Kariz, Z.: Expulsion detection system for resistance spot welding based on a neural network. Measurement Science & Technology 15, 592–598 (2004)
[7] Park, Y., Cho, H.: Quality evaluation by classification of electrode force patterns in the resistance spot welding process using neural networks. Proceedings of the Institution of Mechanical Engineers, Part B (Journal of Engineering Manufacture) 218, 1513 (2004)
[8] El-Banna, M., Filev, D., Chinnam, R.B.: Online Nugget Quality Classification using LVQ Neural Network for Resistance Spot Welding. International Journal of Advanced Manufacturing Technology 36, 237–248 (2008)
[9] Kohonen, T.: Improved versions of learning vector quantization. In: Proc. Int. Joint Conf. on Neural Networks I, 2nd edn., San Diego, pp. 545–550 (1990)
[10] Williams, N.T., Parker, J.D.: Review of resistance spot welding of steel sheets Part 2 - Factors influencing electrode life. International Materials Reviews 49, 77–108 (2004)

[11] Williams, N.T., Holiday, R.J., Parker, J.D.: Current stepping programmes for maximizing electrode campaign life when spot welding coated steels. Science and Technology of Welding and Joining 3, 286–294 (1998)
[12] Messler, R.W., Min, J., Li, C.J.: An intelligent control system for resistance spot welding using a neural network and fuzzy logic. In: IEEE Industry Applications Conference, Orlando, FL, USA, October 8-12 (1995)
[13] Chen, X., Araki, K.: Fuzzy adaptive process control of resistance spot welding with a current reference model. In: Proc. of the IEEE International Conference on Intelligent Processing Systems, ICIPS, Beijing, China (1998)
[14] Lee, S., Choo, Y., Lee, T., Kim, M., Choi, S.: A quality assurance technique for resistance spot welding using a neuro-fuzzy algorithm. Journal of Manufacturing Systems 20, 320–328 (2001)
[15] El-Banna, M., Filev, D., Chinnam, R.B.: Intelligent Constant Current Control for Resistance Spot Welding. In: Proc. of 2006 IEEE World Congress of Computational Intelligence, 2006 IEEE International Conference on Fuzzy Systems, Vancouver, pp. 1570–1577 (2006)
[16] Dickinson, D., Franklin, J., Stanya, A.: Characterization of Spot-Welding Behavior by Dynamic Electrical Parameter Monitoring. Welding Journal, vol 176, S170–S176 (1980)
[17] Hao, M., Osman, K.A., Boomer, D.R., Newton, C.J., Sheasby, P.G.: On-line nugget expulsion detection for aluminum spot welding and weldbonding. SAE Trans. Journal of Materials and Manufacturing 105, 209–218 (1996)
[18] Hasegawa, H., Furukawa, M.: Electric Resistance Welding System, U.S Patent 6130369 (2000)
[19] Yager, R.R., Filev, D.: Essentials of Fuzzy Modeling & Control. John Wiley & Sons, New York (1994)
[20] Kohonen, T.: Self-Organization and Associative Memory, 2nd edn. Springer, Berlin (1987)

Intelligent Control of Mobility Systems

James Albus, Roger Bostelman, Raj Madhavan, Harry Scott, Tony Barbera, Sandor Szabo, Tsai Hong, Tommy Chang, Will Shackleford, Michael Shneier, Stephen Balakirsky, Craig Schlenoff, Hui-Min Huang, and Fred Proctor

Intelligent Systems Division, National Institute of Standards and Technology (NIST), 100 Bureau Drive, Mail Stop 8230, Gaithersburg, MD 20899-8230

1 Introduction

The National Institute of Standards and Technology (NIST) Intelligent Control of Mobility Systems (ICMS) Program provides architectures and interface standards, performance test methods and data, and infrastructure technology needed by the U.S. manufacturing industry and government agencies in developing and applying intelligent control technology to mobility systems to reduce cost, improve safety, and save lives. The ICMS Program is made up of several areas including: defense, transportation, and industry projects, among others. Each of these projects provides unique capabilities that foster technology transfer across mobility projects and to outside government, industry and academia for use on a variety of applications. A common theme among these projects is autonomy and the Four Dimensional (3D+time)/Realtime Control System (4D/RCS) standard control architecture for intelligent systems that has been applied to these projects.

NIST's Intelligent Systems Division (ISD) has been developing the 4D/RCS [1, 2] reference model architecture for over 30 years. 4D/RCS is the standard reference model architecture that ISD has applied to many intelligent systems [3, 4, 5]. 4D/RCS is the most recent version of RCS developed for the Army Research Lab (ARL) Experimental Unmanned Ground Vehicle program. ISD has been applying 4D/RCS to the ICMS Program for defense, industry and transportation applications.

The 4D/RCS architecture is characterized by a generic control node at all the hierarchical control levels. Each node within the hierarchy functions as a goal-driven, model-based, closed-loop controller. Each node is capable of accepting and decomposing task commands with goals into actions that accomplish task goals despite unexpected conditions and dynamic perturbations in the world. At the heart of the control loop through each node is the world model, which provides the node with an internal model of the external world. The fundamental 4D/RCS control loop structure is shown in Figure 1.

The world model provides a site for data fusion, acts as a buffer between perception and behavior, and supports both sensory processing and behavior generation. In support of behavior generation, the world model provides knowledge of the environment with a range and resolution in space and time that is appropriate to task decomposition and control decisions that are the responsibility of that node.

The nature of the world model distinguishes 4D/RCS from conventional artificial intelligence (AI) architectures. Most AI world models are purely symbolic. In 4D/RCS, the world model is a combination of instantaneous signal values from sensors, state

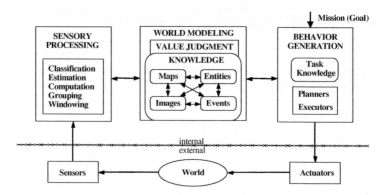

Fig. 1. The fundamental structure of a 4D/RCS control loop

variables, images, and maps that are linked to symbolic representations of entities, events, objects, classes, situations, and relationships in a composite of immediate experience, short-term memory, and long-term memory. Real-time performance is achieved by restricting the range and resolution of maps and data structures to what is required by the behavior generation module at each level. Short range, high resolution maps are implemented in the lower levels, with longer range, lower resolution maps at the higher levels.

A world modeling process maintains the knowledge database and uses information stored in it to generate predictions for sensory processing and simulations for behavior generation. Predictions are compared with observations and errors are used to generate updates for the knowledge database. Simulations of tentative plans are evaluated by value judgment to select the "best" plan for execution. Predictions can be matched with observations for recursive estimation and Kalman filtering. The world model also provides hypotheses for gestalt grouping and segmentation. Thus, each node in the 4D/RCS hierarchy is an intelligent system that accepts goals from above and generates commands for subordinates so as to achieve those goals.

The centrality of the world model to each control loop is a principal distinguishing feature between 4D/RCS and behaviorist architectures. Behaviorist architectures rely solely on sensory feedback from the world. All behavior is a reaction to immediate sensory feedback. In contrast, the 4D/RCS world model integrates all available knowledge into an internal representation that is far richer and more complete than is available from immediate sensory feedback alone. This enables more sophisticated behavior than can be achieved from purely reactive systems.

A high level diagram of the internal structure of the world model and value judgment system is also shown in Figure 1. Within the knowledge database, iconic information (images and maps) is linked to each other and to symbolic information (entities and events). Situations and relationships between entities, events, images, and maps are represented by pointers. Pointers that link symbolic data structures to each other form syntactic, semantic, causal, and situational networks. Pointers that link symbolic data structures to regions in images and maps provide symbol grounding and enable the world model to project its understanding of reality onto the physical world.

Figure 2 shows a 4D/RCS high level diagram duplicated many times, both horizontally and vertically into a hierarchical structure as applied to a single military vehicle (lowest level) through an entire battalion formation (highest level). This structure, now adopted as a reference model architecture for the US Army Future Combat System, among other organizations, could also be applied to civilian on-road single or multiple vehicles as information could be passed from one vehicle to the next or to highway communication and control infrastructure.

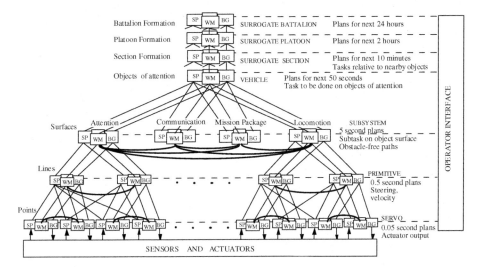

Fig. 2. The 4D/RCS Reference Model Architecture showing multiple levels of hierarchy

This chapter will briefly describe recent project advances within the ICMS Program including: goals, background accomplishments, current capabilities, and technology transfer that has or is planned to occur. Several projects within the ICMS Program have developed the 4D/RCS into a modular architecture for intelligent mobility systems, including: an Army Research Laboratory (ARL) Project currently studying on-road autonomous vehicle control; a Defense Advanced Research Project Agency (DARPA) Learning Applied to Ground Robots (LAGR) Project studying learning within the 4D/RCS architecture with road following application; and an Intelligent Systems Ontology project that develops the description of intelligent vehicle behaviors. Within the standards and performance measurements area of the ICMS program, a Transportation Project is studying components of intelligent mobility systems that are finding their way into commercial crash warning systems (CWS). Also, the ALFUS (Autonomy Levels For Unmanned Systems) project determines the needs for metrics and standard definitions for autonomy levels of unmanned systems. And a JAUS (Joint Architecture for Unmanned Systems) project is working to set a standard for interoperability between components of unmanned robotic vehicle systems. Testbeds and frameworks underway at NIST include the PRIDE (Prediction in Dynamic Environments) framework to provide probabilistic predictions of a moving object's future position to an autonomous vehicle's planning system, as well as the USARSim/MOAST (Urban Search and Rescue Simulation/Mobility Open Architecture

Simulation and Tools) framework that is being developed to provide a comprehensive set of open source tools for the development and evaluation of autonomous agent systems. A NIST Industrial Autonomous Vehicles (IAV) Project provides technology transfer from the defense and transportation projects directly to industry through collaborations with automated guided vehicles manufacturers by researching 4D/RCS control applications to automated guided vehicles inside facilities. These projects are each briefly described in this Chapter followed by Conclusions and continuing work.

2 Autonomous On-Road Driving

2.1 NIST HMMWV Testbed

NIST is implementing the resulting overall 4D/RCS agent architecture on an ARL High-Mobility Multipurpose Wheeled Vehicle (HMMWV) testbed (see Figure 3). Early work has focused on the lower agent control modules responsible for controlling the speed, steering, and real-time trajectory paths based on sensed road features such as curbs. This effort has resulted in sensor-based, on-road driving along dynamically-smooth paths on roadways and through intersection turns [6]. Future work includes the implementation of selected driving and tactical behaviors to further validate the knowledge set.

NIST has put in place several infrastructural elements at its campus to support the intelligent vehicle systems development described above. An aerial survey was completed for the entire site, providing high resolution ground truth data. A GPS base station was installed that transmits differential GPS correction data across the site. Testbed vehicles, equipped with appropriate GPS hardware, can make use of these corrections to determine their location in real-time with an uncertainty a few centimeters. This data can be collected and compared to the ground truth data from the survey to make possible extensive vehicle systems performance measurements of, for example, mobility control and perception components. In addition, lane markings consistent with the DOT Manual on Uniform Traffic Control Devices, have been added to parts of the campus to support development of perception capabilities required for urban driving.

Fig. 3. ARL/NIST HMMWV Testbed showing several sensors mounted above the cab

Perception

The perception in an autonomous mobile robot provides information about the world that enables a mobile robot to autonomously navigate its environment and react to events or changes that influence its task. The goal of Perception is to build and maintain the internal representation of the world (World Model) that the behavior generation components of the mobile robot can use to plan and execute actions that modify the world or the robot's position in the world. Since the world in general is not static, the perception algorithms must update the internal model rapidly enough to allow changes, such as the positions of moving objects, to be represented accurately. This places constraints on the sensors that can be used and on the processing that can be applied.

This section will address the perception aspects of an autonomous mobile robot under the ARL project. Under this project, we developed and demonstrated solutions to a number of perception problems in autonomous navigation. In order to achieve goals in supporting navigation through real-time algorithms, color cameras, FLIR (forward looking infrared), LADARs (laser detection and ranging) (see Figure 4) are used and the following topics are investigated and implemented:

Sensors, Sensor Registration, and Sensor Fusion: Sensors in use include color cameras, FLIR and scanning and imaging LADARs. These sensors have to be registered so that data from the sensors can be combined. Registration involves accurately locating the sensors relative to each other and time- and position-stamping the outputs of each sensor and combined with vehicle motion. Sensor fusion allows a richer description of the world and is particularly useful in associating range information with data from a two-dimensional, monocular color camera.

Obstacle Detection: Obstacles may be features that lie above the ground or holes in the ground. They need to be detected well before the vehicle reaches them and to know the true size of an obstacle to avoid it. Hence a three-dimensional analysis is required. This can be obtained from LADAR or stereo sensors, while a color camera may provide the best data to identify the obstacle.

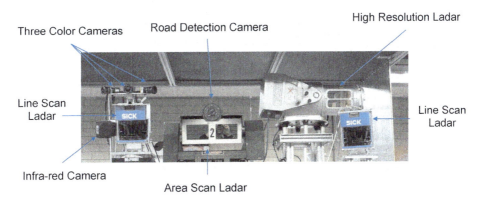

Fig. 4. Sensor suite mounted on the HMMWV cab

Feature Detection: To build up a rich description of the world, a variety of features need to be recognized. These include roads, road signs, water, other vehicles, pedestrians, etc. Special purpose processing is needed for each of these features, which must not only be detected, but tracked while they are within the immediate vicinity of the robotic vehicle.

Performance Evaluation: Sensory processing plays a critical part in keeping the vehicle operating safely. Evaluating the performance of the sensory processing algorithms, which involves algorithm testing in realistic scenarios, provides a way to ensure that they work correctly and robustly enough in the real world.

The world model is used in this project as a basis for temporal fusing the extracted feature from the perception algorithms on different sensors and producing an internal world model. The world model contains a representation of the current state of the world surrounding the vehicle and is updated continually by the sensors. It acts as a bridge between perception and behavior generation by providing a central repository for storing sensory data in a unified representation, and decouples the real-time sensory updates from the rest of the system. The world model therefore constructs and maintains all the information necessary for intelligent path planning.

2.2 4D/RCS Task Decomposition Controller for On-Road Driving

The 4D/RCS design methodology has evolved over a number of years as a technique to capture task knowledge and organize it in a framework conducive to implementation in a computer control system. A fundamental premise of this methodology is that the present state of the task sets the context that identifies the requirements for all of the support processing. In particular, the task activity at any time determines what is to be sensed in the world, what world model states need to be evaluated, which situations need to be analyzed, what plans should be invoked, and what behavior generation knowledge needs to be accessed. This view has resulted in the development of a design methodology that concentrates first and foremost on a clear understanding of the task and all of the possible subtask activities.

The application of this methodology to the on-road driving task has led to the design of a 4D/RCS control system that is formed from a number of agent control modules organized in a hierarchical relationship where each module performs a finer task decomposition of the goal it receives from its supervising module. Each of these control modules receives a goal task/command, breaks it down into a set of simpler subtasks, determines what has to be known from the internal world model to decide on the next course of action, alerts the sensory processing as to what internal world objects have to have their states updated by new sensory readings/measurements, and evaluates the updated state of the world model to determine the next output action.

An on-road driving 4D/RCS control hierarchy is shown in Figure 5, which is an application of multiple levels of the 4D/RCS reference model architecture diagram shown in Figure 2. Here, over 170 intermediate level subtask commands have been identified. These are shown as input commands to the corresponding control module where they execute. A further expansion of these commands is shown at the Vehicle Manager and the Mobility agent control modules. Their commands are underlined in Figure 5 and the corresponding set of possible plans that can be selected are listed under each underlined command. For example, at the Mobility module, the

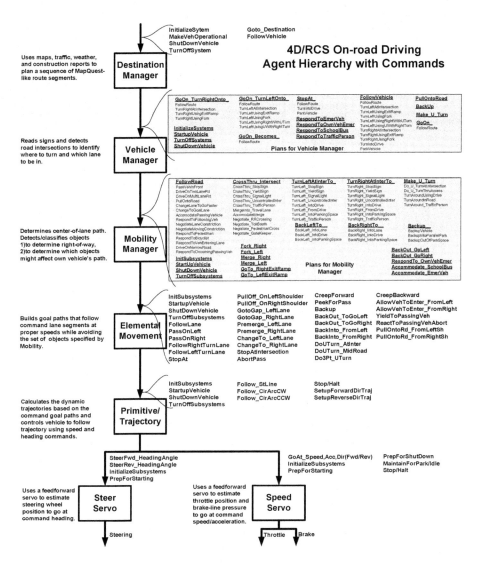

Fig. 5. 4D/RCS on-road driving control hierarchy with over 170 identified intermediate level subtask commands/plans that decompose the **GoTo_Destination** command all the way down to the instant-by-instant steer and speed/acceleration commands to control the vehicle.

FollowRoad command can select a plan to **PassVehInFront**, or **DriveOnTwoLaneRd**, or **DriveOnMultiLaneRd**, or **PullOntoRoad**, etc. depending on the present world state. Each of these plans is a separate state table describing this task behavior with an appropriate set of rules. Each of these rules would have a branching condition/situation from which are derived detail world states, objects, and sensing resolutions. An example of the command for each module at a particular instant in

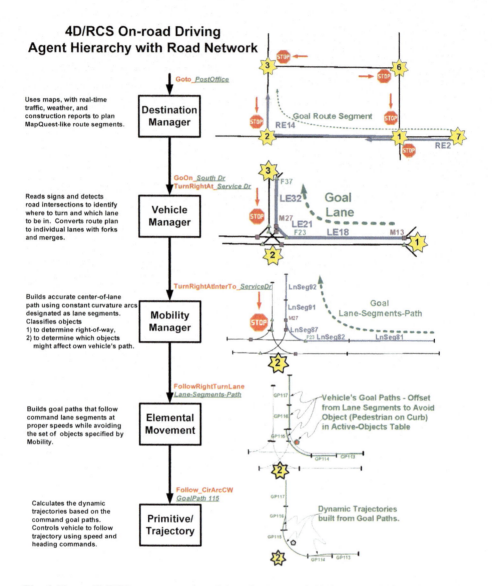

Fig. 6. Upper 4D/RCS agent control modules of an on-road driving control hierarchy summarizing the road network interactions for each control module. Each lower module processes a more detailed representation of the vehicle's goal path.

time is shown in Figure 6: a **GoTo** *PostOffice* command to the Destination Manager might result in a **GoOn** *SouthDr* **TurnLeftOnto** *ServiceDr* command to Vehicle Manager resulting in a **TurnRightAtInterTo_*ServiceDr*** command to Mobility Manager resulting in a **FollowRightTurnLane** command to Elemental Movement resulting in a **Follow_CirArcCW** goal path command to Primitive/Trajectory resulting in a **SteerFwd_*HeadingAngle*** command to the Steer Servo module and a

GoAt_*Speed,Acc,Fwd* command to Speed Servo module. With each of these control modules in Figure 6, a summary view of the road network data structure is also pictured. The commands (goal/action-verbs) in this example are emphasized by a red font with parameters shown in green. There are, of course, many more parameters (not shown here) associated with each of these commands that identify specific road and lane parameters, and objects of interest that can affect right-of-way and own vehicle motion paths. At each control cycle, each of the agent control modules executes as a state machine where its input state is defined by the input command from its supervisor module, status from its subordinate module and the present state of its view of the world model. If this input state has changed from the last cycle, then the control module transitions to a new state and may change the output command to a subordinate module, and/or the status to the supervisor module, along with changes to the world model.

The following outline will detail activities at each of these agent control modules summarizing the level and type of knowledge and data each process. This outline uses a standardized format where each module is described by a five part section consisting of summary of its responsibilities and actions including the world model data generated, its input command from the supervisor control module, its input from the world model, its output command to its subordinate control module, and its output status. Again, each command name will be highlighted in red (with parameters in green). Road network data structure elements will be highlighted in blue.

Destination Manager Control Module

Responsibilities and Actions: *This module's primary responsibility is to plan the route to get to a single commanded destination, which may include passing through specified intermediate control points. This is route planning at the level that determines which road to follow and which intersections to turn at. At any time, real-time inputs concerning weather, traffic, construction, civil situations, etc. can cause a re-planning/re-sequencing of **route elements** used to define the route.*

*Retrieves and builds a sequence of **route elements** to specify a **route segment**.*

This module is commanded to go to a ***destination***, which is described by not only the destination point itself but also by a list of possible intermediate control points so that the supervisor module can control the shape of the solution path. The input command further constrains the route to be planned to arrive at the destination by a specified time and to be prioritized by the shortest distance, or time, or the least number of turns or controlled intersections. As a result, this module has the responsibility to plan the sequence of the **route elements** (see Figure 7) that meet the commanded constraints. It creates multiple sets of possible sequences of **route elements** and evaluates them against the commanded constraints to select a plan set to execute. It is continually updating the attributes of the **route elements** based on the real-time reports of weather, traffic, construction, civil situations, and accidents. Anytime new real-time inputs are available, this module re-plans the sequence of **route elements**. It processes present experiences to create a history of average effects of these parameters versus time of day, day of week, time of year to build more accurate estimates of the **route elements** availability and average speeds. From this, over time, it can improve the accuracy of its

Retrieves and builds a sequence of route elements to specify a route segment.

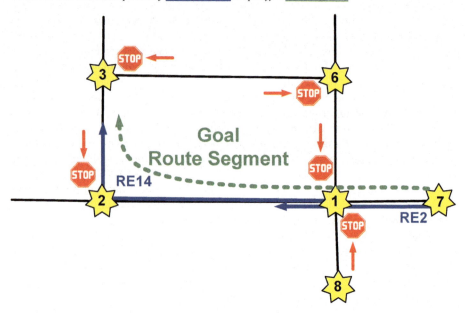

Fig. 7. Route Elements (RE) are route representations that connect two Map Points (which are either intersections or boundary points of the map and are labeled by yellow stars) and indicate a direction away from the end Map Point. Here Route Element 2 (RE2) starts at Map Point 7 and ends at Map Point 1 heading in a direction straight through the intersection at Map Point 1 towards Map Point 2. RE14 starts at Map Point 1 and ends at MapPoint 2 heading in a direction of a right turn at the intersection towards Map Point 3. The Destination Manager control module plans a route as the sequence of these route elements that best meet the constraints of minimum distance, minimum time, or minimum number of turns.

estimations and its planned routes. It commands the Vehicle control module to execute a *route segment*, which is a sequence of **route elements**, all of which pass through each intervening intersection except the last route element, which specifies a turn onto a different road. When the subordinate control module indicates it is nearing completion of its commanded route segment, this module re-plans its set of **route elements** and groups the next sequence of **route elements** that represents the next *route segment* and issues this command to the subordinate module.

Input-Command: A command to **Goto** *Post Office*, which is the form of a ***Goal Destination*** specification. The associated command data includes the parameters of the final goal point along with a sequential list of any intermediate points (control points) to pass through (e.g. **Goto** *Post Office* (from the present position which is the school) by way of the intersection at Service Dr and Research Dr (a control point)) along with a specific time to be at this goal destination (e.g. arrive at Post Office by 10:00am). Additionally, the command specifies that a route is to be planned that prioritizes on the shortest time, or the shortest distance, or the least number of turns, or least number of controlled intersections.

Input-World Model: Present estimate of this module's relevant map of the **road network** – this is a map at the level of **route elements** (an interconnecting single direction of travel between two intersections with a direction pointing out of the second intersection (e.g. turn left, or pass through, or turn right). Each **route element** is a unique possible path for a vehicle to travel. Each is assigned a number of attributes that identify its availability (i.e., is it available for a vehicle to travel on it), time constraints on its availability (not available from 7:30am to 9:00am, such as might apply to a left turn), average travel speeds with time constraints (e.g. 35 kph from 7:00 pm to 7:30 am, 15 kph from 7:30am to 7:00pm), average travel speeds with weather constraints (35 kph clear weather, 20 kph in rain, 15 kph in snow), average travel speeds with traffic constraints (35 kph normal traffic, 25 kph in moderate traffic, 15 kph in heavy traffic) conditions that make it unavailable (e.g. unavailable in heavy rain – has a creek that floods, unavailable in snow or ice because has too steep of a hill, unavailable because of accident report or construction report or civil situation report such as a parade or demonstration). Most of the real-time information is usually received from remote sensing since most of the route is beyond own vehicle's sensing capabilities, but own vehicle's sensing can contribute in situations like observing an accident in the road ahead which has not yet been reported on the radio.

World model road network data includes a set of history attributes that store average times to traverse each route element under the various conditions, which should improve planning estimates.

Output-Command: Goal *Route Segment* which is a Map Quest-like command **GoOn** *SouthDr* **TurnRightAt** *ServiceDr*, which is basically a command to follow a specified road until you have to turn onto a different road.

Associated command data: Sequential list of the **route elements** that combined make up the specified *route segment* command.

Output-Status: Present state of goal accomplishment (ie, the commanded *goal destination*) in terms of executing, done, or error state, and identification of which intermediate control points have been executed along with estimated times to complete the remaining control points.

Vehicle Manager Control Module

Responsibilities and Authority: *This module's primary responsibility is to use the commanded planned* **route elements** *to build out the actual available lanes along these* **route elements**, *identifying wherever a physical branching (fork) exists and selecting the correct branch (such as a right turn as opposed to passing through an intersection). This module assembles the resultant* **lane element** *sequences in groups that define a* **goal lane** *on a roadway between intersections or a* **goal lane** *that defines the maneuver through an intersection. Real-time navigational data input from road name signs, exit signs and lane-use restriction signs are used to assist plan execution.*

Retrieves and builds a sequence of **lane elements** *to specify a* **goal lane**.

This module is commanded to go along a *route segment*, which is described by the corresponding sequence list of **route elements**. This control module converts this higher level of route abstraction into the actual lanes that are available for own vehicle to travel on along these **route elements**. It does this by building out the sequence of **lane elements** that will represent the more detailed route plan for the vehicle and then groups these **lane elements** into commanded *goal lanes* as shown in Figure 8. It further decides, based on navigational concerns at this level, what priority to set on the commanded *goal lane*. If there is an upcoming right hand turn, for instance, on a multi-lane road, it may specify that the *goal lane* (which is the furthest lane to the right) for a **FollowRoad** command is at the priority of only-required-to-be-in-goal-lane-when-reach-end-of-goal-lane so that the subordinate control modules are free to chose adjacent lanes to respond to drive behavior issues but the vehicle must be in that lane at the end of the lane in time for the upcoming right turn.

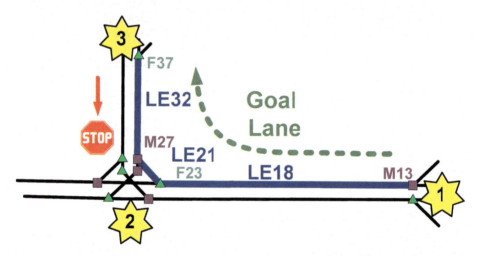

Fig. 8. A Lane Element (LE) is defined single lane section of road that connects two Lane Junctions. Lane Junctions are either Forks (eg, F23 which is a right turn fork) or Merges (eg, M27 which is a merge of a left turn and a right turn). This figure illustrates the lane elements for a single Route Element that turns right at the intersection (defined as Map Point 2). Here, Lane Element 18 (LE18) is one lane of a two lane road that goes from Merge 13 (M13) to Fork 23 (F23). LE21 goes from F23 to M27 and LE32 goes from M27 to F37. The Vehicle Manager Control Module carries out this further refinement of the commanded Route Element 14 from the Destination Manager to build the sequence of these three Lane Elements (LE18, LE21, and LE32) that define the planned Goal Lane for the vehicle.

This module assembles **lane elements** in order to specify either the *goal lanes* between intersections or the *goal lanes* that navigate through intersections (including turns). In this way, these *goal lane* specifications accompany the commands to the subordinate Mobility control module to carry out **Follow Road, Turn Right At Intersection, Pass Through Intersection**, etc. actions.

This module uses those real-time world model data that can affect the choice of **lane elements**, which affects which fork and/or merge **lane junction** to use. Its world data includes the road names of its own road and all crossing roads as well as navigational signs that indicate upcoming intersections, turn lanes and exit ramps, as well as any lane restrictions such as "No Left Turn – 4:30 to 7:00 pm". It also monitors for any situations that might cause changes in the set of **lane elements** selected to execute a commanded **route element**. As an example, a right turn ramp might be closed for construction requiring the vehicle to proceed into the intersection to make the turn, or the left turn lane may be blocked by an accident requiring the vehicle to pass through the intersection and make a U-turn then a left turn. But in all cases, the commanded **route element** is still being followed. However, if the right hand turn road were completely blocked by an accident, this would create a situation requiring a response beyond the level of authority of this Vehicle control module since the commanded **route element** can no longer be followed. This data would feedback through the world model to the Destination manager module to trigger a re-planning of a new sequence of **route elements** to deal with this situation. The Destination manager would then command the Vehicle module with a new *route segment* composed of a new sequence of **route elements**. The Vehicle control module would, in turn, build out its corresponding set of **lane elements** to specify the new *goal lane* and command these to the Mobility control module. All of these responses would happen in a small fraction of a second. This illustrates how each module can only make decisions at its level of responsibility and authority.

Input-Command: A command to **GoOn** *SouthDr* **TurnRightAt** *ServiceDr* which is the form of a *Goal Route Segment* (similar to a MapQuest-like command), which is basically a command to follow a specified road until you have to turn onto a different road.

Associated command data: Sequential list of the **route elements** that when combined, make up the specified route segment command. This is a list of route elements, each of which specifies the route through the next succeeding intersection (these will all be pass-through-the-intersection route elements) until the last **route element** of the list that will specify the turn required to go onto a different road completing the specification of the commanded *route segment*.

Input-World Model: Present estimate of this module's relevant map of the **road network** – this is a map at the level of **lane junctions** and **lane elements**. **Lane junctions** are all of the points along a single travelway where either a physical fork from one lane to two lanes occurs (a Fork Junction) or two lanes come together into one lane (a Merge Junction). The interconnecting stretches of single lane travelways are the **lane elements**. This is analogous to a train track where all of the switches represent the **lane junctions** and the connecting track sections are the **lane elements**. This is a very appropriate analogy since a vehicle traveling on a road is very much constrained to its present lane as a train is to its track and can only change to a different lane at junction/switch points. Every physical fork junction is a decision point where the controller has to decide which of the two lanes to choose. The decision to change lanes along a route with multiple lanes in the same direction is not a navigation issue but a drive behavior issue that will be discussed below. This module's map representation is concerned with navigational decisions at the level of physical road forks and merges. These structures of **lane junctions** and **lane elements** carry many attributes to aid in route planning. Each **lane junction** is classified as a Fork or a Merge. If it is a Fork, it is classified as forming a right or left turn, or adding a lane. For a Merge, it is classified as to which merging lane, right or left, has the right-of-way. If there is a right-of-way control point at the **lane junction**, then whether the control is a stop sign, yield sign, signal light, or own direction of travel has right-of-way is specified.

Each **lane element** is classified as to its type, i.e., right or left turn, entering, exiting, or passing through an intersection, or a roadway. If there is a right-of-way control point within the **lane element**, then whether the control is a stop sign, yield sign, signal light, or own direction of travel has right-of-way. The data structures for the **lane elements** also maintain a large number of relationships, such as, the next following and previous **lane elements**, the adjacent **lane elements**, the type of road (two-lane undivided, four-lane undivided, four-lane divided, etc.) and the **lane element**'s position relative to other **lane elements** in the roadway, whether it starts or ends at an intersection, passes through an intersection or turns, etc. They are also classified by the action taken to enter them (fork right or left, come from merge) and the action at their end (merge-to-left, merge-to-right, etc.) Each **lane element** also references its start and end **lane junction**. All of the **lane elements** carry navigational information in the form of start and end UTM co-ordinates, overall length, speed limits, heading direction, average times to traverse, name of the road, and which **route element** (used by the Destination Manager control module) that they are a part of. In a corresponding cross-referencing, each **route element** maintains a list of **lane elements** that make it up.

World model road network data includes a set of history attributes that store average times to traverse each lane element under various conditions. These should improve the planning estimates.

Output-Command: A command to **Follow Road**, or **Turn Right At Intersection**, or **Cross Through Intersection** along with the data specification of the corresponding *Goal Lane* in the form of a sequential list of **lane elements** along with a specific time to arrive at end of the goal lane. Additionally, in the case of a multi-lane travelway with lanes traveling parallel with the goal lane, the priority of own vehicle being in the goal lane is specified by parameters such as desired-to-be-in-goal-lane, or required-to-be-in-goal-lane, or only-required-to-be-in-goal-lane-when-reach-end-of-goal-lane.

Associated command data: Sequential list of the **lane elements** that identify the goal lane for own vehicle to travel. Additionally, the time of arrival at the end of the specified *GoalLane* is also commanded.

Output-Status: Present state of goal accomplishment (ie, the commanded route segment) in terms of executing, done, or error state, and identification of which intermediate **route elements** have been executed along with estimated time to complete each remaining **route element**.

Mobility Control Module

Responsibilities and Authority: *This module's primary responsibility is to determine own vehicle's present right-of-way based on the road/intersection topography, control devices, rules of the road, and the state of other vehicles and pedestrians. It determines when own vehicle can go and determines the **lane segments** that define its nominal path (see Figure 9). It also determines which objects might affect its path* *motion and places these in an **Objects-of-Interests** table to send to the Elemental Movement control module to plan a path around them.*

Retrieves and builds a sequence of **lane segments** *to specify a **goal lane-segment path**. Builds a table of active objects (with evaluation costs) that might influence own vehicle's path.*

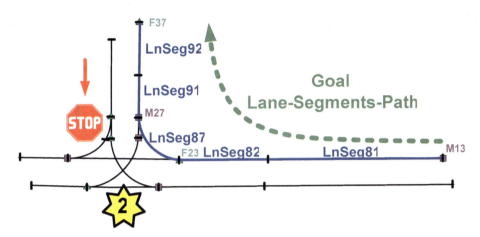

Fig. 9. Lane Segments (LnSeg) are defined as the nominal center-of-lane path representations in the form of constant curvature arcs and straight lines. They pass through the forks and merges that bound the Lane Elements. In this example, the Lane Element 18 references a linked list of lane segments containing LnSeg81 and LnSeg82, LE21 references LnSeg87, and LE32 references LnSeg91 and LnSeg92. Thus, the Lane Element is a data structure to abstract and reference the set of actual center-of-lane paths (Lane Segments). The Mobility control module manages this relationship by constantly updating the list of Lane Segments for a particular Lane Element based on real-time sensing of the lane position and curvature.

This module is commanded to go along a goal lane, which is described by a list of **lane elements**. It derives a set of **lane segments**, which are constant curvature arcs that specify the expected center-of-lane path for the **lane elements**. This module registers/aligns real-time updates of these **lane segments** as derived from sensed position of the road and lanes.

Additionally, this module applies road and intersection topography templates to fill in all of the relevant lane data, specifying such information as to which lanes cross, or merge, or do not intersect own lane (see Figure 10) These structures are then further overlaid with present detected control devices such as stop signs to specify the relative right-of-way considerations (see Figure 11) for all of the lanes nearby own vehicle's goal lane. This world model structure serves as the basis for real-time assessment of own vehicle's right of way when overlaid with other vehicles and their behavior states.

This module uses these observed vehicles' and pedestrians' world states including position, velocity and acceleration vectors, classification of expected behavior type

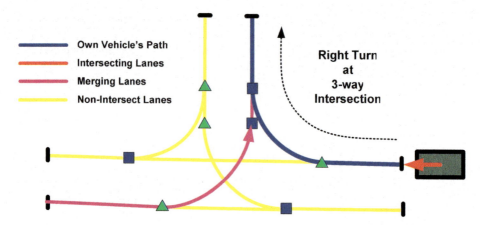

Fig. 10. A Lane Intersection Template for a 3-way intersection describes the type of intersecting interactions between own vehicle goal lane and all of the other lanes in the intersection

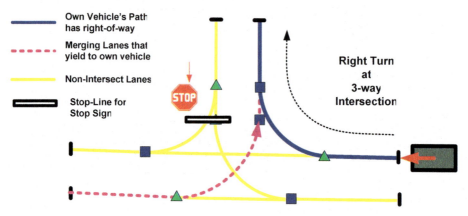

Fig. 11. The specific control devices (here – a Stop Sign) are populated into the template to arrive at the right-of-way behaviors of vehicles in all of the lanes of the intersection

(aggressive, normal, conservative), and intent (stopping at intersection, turning right, asserting right-of-way, etc.) to decide what action own vehicle should do and when to do it.

This module also provides classification of objects relevant to the present commanded driving task (see Figure 12a) by determining which objects are close enough to the identified relevant lanes to have the potential to affect own vehicle's planned path. These objects (see Figure 12b) are placed into an ***Objects-of-Interest*** table along with a number of computed parameters such as required offset distance, passing speed, cost to violate offset or passing speed, cost to collide, as well as dynamic state parameters such as velocity and acceleration vectors and other parameters. This table will serve as part of the command data set to the Elemental Movement control module.

Intelligent Control of Mobility Systems 331

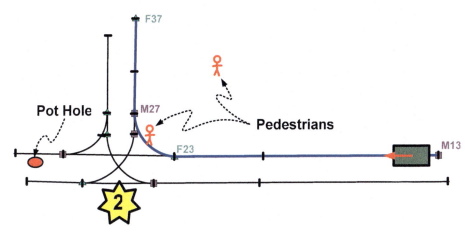

Fig. 12a. Objects relevant to on-road driving (here pedestrians and pot holes) are sensed, classified, and placed into the world model

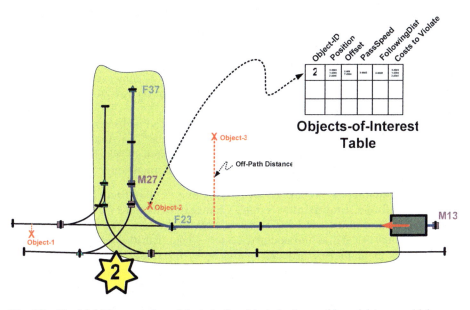

Fig. 12b. The Mobility control module tests the objects in the world model to see which ones are within a specified distance to own vehicle's goal lane (here shown as a shaded green area). In this figure, only Object-2 (a pedestrian) is in this region. Mobility manager places this object into an Objects-of-Interest table along with parameters of minimum offset distance, passing speed, following distance and cost values to exceed these. This is part of the command to the subordinate Elemental Movement control module.

Input-Command: A command to **FollowRoad**, or **TurnRightAtIntersection**, or **Cross ThroughIntersection**, etc. along with the data specification of the corresponding *Goal Lane* in the form of a sequential list of **lane elements**, with a specific time to be at the end of the goal lane. In the case of adjacent lanes in the same direction as the goal lane, the priority of own vehicle being in the goal lane is specified with parameters such as desired, or required, or required-when-reach-end-of-lane.

Input-World Model: Present estimate of this module's relevant map of the **road network** – this is a map at the level of **lane segments** which are the nominal center-of-lane specifications in the form a constant curvature arcs. This module builds out the nominal **lane segments** for each **lane element** and cross-references them with the corresponding **lane elements**. The world model contains estimates of the actual **lane segments** as provided by real-time sensing of roads and lanes and other indicators. This module will register these real-time **lane segments** with its initial nominal set.

This module's world model contains all of the surrounding recognized objects and classifies them according to their relevance to the present commanded driving task. All objects determined to have the potential to affect own vehicle's planned path are placed into an *Objects-of-Interest* table along with a number of parameters such as offset distance, passing speed, cost to violate offset or passing speed, cost to collide as well as dynamic state parameters such as velocity and acceleration and other parameters.

Other objects include detected control devices such as stop signs, yield signs, signal lights and the present state of signal lights. Regulatory signs such as speed limit, slow for school, sharp turn ahead, etc. are included.

The observed other vehicles' and pedestrians' and other animate objects' world states are contained here and include position, velocity and acceleration vectors, classification of expected behavior type (aggressive, normal, conservative), and intent (stopping at intersection, turning right, asserting right-of-way, following-motion-vector, moving-randomly, etc.).

Additionally, this module's world model contains road and intersection topography, intersection, and right-of-way templates for a number of roadway and intersection situations. These are used to aid in the determination of which lanes cross, or merge, or do not intersect own lane and to aid in the determination of right-of-way.

Output-Command: A command to **FollowLane**, **FollowRightTurnLane**, **FollowLeftTurnLane**, **StopAtIntersection**, **ChangeToLeftLane**, etc. along with the data specification of a *Goal Lane-Segment Path* in the form of a sequential list of **lane segments** that define the nominal center-of-lane path the vehicle is to follow. Additionally, the command includes an *Objects-of-Interest* table that specifies a list of objects, their position and dynamic path vectors, the offset clearance distances, passing speeds, and following distances relative to own vehicle, the cost to violate these values, these object dimensions, and whether or not they can be straddled.

Output-Status: Present state of goal accomplishment (i.e., [R: i.e. fixed] the commanded *GoalLane*) in terms of executing, done, or error state, and identification of which **lane elements** have been executed along with estimated time to complete each of the remaining **lane elements**.

Elemental Movement Control Module

Responsibilities and Authority: *This module's primary responsibility is to define the GoalPaths that will follow the commanded lane, slowing for turns and stops, while maneuvering in-lane around the objects in the Objects-of-Interests table.*

*Constructs a sequence of **GoalPaths**.*

This module is commanded to follow a sequence of **lane segments** that define a *goal lane-segment path* for the vehicle. It first generates a corresponding set of goal paths for these lane segments by determining decelerations for turns and stops as well as maximum speeds for arcs both along the curving parts of the roadway and through the intersection turns. This calculation results in a specified enter and exit speed for each goal path. This will cause the vehicle to slow down properly before stops and turns and to have the proper speeds around turns so as not to have too large a lateral acceleration. It also deals with the vehicle's ability to decelerate much faster than it can accelerate.

This module also receives a table of *Objects-of-Interests* that provides cost values to allow this module to calculate how to offset these calculate *GoalPaths* and how to vary the vehicle's speed to meet these cost requirements while being constrained to stay within some tolerance of the commanded *GoalPath*. This tolerance is set to keep the vehicle within its lane while avoiding the objects in the table. If it cannot meet the cost requirements associated with the *Objects-of-Interest* by maneuvering in its lane, it slows the vehicle to a stop before reaching the object(s) unless it is given a command from the Mobility control module allowing it to go outside of its lane. The Elemental Movement module is continually reporting status to the Mobility control module concerning how well it is meeting its goals. If it cannot maneuver around an object while staying in-lane, the Mobility module is notified and immediately begins to evaluate when a change lane command can be issued to Elemental Movement module.

This module will construct one or more *GoalPaths* (see Figure 13) with some offset (which can be zero) for each commanded lane segment based on its calculations of the values in the *Objects-of-Interest* table. It commands one goal path at a time to the Primitive control module but also passes it the complete set of planned *GoalPaths* so the Primitive control module has sufficient look-ahead information to calculate dynamic trajectory values. When the Primitive control module indicates it is nearing completion of its commanded *GoalPath*, the Elemental Movement module re-plans its set of *GoalPaths* and sends the next *GoalPath*. If, at anytime during execution of a *GoalPath*, this module receives an update of either the present commanded lane segments or the present state of any of the *Objects-of-Interest*, it performs a re-plan of the *GoalPaths* and issues a new commanded *GoalPath* to the Primitive control module.

Input-Command: A command to **FollowLane, FollowRightTurnLane, FollowLeftTurnLane, StopAtIntersection, ChangeToLeftLane**, etc. along with the data specification of a *Goal Lane-Segment Path* in the form of a sequential list of **lane segments** that define the nominal center-of-lane path the vehicle is to follow. Additionally, the command includes an *Objects-of-Interest* table that specifies a list of objects, their position and dynamic path vectors, the offset clearance distances, passing speeds, and following distances relative to own vehicle, the cost to violate these values, these object dimensions, and whether or not they can be straddled.

Input-World Model: Present estimate of this module's relevant map of the **road network** – this is a map at the level of present estimated **lane segments**. This includes

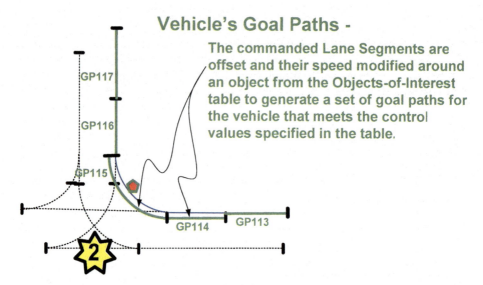

Fig. 13. Elemental Movement control module generates a set of goal paths with proper speeds and accelerations to meet turning, slowing, and stopping requirements to follow the goal lane as specified by the commanded lane segments (center-of-lane paths). However, it will modify these lane segments by offsetting certain ones and altering their speeds to deal with the object avoidance constraints and parameters specified in the Objects-of-Interest table from the Mobility control module. Here, Goal Path 114 (GP114) and GP115 are offset from the original lane segment specifications (LnSeg82 and LnSeg87) to move the vehicle's goal path far enough out to clear the object (shown in red) from the Objects-of-Interest table at the specified offset distance. The speed along these goal paths is also modified according to the values specified in the table.

the **lane segments** that are in the commanded *goal lane segment path* as well as the real-time estimates of nearby **lane segments** such as the adjacent on-coming **lane segments**. This world model also contains the continuously updated states of all of the objects carried in the *Objects-of-Interest* table. Each object's state includes the position, velocity, and acceleration vectors, and history and classification of previous movement and reference model for the type of movement to be expected such.

Output-Command: A command to **Follow_StraightLine, Follow_CirArcCW, Follow_CirArcCCW**, etc. along with the data specification of a single goal path within a sequential list of *GoalPaths* that define the nominal path the vehicle is to follow.

Output-Status: Present state of goal accomplishment (ie, commanded *goal lane-segment path*) in terms of executing, done, or error state, and identification of which **lane segments** have been executed along with estimated time to complete each of the remaining **lane segments**.

Primitive (Dynamic Trajectory) Control Module

Responsibilities and Authority: *This module's primary responsibility is to pre-compute the set of dynamic trajectory path vectors for the sequence of goal paths, and* *to control the vehicle along this trajectory. Constructs a sequence of **dynamic path vectors** which yields the speed parameters and heading vector.*

This module is commanded to follow a *GoalPath* for the vehicle. It has available a number of relevant parameters such as derived maximum allowed tangential and lateral speeds, accelerations, and jerks. These values have rolled up the various parameters of the vehicle, such as engine power, braking, center-of-gravity, wheel base and track, and road conditions such as surface friction, incline, and side slope. This module uses these parameters to pre-compute the set of dynamic trajectory path vectors (see Figure 14) at a much faster than real-time rate (100 to 1), so it always has considerable look-ahead. Each time a new command comes in from Elemental Movement (because its lane segment data was updated or some object changed state), the Primitive control module immediately begins a new pre-calculation of the dynamic trajectory vectors from its present projected position and immediately has the necessary data to calculate the Speed and Steer outputs from the next vehicle's navigational input relative to these new vectors.

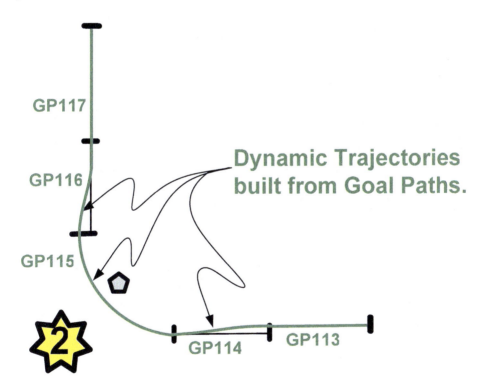

Fig. 14. Primitive/Trajectory control module pre-calculates (at 100x real-time) the set of dynamic trajectory vectors that pass through the specified goal paths while observing the constraints of vehicle-based tangential and lateral maximum speeds, accelerations, and jerks. As seen here, this results in very smooth controlled trajectories that blend across the offset goal paths commanded by the Elemental Movement control module.

On each update of the vehicle position, velocity, and acceleration from the navigation system (every 10 ms), this module projects these values to estimate the vehicle's position at about 0.4 s into the future, finds the closest stored pre-calculated dynamic trajectory path vector to this estimated position, calculates the off-path difference of this estimated position from the vector and derives the next command speed, acceleration, and heading from these relationships.

Input-Command: A command to **Follow_StraightLine**, **Follow_CirArcCW**, **Follow_CirArcCCW**, etc. with the data of a single goal path in the form of a constant curvature arc specification along with the allowed tangential and lateral maximum speeds, accelerations, and jerks. The complete set of constant curvature paths that define all of the planned output *goal paths* from the Elemental Movement control module are also provided.

Input-World Model: Present estimate of this module's relevant map of the **road network** – this is a map at the level of goal paths commanded by the Elemental Movement control module. Other world model information includes the present state of the vehicle in terms of position, velocity, and acceleration vectors. This module's world model also includes a number of parameters about the vehicle such as maximum acceleration, deceleration, weight, allowed maximum lateral acceleration, center-of-mass, present heading, dimensions, wheel base, front and rear overhang, etc.

Output-Command: Commanded maximum speed, present speed, present acceleration, final speed at path end, distance to path end, and end motion state (moving or stopped) are sent to the Speed Servo control module. Commanded vehicle center absolute heading, present arc radius, path type (straight line, arc CW, or arc CCW), the average off-path distance, and the path region type (standard-roadway, beginning-of-intersection-turn, mid-way-intersection-turn, arc-to-straight-line-blend) are sent to the Steer Servo control module.

Output-Status: Present state of goal accomplishment (ie, the commanded *goal path*) in terms of executing, done, or error state, and estimated time to complete present *goal path*. This module estimates time to the endpoint of the present goal path and outputs an advance reach goal point state to give an early warning to the Elemental Movement module so it can prepare to send out the next goal path command.

Speed Servo Control Module

Responsibilities and Authority: *This module's primary responsibility is to use the throttle and brake to cause the vehicle to move at the desired speed and acceleration and to stop at the commanded position.*

Uses a feedforward model-based servo to estimate throttle and brake-line pressure values.

This module is commanded to cause the vehicle to move at a speed with a specific acceleration constrained by a maximum speed and a final speed at the path end, which is known by a distance value to the endpoint that is continuously updated by the Primitive module.

This module basically uses a feedforward servo module to estimate the desired throttle and brake-line pressure values to cause the vehicle to attain the commanded speed and acceleration. An integrated error term is added to correct for inaccuracies in this feedforward model. The parameters for the feedforward servo are the commanded speed and acceleration, the present speed and acceleration, the road and vehicle pitch, and the engine rpm. Some of these parameters are also processed to derive rate of change values to aid in the calculations.

Input-Command: A command to **GoForwardAt***Speed* or **GoBackwardAt***Speed* or **StopAt***Point* along with the parameters of maximum speed, present speed, present acceleration, final speed at path end, distance to path end, and end motion state (moving or stopped) are sent to the Speed Servo control module.

Input-World Model: Present estimate of relevant vehicle parameters – this includes real-time measurements of the vehicle's present speed and acceleration, present vehicle pitch, engine rpm, present normalized throttle position, and present brake line pressure. Additionally, estimates are made for the projected vehicle speed, the present road pitch and the road-in-front pitch. The vehicle's present and projected positions are also utilized.

Output-Command: The next calculated value for the normalized throttle position is commanded to the throttle servo module and the desired brake-line pressure value is commanded to the brake servo module.

Output-Status: Present state of goal accomplishment (i.e., the commanded speed, acceleration, and stopping position) in terms of executing, done, or error state, and an estimate of error if this commanded goal cannot be reached.

Steer Servo Control Module

Responsibilities and Authority: *This module's primary responsibility is to control steering to keep the vehicle on the desired trajectory path.*

Uses a feedforward model-based servo to estimate steering wheel values.

This module is commanded to cause the heading value of the vehicle-center forward pointing vector (which is always parallel to the vehicle's long axis) to be at a specified value at some projected time into the future (about 0.4 seconds for this vehicle). This module uses the present steer angle, vehicle speed and acceleration to estimate the projected vehicle-center heading at 0.4 seconds into the future. It compares this value with the commanded vehicle-center heading and uses the error to derive a desired front wheel steer angle command. It evaluates this new front wheel steer angle to see if it will exceed the steering wheel lock limit or if it will cause the vehicle's lateral acceleration to exceed the side-slip limit. If it has to adjust the vehicle center-heading because of these constraints, it reports this scaling back to the Primitive module and includes the value of the vehicle-center heading it has scaled back to.

This module uses the commanded path region type to set the allowed steering wheel velocity and acceleration which acts as a safe-guard filter on steering corrections. This module uses the average off-path distance to continuously correct its alignment of its internal model of the front wheel position to actual position. It does this by noting the need to command a steer wheel value different than its model for straight ahead when following a straight section of road for a period of time. It uses the average off-path value from the Primitive module to calculate a correction to the internal model and updates this every time it follows a sufficiently long section of straight road.

Input-Command: A **GoForwardAt_*HeadingAngle*** or **GoBackwardAt *Heading Angle*** is commanded along with the parameters of vehicle-center absolute heading, present arc radius, path type (straight line, arc CW, or arc CCW), the average off-path distance, and the path region type (standard-roadway, beginning-of-intersection-turn, mid-way-intersection-turn, arc-to-straight-line-blend).

Input-World Model: Present estimate of relevant vehicle parameters – this includes real-time measurements of vehicle's lateral acceleration as well as the vehicle present heading, speed, acceleration, and steering wheel angle. Vehicle parameters of wheel lock positions, and estimated vehicle maximum lateral acceleration for side-slip calculations, vehicle wheel base and wheel track, and vehicle steering box ratios.

Output-Command: The next commanded value of steering wheel position along with constraints on maximum steering wheel velocity and acceleration are commanded to the steering wheel motor servo module.

Output-Status: Present state of goal accomplishment (i.e., the commanded vehicle-center heading angle) in terms of executing, done, or error state, along with status on whether this commanded value had to be scaled back and what the actual heading value used is.

This concludes the description of the 4D/RCS control modules for the on-road driving example.

2.3 Learning Applied to Ground Robots (DARPA LAGR)

Recently, ISD has been applying 4D/RCS to the DARPA LAGR program [7]. The DARPA LAGR program aims to develop algorithms that enable a robotic vehicle to travel through complex terrain without having to rely on hand-tuned algorithms that only apply in limited environments. The goal is to enable the control system of the vehicle to learn which areas are traversable and how to avoid areas that are impassable or that limit the mobility of the vehicle. To accomplish this goal, the program provided small robotic vehicles to each of the participants (Figure 15). The vehicles are used by the teams to develop software and a separate DARPA team, with an identical vehicle, conducts tests of the software each month. Operators load the software onto an identical vehicle and command the vehicle to travel from a start waypoint to a goal waypoint through an obstacle-rich environment. They measure the performance of the system on multiple runs, under the expectation that improvements will be made through learning.

The vehicles are equipped with four computer processors (right and left cameras, control, and the planner), wireless data and emergency stop radios, GPS receiver, inertial navigation unit, dual stereo cameras, infrared sensors, switch-sensed bumper, front wheel encoders, and other sensors listed later in the Chapter.

4D/RCS Applied to LAGR

The 4D/RCS architecture for LAGR (Figure 16) consists of only two levels. This is because the size of the LAGR test areas is small (typically about 100 m on a side, and the test missions are short in duration (typically less than 4 minutes)). For controlling an entire battalion of autonomous vehicles, there may be as many as five or more 4D/RCS hierarchical levels.

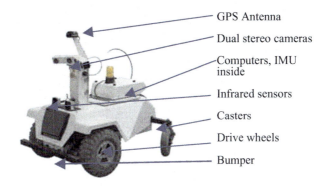

Fig. 15. The DARPA LAGR vehicle

The following sub-sections describe the type of algorithms implemented in sensor processing, world modeling, and behavior generation, as well as a section that describes the application of this controller to road following [38].

Sensory Processing

The sensor processing column in the 4D/RCS hierarchy for LAGR starts with the sensors on board the LAGR vehicle. Sensors used in the sensory processing module include the two pairs of stereo color cameras, the physical bumper and infra-red bumper sensors, the motor current sensor (for terrain resistance), and the navigation sensors (GPS, wheel encoder, and INS). Sensory processing modules include a stereo obstacle detection module, a bumper obstacle detection module, an infrared obstacle detection module, an image classification module, and a terrain slipperiness detection module.

Stereo vision is primarily used for detecting obstacles [8]. We use the SRI Stereo Vision Engine [9] to process the pairs of images from the two stereo camera pairs. For each newly acquired stereo image pair, the obstacle detection algorithm processes each vertical scan line in the reference image and classifies each pixel as GROUND, OBSTACLE, SHORT_OBSTACLE, COVER or INVALID.

A model-based learning process occurs in the SP2 module of the 4D/RCS architecture, taking input from SP1 in the form of labeled pixels with associated (x, y, z) positions from the obstacle detection module. This process learns color and texture models of traversable and non-traversable regions, which are used in SP1 for terrain classification [39]. Thus, there is two-way communication between the levels, with

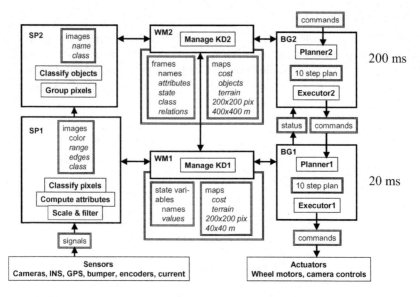

Fig. 16. Two-level instantition of the 4D/RCS hierarchy for LAGR

labeled 3D data passing up, and models passing down. The approach to model building is to make use of the labeled SP1 data including range, color, and position to describe regions in the environment around the vehicle and to associate a cost of traversing each region with its description. Models of the terrain are learned using an unsupervised scheme that makes use of both geometric and appearance information [10].

The system constructs a map of a 40 m by 40 m region of terrain surrounding the vehicle, with map cells of size 0.2 m by 0.2 m and the vehicle in the center of the map. The map is always oriented with one axis pointing north and the other east. The map scrolls under the vehicle as the vehicle moves, and cells that scroll off the end of the map are forgotten. Cells that move onto the map are cleared and made ready for new information. The model-building algorithm takes its input from SP1 as well as the location and pose of the vehicle when the data were collected.

The models are built as a kind of learning by example. The obstacle detection module identifies regions by height as either obstacles or ground. Models associate color and texture information with these labels, and use these examples to classify newly-seen regions. Another kind of learning is also used to measure traversability. This is especially useful in cases where the obstacle detection reports a region to be of one class when it is actually of another, such as when the system sees tall grass that looks like an obstacle but is traversable, perhaps with a greater cost than clear ground. This second kind of learning is learning by experience: Observing what actually happens when the vehicle traverses different kinds of terrain. The vehicle itself occupies a region of space that maps into some neighborhood of cells in the traversability cost map. These cells and their associated models are given an increased traversability weight because the vehicle is traversing them. If the bumper on the vehicle is triggered, the cell that corresponds to the bumper location and its model, if any, are given a decreased traversability weight. We plan to further modify the traversability weights by observing when the wheels on the vehicle slip or the motor has to work harder to traverse a cell.

The models are used in the lower sensory processing module, SP1, to classify image regions and assign traversability costs to them. For this process only color information is available, with the traversability being inferred from that stored in the models. The approach is to pass a window over the image and to compute the same color and texture measures at each window location as are used in model construction. Matching between the windows and the models operates exactly as it does when a cell is matched to a model in the learning stage. Windows do not have to be large, however. They can be as small as a single pixel and the matching will still determine the closest model, although with low confidence (as in the color model method for road detection described below). In the implementation the window size is a parameter, typically set to 16x16. If the best match has an acceptable score, the window is labeled with the matching model. If not, the window is not classified. Windows that match with models inherit the traversability measure associated with the model. In this way large portions of the image are classified.

World Modeling

The world model is the system's internal representation of the external world. It acts as a bridge between sensory processing and behavior generation in the 4D/RCS hierarchy by providing a central repository for storing sensory data in a unified representation. It decouples the real-time sensory updates from the rest of the system. The world model process has two primary functions: To create a knowledge database and keep it current and consistent, and to generate predictions of expected sensory input.

For the LAGR project, two world model levels have been built (WM1 and WM2). Each world model process builds a two dimensional map (200 x 200 cells), but at different resolutions. These are used to temporally fuse information from sensory processing. Currently the lower level (Sensory Processing level one, or SP1) is fused into both WM1 and WM2 as the learning module in SP2 does not yet send its models to WM. Figure 17 shows the WM1 and WM2 maps constructed from the stereo obstacle detection module in SP1. The maps contain traversal costs for each cell in the map. The position of the vehicle is shown as an overlay on the map. The red, yellow, blue, light blue, and green are cost values ranging from high to low cost, and black represents unknown areas. Each map cell represents an area on the ground of a fixed size and is marked with the time it was last updated. The total length and width of the map is 40 m for WM1 and 120 m for WM2. The information stored in each cell includes the average ground and obstacle elevation height, the variance, minimum and maximum height, and a confidence measure reflecting the "goodness" of the elevation data. In addition, a data structure describing the terrain traversability cost and the cost confidence as updated by the stereo obstacle detection module, image classification module, bumper module, infrared sensor module, etc. The map updating algorithm relies on confidence-based mapping as described in [13].

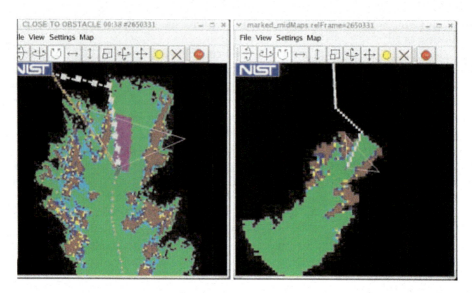

Fig. 17. OCU display of the World Model cost maps built from sensor processing data. WM1 builds a 0.2 m resolution cost map (left) and WM2 builds a 0.6 m resolution cost map (right).

We plan additional research to implement modeling of moving objects (cars, targets, etc.) and to broaden the system's terrain and object classification capabilities. The ability to recognize and label water, rocky roads, buildings, fences, etc. would enhance the vehicle's performance [14-18].

Behavior Generation

Top level input to Behavior Generation (BG) is a file containing the final goal point in UTM (Universal Transverse Mercator) coordinates. At the bottom level in the 4D/RCS hierarchy, BG produces a speed for each of the two drive wheels updated every 20 ms, which is input to the low-level controller included with the government-provided vehicle. The low-level system returns status to BG, including motor currents, position estimate, physical bumper switch state, raw GPS and encoder feedback, etc. These are used directly by BG rather than passing them through sensor processing and world modeling since they are time-critical and relatively simple to process.

Two position estimates are used in the system. Global position is strongly affected by the GPS antenna output and received signal strength and is more accurate over long ranges, but can be noisy. Local position uses only the wheel encoders and inertial measurement unit (IMU). It is less noisy than GPS but drifts significantly as the vehicle moves, and even more if the wheels slip.

The system consists of five separate executables. Each sleeps until the beginning of its cycle, reads its inputs, does some planning, writes its outputs and starts the cycle again. Processes communicate using the Neutral Message Language (NML) in a non-blocking mode, which wraps the shared-memory interface [20]. Each module also posts a status message that can be used by both the supervising process and by developers via a diagnostics tool to monitor the process.

The LAGR Supervisor is the highest level BG module. It is responsible for starting and stopping the system. It reads the final goal and sends it to the waypoint generator. The waypoint generator chooses a series of waypoints for the lowest-cost traversable path to the goal using global position and translates the points into local coordinates. It generates a list of waypoints using either the output of the A* Planner [21] or a previously recorded known route to the goal.

The planner takes a 201 X 201 terrain grid from WM, classifies the grid, and translates it into a grid of costs of the same size. In most cases the cost is simply looked up in a small table from the corresponding element of the input grid. However, since costs also depend on neighboring costs, they are automatically adjusted to allow the vehicle to continue motion. By lowering costs of unknown obstacles near the vehicle, it does not hesitate to move as it would with for example, detected false or true obstacles nearby. Since the vehicle has an instrumented bumper, the choice is to continue vehicle motion.

The lowest level module, the LAGR Comms Interface, takes a desired heading and direction from the waypoint follower and controls the velocity and acceleration, determines a vehicle-specific set of wheel speeds, and handles all communications between the controller and vehicle hardware.

Road and Path Detection in LAGR

In the LAGR environment, roads, tracks, and paths are often preferred over other terrain. A color-based image classification module learns to detect and classify these regions in the scene by their color and appearance, making the assumption that the region directly in front of the vehicle is traversable. A flat world assumption is used to estimate the 3D location of a ground pixel in the image. Our algorithm segments an image of a region by building multiple color models similar to those proposed by Tan et al. [22], who applied the approach to paved road following. For off-road driving, the algorithm was modified to segment an image into traversable and non-traversable regions. Color models are created for each region based on two-dimensional histograms of the colors in selected regions of the image. Previous approaches to color modeling have often made use of Gaussian mixture models, which assumes Gaussian color distributions. Our experiments showed that this assumption did not hold in our domain. Instead, we used color histograms. Many road detection systems have made use of the RGB color space in their methods. However, previous research [23, 24, 25] has shown that other color spaces may offer advantages in terms of robustness against changes in illumination. We found that a 30 x 30 histogram of red (R) and green (G) gave the best results in the LAGR environment.

The approach makes the assumption that the area in front of the vehicle is safe to traverse. A trapezoidal region at the bottom of the image is assumed to be ground. A color histogram is constructed for the points in this region to create the initial ground model. The trapezoidal region is the projection of a 1 m wide by 2 m long area in front of the vehicle under the assumption that the vehicle is on a plane defined by its current pose. In [26] Ulrich and Nourbakhsh addressed the issue of appearance-based obstacle detection using a color camera without range information. Their approach makes the same assumptions that the ground is flat and that the region directly in front of the robot is ground. This region is characterized by Hue and Saturation histograms and used as a model for ground. Ulrich and Nourbakhsh do not model the background, and have only a single ground model (although they observe that more complex environments would call for multiple ground models). Their work was applied to more homogeneous environments than ours, and we found that multiple models of ground are essential for good performance in the kinds of terrain used to test the LAGR vehicles. Substantial work has since been done in the DARPA LAGR program on learning traversability models from stereo, color, and texture (e.g., [14-18]). Other work, such as [19], that makes use of LADAR instead of stereo has also been of growing interest, but is not covered here.

In addition to the ground models, a background model is also built. Construction starts by randomly sampling pixels in the area above the horizon, assuming that they represent non-traversable regions. Because this area might only contain sky pixels, we extend the sampling area to 50 pixels below the horizon. The horizon is the line determined by the points where the line of sight of the cameras stops intersecting the ground plane. Once the algorithm is running, the algorithm randomly samples pixels in the current frame that the previous result identified as background. This enables the background regions to expand below the horizon. These samples are used to update the background color model using temporal fusion. Only one background model is constructed since there is no need to distinguish one type of background from another.

To enable the vehicle to remember traversable terrain with different color characteristics, multiple ground color models are learned. As new data are processed, each color distribution model is updated with new histograms, changing with time to fit changing conditions. Potential new histograms for representing ground are compared to existing models. If the difference is less than a threshold, the histogram is used to upgrade the best matching ground model. Otherwise, if a maximum number of ground models has not yet been reached, the algorithm enters a period known as learning mode. During learning mode, the algorithm monitors new histograms in an attempt to pick out the histogram that is most different from the existing ground models. This is done to avoid picking color models that contain significant amounts of overlap. In the learning mode, if a histogram is found to be more different than a previous histogram, learning mode is extended. Eventually learning mode will end, and the most different histogram is used to create a new color model.

Learning mode is turned off when the region in front of the vehicle assumed to be ground contains obstacles. This is determined by projecting obstacles from the world model map into the image. It is also disabled if the LAGR vehicle is turning faster than 10 degrees per second or if the LAGR vehicle is not moving. The algorithm can also read in a priori models of certain features that commonly appear in the LAGR tests. These include models for a path covered in mulch or in white lime.Figure 18 shows two examples of the output of the color model based classifier. Figure 18a shows a view of an unpaved road, while Figure 18b shows a path laid down in a field that the vehicle is supposed to follow.

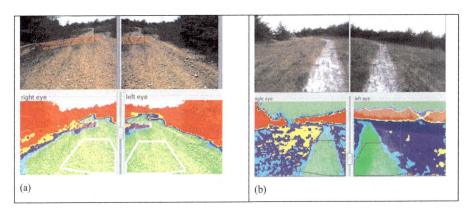

Fig. 18. a. Top: the original images, with the classification images based on the histogram color model shown underneath. Green means ground, the other colors are background regions with higher traversability costs. b. Another scene showing clearly the algorithm's ability to learn to associate traversability with a distinctively-colored path.

Given the models, the algorithm goes through every pixel in an image and calculates a ground probability based upon its color. The end result is a probability map that represents the likelihood that an area is ground. Given the pixel's color, a ground color model and the model for background, ground probability is calculated as:

$$P_{ground} = \frac{N_{ground}}{N_{ground} + N_{background}}$$

where N_{ground} is the count in the ground histogram bin indexed by the pixel, $N_{background}$ is the count in the background histogram bin indexed by the pixel, and P_{ground} is the probability that the pixel is a ground pixel. When there are multiple ground models, all are matched and the largest ground probability is selected. Thus multiple ground probabilities are calculated at each pixel. The largest ground probability is selected as the ground probability for that pixel.

A further step applied to the ground histograms is temporal fusion. This combines ground probabilities across multiple image frames to improve stability and reduce noise. The temporal fusion algorithm can be described as a running average with a parameter to adjust for the influence of new data. The update equation is:

$$P_t = \frac{(w_{t-1} \times P_{t-1}) + (c \times P)}{(w_{t-1} + c)}$$

$$w_t = w_{t-1} + c \text{ if } w_{t-1} < w_{max}$$

Where P is the current probability that the pixel should be labeled ground, P_{t-1} is the probability that the same pixel was labeled ground in the previous frame, P_t is the temporally-fused ground probability of the pixel, and w and c are weighting constants. w_{max} is the maximum number of images used for temporal fusion. The final probability map is used to determine the traversability cost at each pixel as

$$Cost = (1 - P_t) * 250$$

Costs run from 0 being most traversable to 250 being least traversable.

In order to reduce processing requirements, probabilities are calculated on a reduced version of the original image. The original image is resized to 128 x 96 pixels using an averaging filter. This step has the additional benefit of noise reduction. Experiments show that this step does not significantly impact the final segmentation. A noteworthy aspect of this algorithm is that the color models are constructed from the original image for better accuracy, whereas probabilities are calculated on a reduced version of the image for greater speed. The cost of each pixel in the image is sent to the world model with a 3D location determined using the assumption of a flat ground plane.

Summary

The NIST 4D/RCS reference model architecture was implemented on the DARPA LAGR vehicle, which was used to prove that 4D/RCS can learn. Sensor processing, world modeling, and behavior generation processes have been described. Outputs from sensor processing of vehicle sensors are fused with models in WM to update them with external vehicle information. World modeling acts as a bridge between multiple sensory inputs and a behavior generation (path planning) subsystem. Behavior generation plans vehicle paths through the world based on cost maps provided from world modeling. Road following is the example used here to describe how 4D/RCS can be applied in a two level architecture.

Future research will include completion of the sensory processing upper level (SP2) and developing even more robust control algorithms than those described above.

In the second 18 month phase of the LAGR program, NIST was tasked to develop a standard operator control unit color scheme [27] for all performers to use.

The Operator Control Unit (OCU) for a mobile robot needs to display a lot of complex information about the state and planned actions of the vehicle. This includes displays from the robot's sensors, maps of what it knows about the world around it, traces of the path it has already traveled and predictions of the path it is planning to take, and information about obstacles, clear ground, and unseen regions. The information needs to be easy to understand even by people who have no understanding of the way the control system of the robot works, and should enable them to halt the vehicle only if it is about to take an action that will cause damage.

In order to display all the information in an understandable way, it is necessary to use color to represent the different types of region. Figure 19 shows the common color scheme NIST developed that is being used for the LAGR program. The color scheme was distributed to the teams in December, 2006 for use by the January, 2007 test. Use of the color scheme has had the desired effect. The Government evaluation

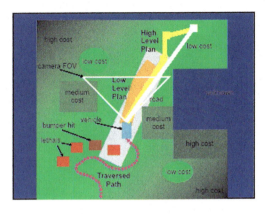 

Fig. 19. (left) A schematic showing the meaning of the colors on the map, (right) A sample NIST high resolution, short-range map

team can more easily understand the OCUs of different teams. It was a useful, if somewhat tedious process to develop the common color scheme. Work needs to be done more broadly to develop common color schemes for different application areas, such as medical images, geographic information systems, and other complex visual displays that require substantial effort to understand.

Also in the second phase, NIST has defined interfaces in a "best-of" LAGR controller for performers to send their best algorithms to NIST for plug-and-play testing against other performers' algorithms. This work is currently ongoing and expected to be completed in 2008.

3 Standards and Performance Measurements

3.1 Autonomy Levels for Unmanned Systems (ALFUS)

The ICMS Program emphasizes to the mobile robotics research community a standard set of related terminology and representation of knowledge. The Autonomy Levels for Unmanned Systems (ALFUS) Ad Hoc Work Group [19], led by NIST, aims at developing a framework to facilitate the characterization of the autonomy capabilities of unmanned systems (UMS). The Group was formed, in 2003, to respond to the user community needs, including many Federal Agencies and private industry. The group has been holding quarterly workshops ever since.

An ALFUS premise is that the autonomous capabilities are characterized with three significant aspects: mission complexity, environmental complexity, and human independence, as shown in Figure 20. Detailed metrics, in turn, further characterize each of the three aspects.

ALFUS is collaborating with the U.S. Army Future Combat System (FCS) (http://www.army.mil/fcs/) and a number of other UMS programs on the issues of autonomy requirements, testing, and evaluation. Interim results have been published [20, 21, 22]. They have been significantly referenced in various public documents,

Fig. 20. ALFUS framework contextual autonomous capability model

including the ASTM Standards E2521-07, F 2395-05, and F 2541-06 for various UMSs as well as the U.S. Army UNMANNED and AUTONOMOUS SYSTEMS TESTING Broad Agency Announcement.

3.2 Joint Architecture for Unmanned Systems (JAUS)

The Joint Architecture for Unmanned Systems (JAUS) is a standard for interoperability between components of unmanned robotic vehicle systems such as unmanned ground, air and underwater vehicles (UGV, UAV and UUV, respectively). JAUS is sponsored by the Office of the Under Secretary of Defense for Acquisition, Technology and Logistics through the Joint Ground Robotics Enterprise (JGRE). It is mandated for use by all JGRE programs. The goals of JAUS are to reduce life cycle costs, reduce development and integration time, provide a framework for technology insertion, and accommodate expansion of existing systems with new capabilities. JAUS is a standard published by the Society of Automotive Engineers (SAE) via their Aerospace Avionics Systems Division AS-4 committee on Unmanned Systems.

JAUS is a component based, message-passing architecture that specifies data formats and methods of communication among computing nodes. It defines messages and component behaviors that are independent of technology, computer hardware, operator use, and vehicle platforms and isolated from mission. It uses the SAE Generic Open Architecture (GOA) framework to classify the interfaces.

JAUS benefits from an active user and vendor community, including Government programs such as FCS, academia such as University of Florida and Virginia Polytechnic Institute and State University (Virginia Tech), and industry such as Applied Perception, Autonomous Solutions, Kairos Autonomi, OpenJAUS, RE2 and TORC Technologies, which could easily be identified with a search on the internet. Users define requirements and sponsor pilot projects. Together with vendors, they participate in testing that helps JAUS evolve toward better performance and to support new technology.

JAUS products include commercially-available ground robots as well as software development kits (SDKs) that help robot vendors put a JAUS-compliant interface on their new or legacy products. Open-source implementations of JAUS exist that serve as reference implementations and help speed deployment.

NIST was one of two teams that participated in an early Red Team/Blue Team interoperability test. The test objective was to determine how well the JAUS specification enabled independent implementors to build JAUS-compliant systems that worked seamlessly together. For this particular test, one team built a JAUS component that continually produced vehicle position data. The other team built a component that displayed the current vehicle position in real time. Neither team knew the identity of the other, and interacted only with the JGRE test sponsor. The sponsor provided information necessary for interoperability but not part of the JAUS specification, such as the particular communication mechanism (in this case, TCP/IP Ethernet). The sponsor also resolved ambiguities in the specification that resulted in different interpretations of how JAUS works. These resolutions were provided to the working groups responsible for the specification for incorporation into the next versions of the relevant documents.

Since those early tests, the JAUS working group has expanded their test activities. These are now conducted by the Experimental Test Group (ETG). The ETG is chartered with implementing and testing proposed changes or extensions of the JAUS specification, and reporting back with recommendations for how these proposals should be formalized into the specification. The ETG is comprised of a core group of JAUS experts from the vendor community who are well equipped to quickly build and test new JAUS capabilities. The JAUS ETG has invested in infrastructure development, such as setting up a distributed virtual private network that allows participants to connect to a JAUS system from any location.

3.3 The Intelligent Systems (IS) Ontology

The level of automation in ground combat vehicles being developed for the Army's objective force is greatly increasing over the Army's legacy force. This automation is taking many forms in emerging ground vehicles, varying from operator decision aides to fully autonomous unmanned systems. The development of these intelligent systems requires a thorough understanding of all of the intelligent behavior that needs to be exhibited by the system so that designers can allocate functionality to humans and/or machines. Traditional system specification techniques focus heavily on the functional description of the major systems of a vehicle and implicitly assume that a well-trained crew would operate these systems in a manner to accomplish the tactical mission assigned to the vehicle. In order to allocate some or all of these intelligent behaviors to machines in future ground vehicles, it is necessary to be able to identify and describe these intelligent behaviors.

The U.S. Army Tank Automotive Research, Development and Engineering Center (TARDEC) has funded NIST to explore approaches to model the ground vehicle domain with explicit representation of intelligent behavior. This exploration has included the analysis of modeling languages (i.e., UML, DAML, OWL) as well as reference architectures. A major component of this effort has been the development of an Intelligent Systems (IS) Ontology.

NIST has taken the view that an IS can be viewed as a multi-agent system, where agents can represent components within the vehicle (e.g., a propulsion system, a lethality system, etc). In addition, an Intelligent Ground Vehicle (IGV), as a whole, can serve as a single agent within a troop, platoon, or section, where multiple IGVs are present. In order for a group of agents to work together to accomplish a common goal, they must be able to clearly and unambiguously communicate with each other without the fear of loss of information or misinterpretation. The IS Ontology has been used to specify a common lexicon and semantics to address this challenge.

The IS Ontology uses that OWL-S upper ontology [23] as the underlying representation to document 4D/RCS in a more open XML (eXtensible Markup Language) format. OWL-S is a service ontology, which supplies a core set of markup language constructs for describing the properties and capabilities of services in an unambiguous, computer-interpretable format. This ontology has been built within the Protégé framework [24], which is an ontology editor, a knowledge-base editor, as well as an open-source, Java tool that provides an extensible architecture for the creation of customized knowledge-based applications.

The IS Ontology is based on the concept of agents, service that the agents can perform, and procedures that the agents follow to perform the services. Figure 21 shows an agent hierarchy for a light cavalry troop. A detailed description of this hierarchy is outside the scope of this chapter. Also specified in the constraints for each class is whom each agent can send external service requests to and who they can receive them from. Any activity that can be called by another agent is considered a service in OWL-S. Any activity that the agent performs internally that cannot be externally requested is called a process. As such, "Conduct A Tactical Road March to an Assembly Area" is modeled as a service that is provided by a Troop agent (and can be called by a Squadron agent). The Troop agent can call services provided by other agents. In this example, a service called "Conduct Route Reconnaissance" is defined and associated with the Scout Platoon agent.

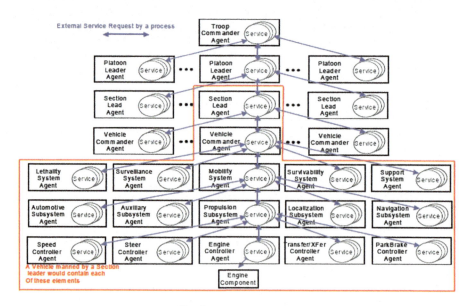

Fig. 21. Agent hierarchy

Process models in OWL-S are used to capture the steps that must be accomplished to carry out the service, and the ordering constraints on those steps. Each step can be performed internally by the agent or could involve making an external service request (a service call) to another agent. OWL-S provides a number of control constructs that allow one to model just about any type of process flow imaginable. Control constructs provided in OWL-S have been sufficient to model all the behaviors explored to date.

Environmental entities and their attributes are a primary output of the 4D/RCS methodology. These include other vehicles, bridges, vegetation, roads, water bodies; anything that is important to perceive in the environment relative to task that is being performed. An environment ontology in OWL-S has been built from the bottom up (i.e., including only entities that prove to be important). Environmental ontologies have started to be explored to see what could be leveraged.

3.4 DOT Integrated Vehicle Based Safety System (IVBSS)

The Transportation Project within the ICMS Program, although not 4D/RCS based, is an important part of the program. While autonomous vehicles on the nations highway are still many years in the future, components of intelligent mobility systems are finding their way into commercial crash warning systems (CWS). The US Department of Transportation, in attempt to accelerate the deployment of CSW, recently initiated the Integrated Vehicle-Based Safety System (IVBSS) program designed to incorporate forward collision, side collision and road-departure warning functions into a single integrated CSW for both light vehicles and heavy trucks [25]. Current analyses estimate that IVBSS has the potential to address 3.6 M crashes annually, of which 27,500 result in one or more fatalities.

NIST's role in the IVBSS program is to assist in the development of objective tests and to evaluate system performance independently during the tests. NIST's approach for measurement-based performance evaluation starts by developing a set of scenarios describing the problem, which in this case evolved from a DOT analysis of the crash database statistics. The scenarios lead to a set of track- and road-based test procedures to determine the system's effectiveness in dealing with the problems. Current plans call for carrying out over 34 track-based tests. Figure 22 provides an example of a multiple threat scenario that forces the system to recognize and respond to both a forward- and side-collision situation.

The pass/fail criteria for the tests include metrics that quantify acceptable performance. Various Crash Prevention Boundary (CPB) equations [26] determine the acceptable time for a warning; a driver cannot respond to late warnings and early

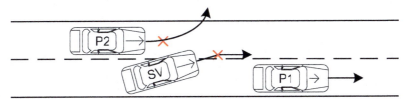

Fig. 22. Multiple threat test scenario. SV (equipped with IVBSS) encounters stopped vehicle (P1) and attempts to avoid collision by changing lanes into second vehicle (P2).

warning may annoy the driver. To promote objectivity further, the tests rely on an independent measurement system (IMS) to provide performance data as opposed to using measurements taken by the system under test. NIST is currently developing a 3rd generation IMS that incorporates calibrated cameras (for lane geometry measurements) and two laser-scanners (for obstacle range measurements). Figure 23 shows the sensor package mounted on the front of the NIST/DOT test bed vehicle. The IMS design allows for quick installation on cars and trucks. Data is collected in real time and a suite of software exists for post-process analysis.

NIST is developing static and dynamic accuracy tests to verify the IMS meets requirements that range error be less than 5% of the actual range. Static tests evaluate sensor performance from a stationary position with targets at different ranges, reflectivity and orientations. The mean errors (mean of differences between measured range and actual range) serve as a calibration factor and the standard deviations of the errors define the uncertainty. Dynamic tests yield insight into sensor performance when operated from a fast moving vehicle (e.g., highway speeds). No standard procedure exists for dynamic testing and NIST is working on a novel approach involving precise time measurement, using a GPS receiver, of when a range reading takes place. For a dynamic test, an optical switch mounted on the vehicle senses when the vehicle crosses over reflectors placed on a track at known distances from a target. The switch causes a GPS receiver to latch the GPS time. The approach requires that all measurements, for example video streams, laser-scanner range measurements, etc., be stamped using GPS time (which is the case with the IVBSS warning system and the

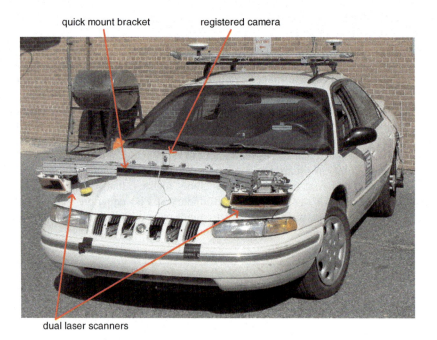

Fig. 23. IMS consisting of calibrated cameras and laser-scanners mounted on the NIST/DOT testbed vehicle

IMS). The range error comes from the difference (at the GPS time the vehicle crosses the reflector) between the laser-scanner's measured range to the target and the known range. We compute a mean error and uncertainty from measurements taken over several laps around a track. Current results indicate the laser scanner uncertainty is approximately 1% of the actual range.

Program goals for this year include track testing and on-road testing of an IVBSS developed for light vehicles and a second IVBSS developed for heavy truck.

4 Testbeds and Frameworks

4.1 USARSim/MOAST Framework

Many of the systems described in this chapter may be viewed as intelligent embodied agents. These agents require an environment to operate in, an embodiment that allows them to affect and move in the environment, and intelligence that allows them to execute useful behaviors that have the desired outcome in the environment. There are many aspects of the development of these agents in which simulations can play a useful role. If correctly implemented, simulation can be an effective first step in the development and deployment of new algorithms. Simulation environments enable researchers to focus on the algorithm development without having to worry about hardware aspects of the robots such as maintenance, availability, and operating space. Simulation provides extensive testing opportunities without the risk of harm to personnel or equipment. Major components of the robotic architecture (for example, advanced sensors) that may be too expensive for an institution to purchase can be simulated and enable the developers to focus algorithm development. The remainder of this section will present an overview of a control framework that provides all three aspects of the embodied agent. Urban Search and Rescue Simulation (USARSim) provides the environment and embodiment, while Mobility Open Architecture Simulation and Tools (MOAST) provides the intelligence.

Urban Search and Rescue Simulation (USARSim)
The current version of Urban Search and Rescue Simulation (USARSim) [2] is based on the UnrealEngine2[1] game engine that was released by Epic Games as part of UnrealTournament 2004. This engine may be inexpensively obtained by purchasing the Unreal Tournament 2004 game. The USARSim extensions may then be freely downloaded from *sourceforge.net/projects/usarsim*. The engine handles most of the basic mechanics of simulation and includes modules for handling input, output (3D rendering, 2D drawing, and sound), networking, physics and dynamics. USARSim uses these features to provide controllable camera views and the ability to control multiple robots. In addition to the simulation, a sophisticated graphical development environment and a variety of specialized tools are provided with the purchase of Unreal Tournament.

---

[1] Certain commercial software and tools are identified in this paper in order to explain our research. Such identification does not imply recommendation or endorsement by the authors, nor does it imply that the software tools identified are necessarily the best available for the purpose.

The USARSim framework builds on this game engine and consists of:

- standards that dictate how agent/game engine interaction is to occur,
- modifications to the game engine that permit this interaction
- an Application Programmer's Interface (API) that defines how to utilize these modifications to control an embodied agent in the environment
- 3-D immersive test environments
- models of several commercial and laboratory robots and effectors
- models of commonly used robotic sensors

The USARSim interaction standards consist of items such as robot coordinate frame definitions and unit declarations while the API specifies the command vocabulary for robot/sensor control and feedback. Both of these items have become the *de facto* standard interfaces for use in the RoboCup Rescue Virtual Competition which utilizes USARSim to provide an annual Urban Search and Rescue competition. In 2007 this competition had participation from teams representing 5 countries.

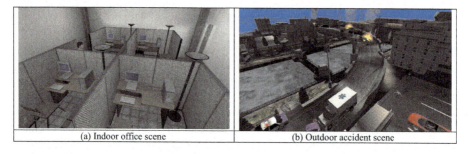

(a) Indoor office scene (b) Outdoor accident scene

Fig. 24. Sample USARSim Environments

Fig. 25. Simulated Talon Robot

Highly realistic environments are also provided with the USARSim release. Example indoor and outdoor environments may be seen in 24(a) and 24(b). In addition, a provided editor and the ability to import models simplifies the creation of additional worlds. In addition to environments, USARSim provides numerous robot and sensor models. Figure 25 shows the virtual version of the Talon robot. This robot features a simulated track and arm with gripper. In addition to laboratory robots, aerial, road vehicles, commercial robots (both for manufacturing and bomb disposal), and robotic arms are modeled.

USARSim does not provide a robot controller. However, several open source controllers may be freely downloaded. These include the community developed Mobility Open Architecture Simulation and Tools (MOAST) controller (sourceforge.net/ projects/moast), the player middle-ware (sourceforge.net/projects/playerstage), and any of the winning controllers from previous year's competitions (2006's and 2007's winning controllers may be found on the robocuprescue wiki (www.robocuprescue.org/ wiki). A description of the winning algorithms may be found in [1].

Mobility Open Architecture Simulation and Tools (MOAST)
MOAST is a framework that provides a baseline infrastructure for the development, testing, and analysis of autonomous systems that is guided by three principles:

- Create a multi-agent simulation environment and tool set that enables developers to focus their efforts on their area of expertise without having to have the knowledge or resources to develop an entire control system.
- Create a baseline control system which can be used for the performance evaluation of the new algorithms and subsystems.
- Create a mechanism that provides a smooth gradient to migrate a system from a purely virtual world to an entirely real implementation.
- MOAST has the 4D/RCS architecture at its core (described elsewhere in this chapter) and consists of the additional components of control modules, interface specs, tools, and data sets. MOAST is fully integrated with the USARSim simulation system and may communicate with real hardware through the Player interface.

MOAST provides an implementation of primitive echelon through the section echelon of the 4D/RCS reference model architecture. This implementation is not designed to be complete, but rather is designed to provide examples of control strategies and a starting point for further research. The framework provides methods of mobility control for vehicles including Ackerman steered, skid steered, omni-drive, helicopter-type flying machines, and water craft. Control modalities increase in complexity as one moves up the hierarchy and include velocity/steering angle control, waypoint following, and an exploration behavior.

All of the control modules are designed to be self-contained and fully accessible through well-defined interfaces. This allows a developer to create a module that conforms to the specifications and replace any MOAST provided system. The idea is to allow a researcher to utilize all of the architecture except for the modules that are in the area of their expertise. These modules would be replaced with their own research code.

In order to simplify this replacement, several debug and diagnostic tools are also provided. These allow for unit testing of any module by providing a mechanism to send any command, status, and data into a module that is under test. In this way, a module may be fully and repeatably tested. Once the system is debugged in simulation, the standardized interfaces allow the user to slowly move systems from simulation to actual hardware. For example, planning algorithms may be allowed to control the actual vehicle while sensing may be left in simulation.

4.2 PRediction In Dynamic Environments (PRIDE) Framework

There have been experiments performed with autonomous vehicles during on-road navigation. Perhaps the most successful was that of Prof. Dr. Ernst Dickmanns [31] as part of the European Prometheus project in which the autonomous vehicle performed a trip from Munich to Odense (>1600 km) at a maximum velocity of 180 km/h. Although the vehicle was able to identify and track other moving vehicles in the environment, it could only make basic predictions of where those vehicles were expected to be at points in the future, considering the vehicle's current velocity and acceleration.

What is missing from all of these experiments is a level of situation awareness of how other vehicles in the environment are expected to behave considering the situation in which they find themselves. To date, the authors are not aware of any autonomous vehicle efforts that account for this information when performing path planning. To address this need, a framework, called PRIDE (PRediction in Dynamic Environments) was developed that provides an autonomous vehicle's planning system with information that it needs to perform path planning in the presence of moving objects [32]. The underlying concept is based upon a multi-resolutional, hierarchical approach that incorporates multiple prediction algorithms into a single, unifying framework. This framework supports the prediction of the future location of moving objects at various levels of resolution, thus providing prediction information at the frequency and level of abstraction necessary for planners at different levels within the hierarchy. To date, two prediction approaches have been applied to this framework.

At the lower levels, estimation theoretic short-term predictions is used via an extended Kalman filter-based algorithm using sensor data to predict the future location of moving objects with an associated confidence measure [33]. At the higher levels of the framework, moving object prediction needs to occur at a much lower frequency and a greater level of inaccuracy is tolerable. At these levels, moving objects are identified as far as the sensors can detect, and a determination is made as to which objects should be classified as "objects of interest". In this context, an object of interest is an object that has a possibility of affecting the path in the planning time horizon. Once objects of interest are identified, a moving object prediction approach based on situation recognition and probabilistic prediction algorithms is used to predict where object will be at various time steps into the future. Situation recognition is performed using spatio-temporal reasoning and pattern matching with an *a priori* database of situations that are expected to be seen in the environment.

The algorithms are used to predict the future location of moving objects in the environment at longer time planning horizons on the order of tens of seconds into the future with plan steps at about one second intervals.

The steps within the algorithm (shown in Figure 26) are:

- For each vehicle on the road (α), the algorithm gets the current position and velocity of the vehicle by querying external programs/sensors (β).
- For each set of possible future actions (δ), the algorithm creates a set of next possible positions and assigns an overall cost to each action based upon the cost incurred by performing the action and the cost incurred based upon the vehicle's proximity to static objects. An underlying cost model is developed to represent these costs.
- Based upon the costs determined in Step 2, the algorithm computes the probability for each action the vehicle may perform (ε).
- Predicted Vehicle Trajectories (PVT) (ξ) are built for each vehicle which will be used to evaluate the possibility of collision with other vehicles in the environment. PVTs are a vector that indicates the possible paths that a vehicle will take within a predetermined number of time steps into the future.
- For each pair of PVTs (η), the algorithm checks if a possible collision will occur (where PVTs intersect) and assigns a cost if collision is expected. In this step, the probabilities of the individual actions (θ) are recalculated, incorporating the risk of collision with other moving objects, as shown in Figures 15a-c.

At the end of the main loop, the future positions with the highest probabilities for each vehicle represent the most likely location of where the vehicles will be in the future. More information about the cost-based probabilistic prediction algorithms can be found in [33].

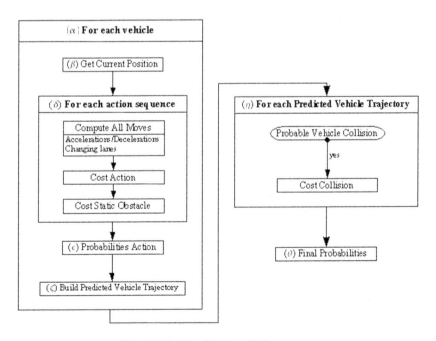

Fig. 26. Moving object prediction process

4.3 Industrial Automated Guided Vehicles

Study of Next Generation Manufacturing Vehicles
This effort, called the Industrial Autonomous Vehicles (IAV) Project, aims to provide industries with standards, performance measurements, and infrastructure technology needs for the material handling industry. The NIST ISD has been working with the material handling industry, specifically on automated guided vehicles (AGV's), to develop next generation vehicles. A few example accomplishments in this area include: determining the high impact areas according to the AGV industry, partnering with an AGV vendor to demonstrate pallets visualization using LADAR towards autonomous truck unloading, and demonstrating autonomous vehicle navigation through unstructured facilities. Here, we briefly explain each of these points.

Generation After Next AGV
NIST recently sponsored a survey of AGV manufacturers in the US, conducted by Richard Bishop Consulting, to help determine their "generation-after-next" technology needs. Recognizing that basic engineering issues to enhance current AGV systems and reduce costs are being addressed by AGV vendors, the study looks beyond today's issues to identify needed technology breakthroughs that could open new markets and improve US manufacturing productivity. Results of this study are described in [34].

Within the survey and high on the list, AGV vendors look to the future for: reduced vehicle costs, navigation in unstructured environments, onboard vehicle processing, 3D imaging sensors, and transfer of advanced technology developed for Department of Defense. Current AGVs are "guided" by wire, laser or other means, operate in structured environments tailored to the vehicle, have virtually no 3D sensing and operate from a host computer with limited onboard-vehicle control.

Visualizing Pallets
Targeting the high impact are of using 3D imaging sensors on AGV's, NIST ISD teamed with Transbotics, an AGV vendor, to visualize pallets using panned line-scan LADAR towards autonomous truck unloading. [35] A cooperative agreement between NIST and Transbotics allowed NIST to: 1) set up mock pallets, conveyer and truck loading on a loading dock, 2) to develop software to visualize pallets, the conveyer and the truck in 3D space, and 3) verify if the pallet, conveyor and truck are in their expected location with respect to the AGV. The project was successful on mock components used at NIST and software was transferred to Transbotics for implementation on their AGV towards use in a production facility.

Navigation Through Unstructured Facilities
Also targeting a high impact AGV industry requested area, the ICMS Program has been transferring technology from defense mobility projects through its' IAV Project to the AGV industry. By focusing on AGV industry related challenges, for example autonomous vehicle navigation through unstructured facilities [36], the IAV project attempts to provide improved AGV capabilities to do more than point–to–point, part pick-up/delivery operations. For example, AGV's could: avoid obstacles and people in the vehicle path, adapt to facilities instead of vice versa, navigate both indoors and

outdoors using the same adaptable absolute vehicle position software modules - all towards doing more with end users' vehicle capital investments and developing niche markets.

A number of changes were made to the LAGR control system software in order to transfer the military outdoor vehicle application to an indoor industrial setting. Two RFID sensors, batteries, laptop, and network hub were added. Active RFID sensors were integrated into the vehicle position estimate. Also, a passive RFID system was used including tags that provide a more accurate vehicle position to within a few centimeters. RFID systems updates replaced the outdoor GPS positioning system updates in the controller.

The control system also needed to be less aggressive for safety of people and equipment, use stereo vision indoors, negotiate tighter corners than are typically encountered outdoors, display facility maps and expected paths (see Figure 27), and many other modifications detailed in [36]. The demonstration was successful and allowed the AGV industry to view how vehicles could adapt to a more cluttered facility than AGVs typically navigate.

Future research will include integration of a 2D safety sensor to eliminate false positives on obstacles near ground level caused by low stereo disparity. Demonstration of

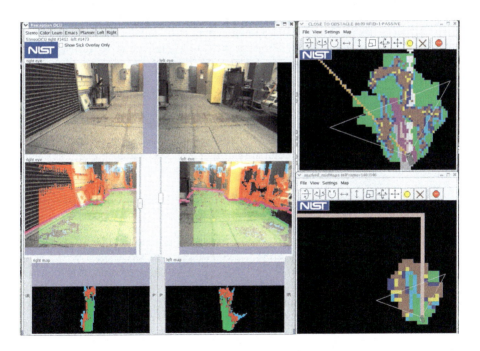

Fig. 27. LAGR AGV Graphical Displays – right and left stereo images (upper left); images overlaid with stereo obstacle (red) and floor (green) detection and 2D scanner obstacle detection (purple) (middle left); right and left cost maps (lower left); low level map (upper right); and high level map (lower right)

controlling more than one intelligent vehicle at a time in the same unstructured environment along with other moving obstacles is also planned.

5 Conclusions and Continuing Work

The field of autonomous vehicles has grown tremendously over the past few years. This is perhaps most evident by the performance of these vehicles in the DARPA-sponsored Grand Challenge events which occurred in 2204, 2005 and most recently in 2007 [37]. The purpose of the DARPA Grand Challenge was to develop autonomous vehicle technologies that can be applied to military requirements. The Grand Challenge courses gradually got harder, with the most recent event incorporating moving objects in the environment and on-road driving. The performance of the vehicles continuously improved, even as the environment got more difficult. This is in part to the advancement of technologies that are being explored as part of the ICMS program. The ICMS Program and its development of 4D/RCS have been ongoing for nearly 30 years with the goal to provide architectures and interface standards, performance test methods and data, and infrastructure technology needed by US manufacturing industry and government agencies in developing and applying intelligent control technology to mobility systems to reduce cost, improve safety, and save lives.

The 4D/RCS has been the standard intelligent control architecture on many of the Defense, Learning, and Industry Projects providing application to respective real world issues. The Transportation Project provides performance analysis of the latest mobile system sensor advancements. And the Research and Engineering Projects allow autonomy capabilities to be defined along with simulation and prediction efforts for mobile robots.

Future ICMS efforts will focus deeper into these projects with even more autonomous capabilities expected. As well, a broader application to robots supporting humans in manufacturing, construction, and farming are expected once major key intelligent mobility elements in perception and control are solved.

References

[1] Albus, J.S., Huang, H.-M., Messina, E., Murphy, K., Juberts, M., Lacaze, A., Balakirsky, S., Shneier, M.O., Hong, T., Scott, H., Horst, J., Proctor, F., Shackleford, W., Szabo, S., Finkelstein, R.: 4D/RCS Version 2.0: A Reference Model Architecture for Unmanned Vehicle Systems. NIST, Gaithersburg, MD, NISTIR 6912 (2002)

[2] Balakirsky, S., Messina, E., Albus, J.S.: Architecting a Simulation and Development Environment for Multi-Robot Teams. In: Proceedings of the International Workshop on Multi Robot Systems (2002)

[3] Balakirsky, S.B., Chang, T., Hong, T.H., Messina, E., Shneier, M.O.: A Hierarchical World Model for an Autonomous Scout Vehicle. In: Proceedings of the SPIE 16th Annual International Symp. on Aerospace/Defense Sensing, Simulation, and Controls, Orlando, FL, April 1-5 (2002)

[4] Albus, J.S., Juberts, M., Szabo, S.: RCS: A Reference Model Architecture for Intelligent Vehicle and Highway Systems. In: Proceedings of the 25th Silver Jubilee International Symposium on Automotive Technology and Automation, Florence, Italy, June 1-5 (1992)

[5] Bostelman, R.V., Jacoff, A., Dagalakis, N.G., Albus, J.S.: RCS-Based RoboCrane Integration. In: Proceedings of the International Conference on Intelligent Systems: A Semiotic Perspective, Gaithersburg, MD, October 20-23 (1996)

[6] Madhavan, R., Messina, E., Albus, J. (eds.): Low-Level Autonomous Mobility Implementation part of Chapter 3: Behavior Generation in the book, Intelligent Vehicle Systems: A 4D/RCS Approach (2007)

[7] Jackel, Larry: LAGR Mission, DARPA Information Processing Technology Office, http://www.darpa.mil/ipto/programs/lagr/index.htm

[8] Konolige, K.: SRI Stereo Engine, http://www.ai.sri.com/~konolige/svs/

[9] Tan, C., Hong, T., Shneier, M., Chang, T.: Color Model-Based Real-Time Learning for Road Following. In: Proc. of the IEEE Intelligent Transportation Systems Conference, Toronto, Canada (submitted, 2006)

[10] Oskard, D., Hong, T., Shaffer, C.: Real-time Algorithms and Data Structures for Underwater Mapping. In: Proceedings of the SPIE Advances in Intelligent Robotics Systems Conference, Boston, MA (November 1988)

[11] Shackleford, W.: The NML Programmer's Guide (C++ Version), http://www.isd.mel.nist.gov/projects/rcslib/NMLcpp.html

[12] Heyes-Jones, J.: A* algorithm tutorial, http://us.geocities.com/jheyesjones/astar.html

[13] Tan, C., Hong, T., Shneier, M., Chang, T.: Color Model-Based Real-Time Learning for Road Following. In: Proceedings of the IEEE Intelligent Transportation Systems Conference (2006)

[14] He, Y., Wang, H., Zhang, B.: Color-based road detection in urban traffic scenes. IEEE Transactions on Intelligent Transportation Systems 5(4), 309–318 (2004)

[15] Kristensen, D.: Autonomous Road Following. PhD thesis, KTH Royal Institute of Technology, Stockholm, Sweden (2004)

[16] Lin, X., Chen, S.: Color image segmentation using modified HSI system for road following. In: IEEE International Conference on Robotics and Automation, pp. 1998–2003 (1991)

[17] Ulrich, I., Nourbakhsh, I.: Appearance-Based Obstacle Detection with Monocular Color Vision. In: Proceedings of the AAAI National Conference on Artificial Intelligence (2000)

[18] Shneier, M., Bostelman, R., Albus, J.S., Shackleford, W., Chang, T., Hong, T.: A Common Operator Control Unit Color Scheme for Mobile Robots. National Institute of Standards and Technology, Gaithersburg, MD (August 2007)

[19] http://www.isd.mel.nist.gov/projects/autonomy_levels/

[20] Huang, H.-M.: The Autonomy Levels for Unmanned Systems (ALFUS) Framework–Interim Results. In: Performance Metrics for Intelligent Systems (PerMIS) Workshop, Gaithersburg, Maryland (2006)

[21] Huang, H.-M., et al.: Characterizing Unmanned System Autonomy: Contextual Autonomous Capability and Level of Autonomy Analyses. In: Proceedings of the SPIE Defense and Security Symposium (April 2007)

[22] Huang, H.-M. (ed.): Autonomy Levels for Unmanned Systems (ALFUS) Framework, vol. I, Terminology, NIST Special Publication 1011, National Institute of Standards and Technology, Gaithersburg (2004)

[23] The OWL Services Coalition, OWL-S 1.0 Release (2003), http://www.daml.org/services/owl-s/1.0/owl-s.pdf

[24] Schlenoff, C., Washington, R., Barbera, T.: Experiences in Developing an Intelligent Ground Vehicle (IGV) Ontology in Protege. In: Proceedings of the 7th International Protege Conference, Bethesda, MD (2004)
[25] http://www.its.dot.gov/ivbss
[26] Szabo, S., Wilson, B.: Application of a Crash Prevention Boundary Metric to a Road Departure Warning System. In: Proceedings of the Performance Metrics for Intelligent Systems (PerMIS) Workshop, NIST, Gaithersburg, MD, August 24-26 (2004), http://www.isd.mel.nist.gov/documents/szabo/PerMIS04.pdf
[27] Balakirsky, S., Scrapper, C., Carpin, S., Lewis, M.: USARSim: Providing a Framework for Multi-robot Performance Evaluation. In: Proceedings of the Performance Metrics for Intelligent Systems (PerMIS) Workshop (2006)
[28] Scrapper, C., Balakirsky, S., Messina, E.: MOAST and USARSim - A Combined Framework for the Development and Testing of Autonomous Systems. In: SPIE 2006 Defense and Security Symposium (2006)
[29] USARSim Homepage (2007), http://usarsim.sourceforge.net/
[30] MOAST Homepage (2007), http://sourceforge.net/projects/moast/
[31] Dickmanns, E.D.: A General Dynamic Vision Architecture for UGV and UAV. Journal of Applied Intelligence 2, 251 (1992)
[32] Schlenoff, C., Ajot, J., Madhavan, R.: PRIDE: A Framework for Performance Evaluation of Intelligent Vehicles in Dynamic, On-Road Environments. In: Proceedings of the Performance Metrics for Intelligent Systems (PerMIS) 2004 Workshop (2004)
[33] Madhavan, R., Schlenoff, C.: The Effect of Process Models on Short-term Prediction of Moving Objects for Autonomous Driving. International Journal of Control, Automation and Systems 3, 509–523 (2005)
[34] Bishop, R.: Industrial Autonomous Vehicles: Results of a Vendor Survey of Technology Needs. Bishop Consulting (February 16, 2006)
[35] Bostelman, R., Hong, T., Chang, T.: Visualization of Pallets. In: Proc. SPIE Optics East 2006 Conference, Boston, MA, USA, October 1-4 (2006)
[36] Bostelman, R., Hong, T., Chang, T., Shackleford, W., Shneier, M.: Unstructured Facility Navigation by Applying the NIST 4D/RCS Architecture. In: CITSA 2006 Conference Proceedings, July 20-23 (2006)
[37] Iagnemma, K., Buehler, M.: Special Issues on the DARPA Grand Challenge. Journal of Field Robotics 23(8&9), 461–835 (2006)
[38] Albus, J., Bostelman, R., Chang, T., Hong, T., Shackleford, W., Shneier, M.: Learning in a Hierarchical Control System: 4D/RCS in the DARPA LAGR Program. Journal of Field Robotics, Special Issue on Learning in Unstructured Environments 23(11/12), 975–1003 (2006)
[39] Shneier, M., Chang, T., Hong, T., Shackleford, W., Bostelman, R., Albus, J.S.: Learning traversability models for autonomous mobile vehicles. Autonomous Robots 24(1), 69–86 (2008)

Index

Active Appearance Models   45
adaptive boosting (AdaBoost)   13, 79, 89, 92, 278
adaptive DWE   21
air-intake system   261
air intake subsystem (AIS)   257, 262
airpath
   control   173, 174
   in-cylinder air mass observer   177
   manifold pressure   180
   recirculated gas mass   175, 176
   residual gases   182
   volumetric efficiency   177
ALOPEX   151
analysis of variance   13
annotation tool   54, 57, 58, 64, 76
anti-lock braking system (ABS)   257, 283
automotive suspension system   257, 270

backpropagation through time (BPTT)
   truncated   145
Bayesian network   105, 109, 110
blink frequency   29, 35, 36, 39
branch and bound   117, 118, 124, 125, 128, 129

Chamfer matching   81–83
Classifier Fusion Techniques   276
cognitive workload   1
computer vision   28, 45
Constrained Local Models   47
cross validation   14

decision tree   7, 11, 14–17, 19, 59–62, 278
decision tree learning   11, 21
diagnostic matrix   265
diagnostic matrix (D-matrix)   263
diagnostic tree   264
distraction   26
   cognitive distraction   26, 35, 38, 44
   visual distraction   26, 35
driver assistance   53, 54, 67, 75, 76
driver inattention   25, 26, 35, 54, 67–70, 75, 76
   detection   67–69, 75
   driver inattentiveness level   39
driver workload estimation (DWE)   2, 4, 13
driving/driver activity   53, 57, 75
driving activity   65
driving pattern   243
Dynamic fusion   279
Dynamic Programming   248
dynamic resistance approach   295
Dynamic resistance profile   296
dynamic resistance profile   296, 298, 300, 301, 303, 311

engine
   actuators   166
   control   165, 173
      common features   166
      development cycle   168
   downsizing   173
   Spark Ignition (SI) —   173
   turbocharging   173

Error-Correcting Output Codes (ECOC) 279
extended Kalman filter (EKF) 261
eye closure duration 39

face pose 29, 34, 37, 39, 48
fatigue 27–29, 35, 37, 38, 40, 42
fault detection and diagnosis (FDD) 254
fixed gaze 35, 38, 39, 42, 43, 48
fuzzy logic 292, 301–304, 311
fuzzy rule 240
fuzzy rules 240, 241, 305
fuzzy system 29, 38

graphical models 103, 104, 114
grey box approach 168

hardware-in-the-loop simulations (HILS) 254, 261
Hessian matrix 148
hypervariate 54

Image acquisition 29
Intelligent Constant Current Control 293, 301, 302, 306, 307, 309–312

K-Nearest Neighbor (KNN) 275
Kalman filter 29, 33, 37, 38, 43
 extended (EKF) 150, 151, 153
 multi-stream 150
 non-differential or nonlinear 153
kernel function 171
knowledge discovery 117

lane tracking 49
learning machines 165
learning rate 150
Learning vector quantization (LVQ) 246, 292, 296, 297, 299, 300, 304, 311
learning-based DWE design process 5
linear parameter varying (LPV) system 179

maneuver 1, 6, 53, 54, 57, 59, 62–64, 70, 76
 classification 63, 64
 detection 57, 62, 64
Markov network 105, 106

micro-camera 29
Multi-way partial least squares (MPLS) 274
multi-way partial least squares (MPLS) 274
multilayer perceptron (MLP) 135, 136, 167, 170

near-IR 29, 33, 45
neural network 236, 246
 controller 140, 143, 146, 148
 in engine control 167
 models 135, 136, 153, 169
neuro-fuzzy inference system 295
nodding 27, 39

observer 136, 167, 174–177, 179
 polytopic 165, 176, 179
Observers 259
Output error (prediction-error) method 258

partial least squares (PLS) 280
particle swarm optimization (PSO) 151, 152
PERCLOS 28, 35–37, 39, 41, 42, 44, 48
prior knowledge 168, 182
process capability index 117, 118, 122, 124, 127, 130
prognostic model 267
Prognostics 266
Pupil detection 32

radial basis function (RBF)
 kernel 171, 183
 networks 136, 137, 167, 170, 172
Random Forest 61–65, 72, 74–76
real-rime recurrent learning (RTRL) 145
recurrent neural network (RNN) 134, 138, 143, 145, 147, 153
remaining useful life (RUL) 256
Residual 263
residual 264–266, 269
Resistance Spot Welding 293
resistance spot welding 293, 295, 311
Resistance spot welding (RSW) 291

Index   365

root cause analysis   117, 118, 124, 126, 127
rule extraction   117, 118, 122

sensor selection   54, 64, 65, 67
Simultaneous Perturbation Stochastic Approximation (SPSA)   152, 153
soft (indirect) sensor   301
soft sensing   296, 303, 306, 307, 311
soft sensor   136, 167, 292, 301, 302, 311
Stochastic Meta-Descent (SMD)   151, 153
support vector machine (SVM)   171, 275, 280
support vector machines   280

support vector regression (SVR)   170, 182

traffic accidents   25

variable camshaft timing (VCT)   173, 174, 182
vehicle power management   239, 243, 245, 248
virtual or soft (indirect) sensor   292
virtual sensor   133, 134, 136–139
visual behaviors   27–29, 34, 37, 38, 43
visualization   104, 110, 111, 113

weight update method   146, 147
  first-order   148
  second-order   149
workload management   1, 53, 54

Author Index

Albus, James 315
Anuradha, Kodali 253
Arsie, Ivan 191

Balakirsky, Stephen 315
Barbera, Tony 315
Barea, Rafael 25
Bergasa, Luis M. 25
Bloch, Gérard 165
Bostelman, Roger 315

Chaitanya, Sankavaram 253
Chang, Tommy 315
Chinnam, Ratna Babu 291
Colin, Guillaume 165

El-Banna, Mahmoud 291

Filev, Dimitar 291

Gandhi, Tarak 79
Gardner, Mike 53

Hong, Tsai 315
Hrycej, Tomas 117
Huang, Hui-Min 315

Jianhui, Luo 253

Kihoon, Choi 253
Krishna, Pattipati 253
Kruse, Rudolf 103

Lauer, Fabien 165
Leivian, Bob 53
Liu, Qiao 253
Lopez, Elena 25

Madhavan, Raj 315
Madhavi, Namburu Setu 253
Murphey, Yi L. 223

Nuevo, Jesús 25

Owechko, Yuri 1

Pianese, Cesare 191
Proctor, Fred 315
Prokhorov, Danil 133

Rügheimer, Frank 103

Satnam, Singh 253
Schlenoff, Craig 315
Schreiner, Chris 53
Scott, Harry 315
Shackleford, Will 315
Shneier, Michael 315
Shunsuke, Chigusa 253
Sorrentino, Marco 191
Sotelo, Miguel A. 25
Steinbrecher, Matthias 103
Strobel, Christian Manuel 117
Summers, John 53
Suvasri, Mandal 253
Szabo, Sandor 315

Torkkola, Kari 53
Trivedi, Mohan Manubhai 79

William, Donat 253

Zhang, Harry 53
Zhang, Jing 1
Zhang, Keshu 53
Zhang, Yilu 1

CPSIA information can be obtained
at www.ICGtesting.com
Printed in the USA
LVOW01s0832231115
463691LV00002BC/2/P